Polymer Processing

Polymer Processing

by James M. McKelvey

Professor of Chemical Engineering

Washington University

St. Louis, Missouri

John Wiley and Sons, Inc.

New York

London

Library of Congress Catalog Card Number: 62-8780

Printed in the United States of America

To Edith

Preface

The last few decades have seen the development of a new and important industry based on the *synthesis* of high polymers and their *conversion* to useful products. The conversion operations—molding, extrusion, blending, and calendering, to name but a few—constitute a challenging branch of technology called, here, polymer processing. Polymer conversion operations are today carried out on a large scale and involve an increasingly complex technology fraught with many unusual engineering problems.

My introduction to polymer processing occurred at the du Pont Experimental Station where, in the early 1950's, it became apparent that empirical studies in polymer processing were often inefficient and yielded limited results. Furthermore, it appeared that some polymer conversion operations might be amenable to mathematical analysis, which would yield results of a more general nature. With this in mind, the research group with which I was associated (which included W. L. Gore, Carl Jepson, R. S. Mallouk, J. F. Carley, and E. C. Bernhardt) undertook the study of thermoplastic extrusion. The work culminated in the Extrusion Symposium presented at the American Chemical Society meeting in the fall of 1952. In a sense this textbook is an outgrowth of that work.

It is now clear that rheology, the science of deformation and flow, is the key to understanding many of the phenomena encountered in polymer processing. Today non-Newtonian fluid mechanics is particularly useful in the analysis of most polymer conversion operations. It is almost certain that before long the concepts of viscoelasticity will become an important part of process engineering in the high polymer industry. Hence a substantial portion of this textbook is concerned with rheology and with applications in non-Newtonian flow.

My aim has been to write a textbook for senior and graduate engineering courses that would be an introductory (*but analytical rather than descriptive*) presentation of polymer processing. The book should also be useful to engineers practicing in the high polymer industry. To facilitate independent study, derivations have been given in detail, and examples showing numerical computations appear throughout the text.

When used as a textbook for a course in polymer processing, it should be possible to cover the material in a one-semester course. The prime prerequisite would be a working knowledge of differential equations—the language of modern engineering—and some previous exposure to the subject of transport phenomena. Although mathematical techniques more advanced than differential equations—vector analysis, for example—are used in some sections, the main theme can be followed without a complete understanding of the mathematical details of these sections. The instructor may find it desirable to provide certain introductory material concerning high polymer technology.

The presentation of the material relating to transport phenomena, particularly non-Newtonian fluid mechanics, has been strongly influenced by the work of Professor R. B. Bird (University of Wisconsin), who also reviewed the manuscript and made many important suggestions regarding the organization. I am also indebted to Professor A. T. DiBenedetto (University of Wisconsin), who used the material in manuscript form for his course in high polymer engineering and who commented most helpfully, and to Professor Eric Baer (University of Illinois), who, when associated with the du Pont Company, wrote Chapter 5 on Crystallization of Polymers. I also wish to express my appreciation to Professor L. E. Stout, head of Chemical Engineering at Washington University, for his encouragement and help during the time the book was being written.

JAMES M. MCKELVEY

St. Louis, Missouri
February, 1962

Contents

Polymer Processing

chapter 1

Introduction to Polymer Processing

The modern high polymer industry—a development of the past century—began with the discovery of techniques for the chemical modification of natural polymers. Examples of important modification processes include rubber vulcanization and cellulose acetylation. These processes greatly extended the utility of natural rubber and cellulose and opened many new fields of application.

Within the past few decades, synthetic polymers have assumed a commanding position in the industry. A wide variety of them, having an enormous range of physical and chemical properties, is now manufactured. They have found such wide application that their synthesis is now one of the most important activities of the chemical industry.

Along with the development of synthetic and modified natural polymers, there has been, of necessity, a parallel development of techniques for the conversion of polymeric materials (resins) of all kinds into useful products, such as textile fibers, packaging films, and surface coatings, to name but a few. The term *polymer processing* is used to describe these conversion operations of the high polymer industry.

1-1 Definition and Scope of Polymer Processing

The scope of polymer processing can be deduced from the following definition:[1]

> Polymer processing is an engineering specialty concerned with the operations carried out on polymeric materials or systems to increase their utility. These operations produce one or more of the following effects: chemical reaction, flow, or a permanent change in a physical property. Specifically excluded are the chemical reactions involved in the manufacture of resins.

[1] E. C. Bernhardt and J. M. McKelvey, *Modern Plastics*, p. 154, July, 1958.

A few typical polymer processing operations coming within the scope of this definition include injection molding, calendering, dispersion of pigments in polymers, and surface activation of plastic film.

Injection molding, one of the most important operations of polymer processing, involves flow but neither chemical reaction nor a permanent change in a physical property. This is also true of the operations of sheet thermoforming, extrusion, and simple mixing of molten polymers.

Operations involving chemical reaction include flame and electronic activation of plastic film surfaces and compression molding of thermosetting resins. The control of crystallization and orientation in textile fibers and packaging films to improve mechanical properties is an example of an operation effecting permanent changes in physical properties without chemical reaction or flow in the ordinary sense.

The technical aspects of the polymer conversion industry include machine and equipment design (primarily a mechanical engineering problem); plant design and operations (problems such as quality control and time and motion study, which are primarily industrial engineering); product development (work concerned with the physical and chemical evaluation of materials for specific applications); and the subject with which this textbook is concerned, analysis of polymer conversion operations.

In engineering analysis, the principles of physical science are used to establish the dependent and independent variables of an operation and to determine the functional relationships between them. Because deformation and flow are involved in most polymer processing operations, the concepts of the science of rheology are widely used in this textbook. Other subjects of importance in the analysis of polymer conversion operations include crystallization, dielectric behavior of polymers, *P-V-T* behavior of polymers, surface chemistry, and heat transfer.

The analogy between the polymer conversion operations and the chemical engineering unit operations is close. The unique properties of high polymers, however, raise certain unusual engineering problems that are ordinarily not considered in the chemical engineering unit operations. For this reason polymer processing is treated as a separate field of specialization.

1-2 Classification of Polymer Conversion Operations

One method of classifying polymer conversion operations is based on function. It groups the operations into three broad categories. The function of operations in the first category is to form a mass of the polymer into a useful shape. Operations in the second category have the function of bonding whereas those in the third have the function of modifying the structure of the polymeric system. Table 1-1 lists some of the more important operations in each of these categories.

Molding and extrusion (calendering is considered as a special case of extrusion) are the basic forming techniques of polymer processing. Since a mold can fix three dimensions (height, width, and length) of an object while an extrusion die can fix only two (height and width), extrusion and molding are usually complementary rather than competitive operations.

Forming operations produce a wide variety of products including such diverse items as buttons, refrigerator liners, and packaging film. Although the details of the operations differ greatly, they have at least one common feature. In all forming operations, the substance being formed must *flow* in order to be shaped and then must *solidify* while it still retains the shape of the mold or die.

Table 1-1

Polymer Conversion Operations

Forming Operations	Bonding Operations	Modification Operations
Molding	Heat sealing	Mixing
Extrusion	Laminating	Surface activation
Calendering	Coating	Polymer modification

There are many different types of molding operations. In injection molding, a thermoplastic material is heated above its softening point to obtain flow and then cooled below its softening point to obtain solidification, a purely physical process. Other molding operations are based on the chemical reactions of polymerization and cross-linking. Still others involve a combination of both physical and chemical changes. In each case a different method is used to obtain the flow-solidification sequence.

Molding operations of the first type are those that involve physical changes only. Examples include injection molding of thermoplastics, forming of thermoplastic sheets, and plastisol molding. Molding operations of the second type are those in which chemical change is the basis of the flow-solidification sequence. An example of an operation of this type is the casting of liquid monomers. Solidification starts when the polymerization reaction is initiated. In molding operations of the third type, both physical and chemical changes occur. Compression molding of thermosetting resins is a good example of this type of operation. Here flow is obtained while the resin is still in the thermoplastic state, and solidification is obtained through cross-linking reactions that transform the resin into an infusible mass. Table 1-2 shows various molding operations classified according to this method.

In extrusion, the polymer is shaped by being forced through a die in the fluid state and then solidified while it retains the shape imparted to it by the die. Usually fluidity is obtained by melting the polymer and solidification by freezing. In some fiber spinning processes, however, fluidity is achieved by putting the polymer in solution and solidification by solvent evaporation or chemical reaction. Although an extrusion die can fix only two dimensions (height and width), it offers the advantage, when

Table 1-2

Classification of Molding Operations

Type I Operation Involves Physical Changes Only	Type II Operation Involves Chemical Changes Only	Type III Operation Involves Both Types of Changes
Injection molding	Monomer casting	Compression molding
Thermoforming	Slush molding	Transfer molding
Blow molding		
Plastisol molding		

coupled with a pump, of being a continuous low cost forming technique that can be operated at high speed.

Extrusion operations are used in the manufacture of film, fibers, pipe and other shapes, in coating paper, metal foils, wire, and in many other ways. In all extrusion operations, shear stresses of sufficient magnitude must be generated to cause the polymer to flow through the die at the desired rate. In some cases, when the polymer is initially in the molten state or in solution, screw pumps or gear pumps are used to develop the required pressure. In other cases, where the polymer must also be melted, plasticating extruders are commonly used. In calendering—considered here to be a special case of extrusion—the extrusion die is formed by the gap between the rolls, and it is the motion of the rolls that generates the forces causing the polymer to flow through the die.

Adhesives and adhesive bonding form a large field of technology that is essentially separate and distinct from polymer processing and is outside the scope of this book. The discussion here is limited to those operations in which bonding occurs with the solidification of a molten thermoplastic.

Two kinds of bonding operations are considered. The first kind is that in which both sides of the interface become molten during the operation. A typical example is the heat sealing of thermoplastics. The second kind includes operations in which only one side of the interface becomes molten. The extrusion coating of molten polyethylene on aluminum foil is an example of a bonding operation of the second kind.

The adhesion processes responsible for bonding can be either specific or mechanical. Specific adhesion results from intermolecular attractive forces of various kinds. It is solely responsible for bonding when one side of the interface is a dense, smooth material, such as glass or metal.

Mechanical adhesion occurs when one side of the interface has a porous structure that is penetrated by the adhesive when it is in the fluid state. Upon solidification, mechanical interlocks are formed that account for the gripping action. Mechanical adhesion may, in some cases, assist specific adhesion, but by itself it is probably not a useful means of bonding.

With bonding processes of the first kind where both sides of the interface become fluid, there is probably an additional mechanism that must be considered. This is the interdiffusion of polymer chains across the interface. This process, if continued long enough, would cause the interface to disappear and would obviously result in a bond having the same strength as the material itself.

The object of the polymer processing operations in the third category is to modify the structure of the polymeric system. In surface activation processes, structural modifications are limited to the surface of the polymer where certain chemical and physical changes are induced by exposure to a gas flame or a corona discharge.

Mixing processes of all kinds change the structure of the system. The object of simple mixing is to randomize the spatial distribution of the components of the mixture. The object of dispersive mixing is to reduce the size of clumps, agglomerates, or particles of one or more of the mixture's components. Both types of mixing are important in polymer processing.

Other modification operations result in changes in the microstructure throughout the bulk of the polymeric system. Examples include crystallization and orientation. Additional modification operations of this type induce changes in the polymer molecules through chemical and physical processes, such as vulcanization, and irradiation with high energy electrons.

chapter 2
Rheology

This chapter is an introductory treatment of rheology—the science of deformation and flow. Because flow is a key factor in most polymer processing operations, the chapter deals primarily with methods for describing the behavior of materials that, rheologically speaking, are fluids. Although inelastic fluids are emphasized, an introduction to viscoelasticity is also included.

2-1 The Equations of Continuity, Momentum, and Energy

The equations of continuity, momentum, and energy are mathematical formulations of fundamental physical principles and are independent of the nature of the fluid. One or more of them, together with the rheological equation of the fluid, are always involved in the mathematical formulation of flow problems. They are presented in this section without proof[1] and in a manner intended to minimize the mathematical aspects of formulating flow problems.

Consider a region in space containing a viscous fluid in motion. Its density ρ, pressure P, and temperature T are scalars; its velocity $\boldsymbol{v}$ is a vector; while $\boldsymbol{\tau}$, the stress resulting from viscous forces, is a symmetrical second order tensor.[2] In the following discussion these quantities are considered functions of time and position.

Now, consider an arbitrary element of fluid contained within an imaginary closed surface that follows the motion of the element but exerts no forces upon it. The fluid element constitutes a closed system, that is, a system that can exchange energy, but not matter, with its surroundings.

[1] Appendix A presents a derivation of the equations of continuity, momentum, and energy.

[2] Vector notation and operations are summarized in Appendix A.

The principle of conservation of matter requires that the mass of a closed system be constant. The *continuity equation*

$$\frac{D\rho}{Dt} = -\rho(\nabla \cdot \boldsymbol{v}) \tag{2-1}$$

where t represents time, is a mathematical statement of this principle.

Newton's second law of motion requires that the rate of increase of momentum of the fluid element be equal to the sum of all the forces acting upon it. This is expressed mathematically by the *momentum equation*

$$\rho \frac{D\boldsymbol{v}}{Dt} = -\nabla P + [\nabla \cdot \boldsymbol{\tau}] + \rho \boldsymbol{g} \tag{2-2}$$

where $\boldsymbol{g}$ is a vector representing the resultant of the body forces (per unit mass) acting on the fluid at a point.

The first law of thermodynamics, a statement of the principle of conservation of energy, applied to the fluid element yields the *energy equation*

$$\rho c_v \frac{DT}{Dt} = -(\nabla \cdot \boldsymbol{q}) - T\left(\frac{\partial P}{\partial T}\right)_\rho (\nabla \cdot \boldsymbol{v}) + (\boldsymbol{\tau} : \nabla \boldsymbol{v}) \tag{2-3}$$

where c_v is the specific heat at constant volume of the fluid and $\boldsymbol{q}$ is the conductive heat flux vector. It is related to the temperature gradient, in an isotropic medium, by the equation

$$\boldsymbol{q} = -k \, \nabla T \tag{2-4}$$

which is Fourier's law of heat conduction. The coefficient k is the thermal conductivity of the fluid. The components of the conductive heat flux vector in rectangular, cylindrical, and spherical coordinates are given in Table 2-1.

Table 2-1*

Components of the Conductive Heat Flux Vector

Rectangular	Cylindrical	Spherical
$q_x = -k \frac{\partial T}{\partial x}$	$q_r = -k \frac{\partial T}{\partial r}$	$q_r = -k \frac{\partial T}{\partial r}$
$q_y = -k \frac{\partial T}{\partial y}$	$q_\theta = -k \frac{1}{r} \frac{\partial T}{\partial \theta}$	$q_\theta = -k \frac{1}{r} \frac{\partial T}{\partial \theta}$
$q_z = -k \frac{\partial T}{\partial z}$	$q_z = -k \frac{\partial T}{\partial z}$	$q_\phi = -k \frac{1}{r \sin \theta} \frac{\partial T}{\partial \phi}$

* Reprinted with permission from R. B. Bird, W. E. Stewart, and E. N. Lightfoot, *Transport Phenomena*, John Wiley, (1960).

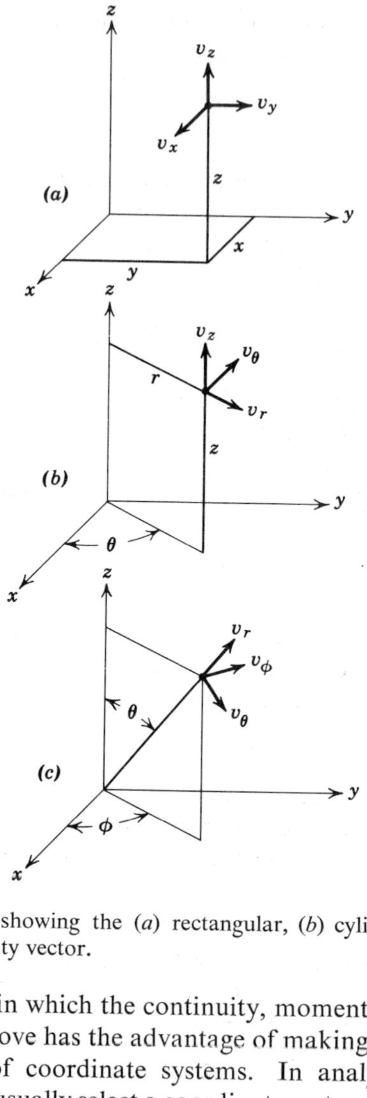

Figure 2-1. Diagrams showing the (*a*) rectangular, (*b*) cylindrical, and (*c*) spherical components of a velocity vector.

The vector form in which the continuity, momentum, and energy equations are written above has the advantage of making the equations concise and independent of coordinate systems. In analyzing flow problems, however, one must usually select a coordinate system and resolve the vector and tensor quantities into components. The choice of a coordinate system depends primarily upon the geometry of the boundaries of the fluid. Rectangular, cylindrical, and spherical coordinates are most often used, and in Table 2-2 the components of the above equations are presented in these coordinates.

Figure 2-1 shows how the location of a point in space is specified in each of the three systems. It also shows the orientation of the components of a

Table 2-2*

Equations of Continuity, Momentum, and Energy in Rectangular, Cylindrical, and Spherical Coordinates

Continuity Equation

Rectangular coordinates (x, y, z):

$$\frac{\partial \rho}{\partial t} + \frac{\partial}{\partial x}(\rho v_x) + \frac{\partial}{\partial y}(\rho v_y) + \frac{\partial}{\partial z}(\rho v_z) = 0$$

Cylindrical coordinates (r, θ, z):

$$\frac{\partial \rho}{\partial t} + \frac{1}{r}\frac{\partial}{\partial r}(\rho r v_r) + \frac{1}{r}\frac{\partial}{\partial \theta}(\rho v_\theta) + \frac{\partial}{\partial z}(\rho v_z) = 0$$

Spherical coordinates (r, θ, ϕ):

$$\frac{\partial \rho}{\partial t} + \frac{1}{r^2}\frac{\partial}{\partial r}(\rho r^2 v_r) + \frac{1}{r \sin\theta}\frac{\partial}{\partial \theta}(\rho v_\theta \sin\theta) + \frac{1}{r \sin\theta}\frac{\partial}{\partial \phi}(\rho v_\phi) = 0$$

Momentum Equation in Rectangular Coordinates

x-component

$$\rho\left(\frac{\partial v_x}{\partial t} + v_x\frac{\partial v_x}{\partial x} + v_y\frac{\partial v_x}{\partial y} + v_z\frac{\partial v_x}{\partial z}\right) = -\frac{\partial P}{\partial x} + \left(\frac{\partial \tau_{xx}}{\partial x} + \frac{\partial \tau_{yx}}{\partial y} + \frac{\partial \tau_{zx}}{\partial z}\right) + \rho g_x$$

y-component

$$\rho\left(\frac{\partial v_y}{\partial t} + v_x\frac{\partial v_y}{\partial x} + v_y\frac{\partial v_y}{\partial y} + v_z\frac{\partial v_y}{\partial z}\right) = -\frac{\partial P}{\partial y} + \left(\frac{\partial \tau_{xy}}{\partial x} + \frac{\partial \tau_{yy}}{\partial y} + \frac{\partial \tau_{zy}}{\partial z}\right) + \rho g_y$$

z-component

$$\rho\left(\frac{\partial v_z}{\partial t} + v_x\frac{\partial v_z}{\partial x} + v_y\frac{\partial v_z}{\partial y} + v_z\frac{\partial v_z}{\partial z}\right) = -\frac{\partial P}{\partial z} + \left(\frac{\partial \tau_{xz}}{\partial x} + \frac{\partial \tau_{yz}}{\partial y} + \frac{\partial \tau_{zz}}{\partial z}\right) + \rho g_z$$

Momentum Equation in Cylindrical Coordinates

r-component

$$\rho\left(\frac{\partial v_r}{\partial t} + v_r\frac{\partial v_r}{\partial r} + \frac{v_\theta}{r}\frac{\partial v_r}{\partial \theta} - \frac{v_\theta^2}{r} + v_z\frac{\partial v_r}{\partial z}\right) = -\frac{\partial P}{\partial r} + \left(\frac{1}{r}\frac{\partial}{\partial r}(r\tau_{rr}) + \frac{1}{r}\frac{\partial \tau_{r\theta}}{\partial \theta} - \frac{\tau_{\theta\theta}}{r} + \frac{\partial \tau_{rz}}{\partial z}\right) + \rho g_r$$

θ-component

$$\rho\left(\frac{\partial v_\theta}{\partial t} + v_r\frac{\partial v_\theta}{\partial r} + \frac{v_\theta}{r}\frac{\partial v_\theta}{\partial \theta} + \frac{v_r v_\theta}{r} + v_z\frac{\partial v_\theta}{\partial z}\right) = -\frac{1}{r}\frac{\partial P}{\partial \theta} + \left(\frac{1}{r^2}\frac{\partial}{\partial r}(r^2\tau_{r\theta}) + \frac{1}{r}\frac{\partial \tau_{\theta\theta}}{\partial \theta} + \frac{\partial \tau_{\theta z}}{\partial z}\right) + \rho g_\theta$$

* Reprinted with permission from Bird, Stewart, and Lightfoot, *Transport Phenomena*.

Table 2-2 (continued)

z-component

$$\rho\left(\frac{\partial v_z}{\partial t} + v_r \frac{\partial v_z}{\partial r} + \frac{v_\theta}{r}\frac{\partial v_z}{\partial \theta} + v_z \frac{\partial v_z}{\partial z}\right) = -\frac{\partial P}{\partial z} + \left(\frac{1}{r}\frac{\partial}{\partial r}(r\tau_{rz}) + \frac{1}{r}\frac{\partial \tau_{\theta z}}{\partial \theta} + \frac{\partial \tau_{zz}}{\partial z}\right) + \rho g_z$$

Momentum Equation in Spherical Coordinates

r-component

$$\rho\left(\frac{\partial v_r}{\partial t} + v_r \frac{\partial v_r}{\partial r} + \frac{v_\theta}{r}\frac{\partial v_r}{\partial \theta} + \frac{v_\phi}{r\sin\theta}\frac{\partial v_r}{\partial \phi} - \frac{v_\theta{}^2 + v_\phi{}^2}{r}\right)$$

$$= -\frac{\partial P}{\partial r} + \left(\frac{1}{r^2}\frac{\partial}{\partial r}(r^2\tau_{rr}) + \frac{1}{r\sin\theta}\frac{\partial}{\partial \theta}(\tau_{r\theta}\sin\theta) + \frac{1}{r\sin\theta}\frac{\partial \tau_{r\phi}}{\partial \phi} - \frac{\tau_{\theta\theta} + \tau_{\phi\phi}}{r}\right) + \rho g_r$$

θ-component

$$\rho\left(\frac{\partial v_\theta}{\partial t} + v_r \frac{\partial v_\theta}{\partial r} + \frac{v_\theta}{r}\frac{\partial v_\theta}{\partial \theta} + \frac{v_\phi}{r\sin\theta}\frac{\partial v_\theta}{\partial \phi} + \frac{v_r v_\theta}{r} - \frac{v_\phi{}^2\cot\theta}{r}\right)$$

$$= -\frac{1}{r}\frac{\partial P}{\partial \theta} + \left(\frac{1}{r^2}\frac{\partial}{\partial r}(r^2\tau_{r\theta}) + \frac{1}{r\sin\theta}\frac{\partial}{\partial \theta}(\tau_{\theta\theta}\sin\theta) + \frac{1}{r\sin\theta}\frac{\partial \tau_{\theta\phi}}{\partial \phi} + \frac{\tau_{r\theta}}{r} - \frac{\cot\theta}{r}\tau_{\phi\phi}\right) + \rho g_\theta$$

φ-component

$$\rho\left(\frac{\partial v_\phi}{\partial t} + v_r \frac{\partial v_\phi}{\partial r} + \frac{v_\theta}{r}\frac{\partial v_\phi}{\partial \theta} + \frac{v_\phi}{r\sin\theta}\frac{\partial v_\phi}{\partial \phi} + \frac{v_\phi v_r}{r} + \frac{v_\theta v_\phi}{r}\cot\theta\right)$$

$$= -\frac{1}{r\sin\theta}\frac{\partial P}{\partial \phi} + \left(\frac{1}{r^2}\frac{\partial}{\partial r}(r^2\tau_{r\phi}) + \frac{1}{r}\frac{\partial \tau_{\theta\phi}}{\partial 0} + \frac{1}{r\sin\theta}\frac{\partial \tau_{\phi\phi}}{\partial \phi} + \frac{\tau_{r\phi}}{r} + \frac{2\cot\theta}{r}\tau_{\theta\phi}\right) + \rho g_\phi$$

The Energy Equation

Rectangular coordinates:

$$\rho c_v\left(\frac{\partial T}{\partial t} + v_x \frac{\partial T}{\partial x} + v_y \frac{\partial T}{\partial y} + v_z \frac{\partial T}{\partial z}\right) = -\left[\frac{\partial q_x}{\partial x} + \frac{\partial q_y}{\partial y} + \frac{\partial q_z}{\partial z}\right]$$

$$- T\left(\frac{\partial P}{\partial T}\right)_\rho \left(\frac{\partial v_x}{\partial x} + \frac{\partial v_y}{\partial y} + \frac{\partial v_z}{\partial z}\right) + \left\{\tau_{xx}\frac{\partial v_x}{\partial x} + \tau_{yy}\frac{\partial v_y}{\partial y} + \tau_{zz}\frac{\partial v_z}{\partial z}\right\}$$

$$+ \left\{\tau_{xy}\left(\frac{\partial v_x}{\partial y} + \frac{\partial v_y}{\partial x}\right) + \tau_{xz}\left(\frac{\partial v_x}{\partial z} + \frac{\partial v_z}{\partial x}\right) + \tau_{yz}\left(\frac{\partial v_y}{\partial z} + \frac{\partial v_z}{\partial y}\right)\right\}$$

Table 2-2 (continued)

Cylindrical coordinates:

$$\rho c_v\left(\frac{\partial T}{\partial t} + v_r\frac{\partial T}{\partial r} + \frac{v_\theta}{r}\frac{\partial T}{\partial \theta} + v_z\frac{\partial T}{\partial z}\right) = -\left[\frac{1}{r}\frac{\partial}{\partial r}(rq_r) + \frac{1}{r}\frac{\partial q_\theta}{\partial \theta} + \frac{\partial q_z}{\partial z}\right]$$

$$- T\left(\frac{\partial P}{\partial T}\right)_\rho\left(\frac{1}{r}\frac{\partial}{\partial r}(rv_r) + \frac{1}{r}\frac{\partial v_\theta}{\partial \theta} + \frac{\partial v_z}{\partial z}\right) + \left\{\tau_{rr}\frac{\partial v_r}{\partial r} + \tau_{\theta\theta}\frac{1}{r}\left(\frac{\partial v_\theta}{\partial \theta} + v_r\right)\right.$$

$$\left. + \tau_{zz}\frac{\partial v_z}{\partial z}\right\} + \left\{\tau_{r\theta}\left[r\frac{\partial}{\partial r}\left(\frac{v_\theta}{r}\right) + \frac{1}{r}\frac{\partial v_r}{\partial \theta}\right] + \tau_{rz}\left(\frac{\partial v_z}{\partial r} + \frac{\partial v_r}{\partial z}\right)\right.$$

$$\left. + \tau_{\theta z}\left(\frac{1}{r}\frac{\partial v_z}{\partial \theta} + \frac{\partial v_\theta}{\partial z}\right)\right\}$$

Spherical coordinates:

$$\rho c_v\left(\frac{\partial T}{\partial t} + v_r\frac{\partial T}{\partial r} + \frac{v_\theta}{r}\frac{\partial T}{\partial \theta} + \frac{v_\phi}{r\sin\theta}\frac{\partial T}{\partial \phi}\right) = -\left[\frac{1}{r^2}\frac{\partial}{\partial r}(r^2 q_r)\right.$$

$$\left. + \frac{1}{r\sin\theta}\frac{\partial}{\partial \theta}(q_\theta \sin\theta) + \frac{1}{r\sin\theta}\frac{\partial q_\phi}{\partial \phi}\right] - T\left(\frac{\partial P}{\partial T}\right)_\rho\left(\frac{1}{r^2}\frac{\partial}{\partial r}(r^2 v_r)\right.$$

$$\left. + \frac{1}{r\sin\theta}\frac{\partial}{\partial \theta}(v_\theta \sin\theta) + \frac{1}{r\sin\theta}\frac{\partial v_\phi}{\partial \phi}\right) + \left\{\tau_{rr}\frac{\partial v_r}{\partial r} + \tau_{\theta\theta}\left(\frac{1}{r}\frac{\partial v_\theta}{\partial \theta} + \frac{v_r}{r}\right)\right.$$

$$\left. + \tau_{\phi\phi}\left(\frac{1}{r\sin\theta}\frac{\partial v_\phi}{\partial \phi} + \frac{v_r}{r} + \frac{v_\theta \cot\theta}{r}\right)\right\} + \left\{\tau_{r\theta}\left(\frac{\partial v_\theta}{\partial r} + \frac{1}{r}\frac{\partial v_r}{\partial \theta} - \frac{v_\theta}{r}\right)\right.$$

$$\left. + \tau_{r\phi}\left(\frac{\partial v_\phi}{\partial r} + \frac{1}{r\sin\theta}\frac{\partial v_r}{\partial \phi} - \frac{v_\phi}{r}\right) + \tau_{\theta\phi}\left(\frac{1}{r}\frac{\partial v_\phi}{\partial \theta} + \frac{1}{r\sin\theta}\frac{\partial v_\theta}{\partial \phi} - \frac{\cot\theta}{r}v_\phi\right)\right\}$$

---

velocity vector $\boldsymbol{v}$ at the point. For example, in cylindrical coordinates, the velocity component v_θ is perpendicular to a plane containing r and z.

The stress $\boldsymbol{\tau}$, being a second order tensor, has 9 components of the form τ_{ij}, where i and j take on, in turn, the values x, y, and z. The second subscript indicates the direction in which the stress acts and the first subscript the orientation of the surface upon which it acts. For example, τ_{xy} represents the stress component acting in the y-direction upon a plane perpendicular to the x-axis. Because $\boldsymbol{\tau}$ is a symmetrical tensor

$$\tau_{ij} = \tau_{ji} \tag{2-5}$$

and only 6 of the stress components are independent.

In general, a flow problem cannot be formulated completely by using only the equations of continuity, momentum, and energy. A *rheological equation* will be required to relate the stress components to the rate of strain of the fluid, and an *equation of state* will be required for all problems where the density is not assumed constant. In addition, equations giving the temperature and pressure dependence of the other fluid properties may be required.

Example 2-1.

This example considers steady laminar flow between parallel plates, shown schematically in Figure 2-2. The plates are separated by distance H, which is very small in comparison to the length and width of the plates. The lower plate is stationary and the upper plate moves at a constant speed V_x in the x-direction.

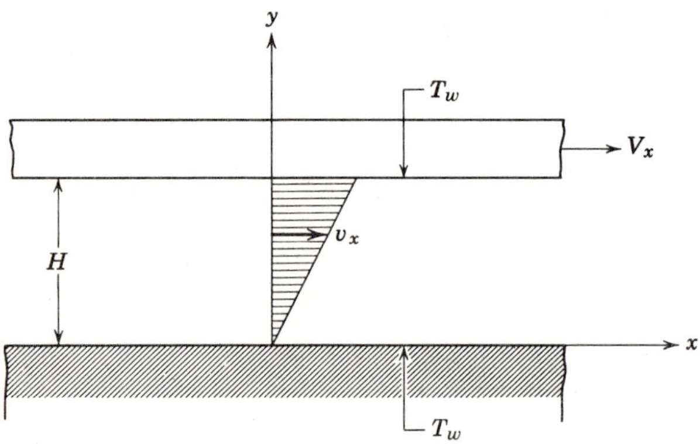

Figure 2-2. Laminar flow between parallel plates. Lower plate is stationary, and upper plate moves at constant speed in the x-direction.

The fluid is open to the atmosphere at the edges so that the hydrostatic pressure is the same at all points in the fluid. Both plates are maintained at a constant temperature T_w. It is assumed that all fluid properties, including viscosity and density, are constant and that there are no body forces acting on the fluid.

The dependent variables in this problem are T, $\mathbf{v}$, and $\boldsymbol{\tau}$ and they are functions only of one independent variable, the y-position coordinate. Therefore, all derivatives of these quantities or their components with respect to x, z, and t are zero. Furthermore, v_x is the only nonzero component of the velocity vector and q_y the only nonzero component of the conductive heat flux vector. For this special case, Table 2-2 shows that the momentum and energy equations, in rectangular coordinates, reduce to

$$0 = \left(\frac{\partial \tau_{yx}}{\partial y}\right) \tag{2-6}$$

$$0 = -\left(\frac{\partial q_y}{\partial y}\right) + \tau_{xy}\left(\frac{\partial v_x}{\partial y}\right) \tag{2-7}$$

The second term of equation (2-7) represents the rate (per unit volume of fluid) of dissipation of mechanical energy into heat at a given point in the fluid.

The first term of (2-7) represents the rate (per unit volume of fluid) at which heat is lost by conduction at the same point. Under steady state conditions the two rates are equal, as is shown by the equation.

The boundary conditions for the problem are written

$$\begin{aligned} v_x(0) &= 0 \qquad & T(0) &= T_w \\ v_x(H) &= V_x \qquad & T(H) &= T_w \end{aligned} \tag{2-8}$$

The y-component of the conductive heat flux vector, obtained from Table 2-1, is

$$q_y = -k\left(\frac{\partial T}{\partial y}\right)$$

and because the order of the subscripts on the stress component can be interchanged, equation (2-7) becomes

$$-k\left(\frac{\partial^2 T}{\partial y^2}\right) = \tau_{yx}\left(\frac{\partial v_x}{\partial y}\right) \tag{2-9}$$

The flow problem as now formulated consists of three independent variables and two equations. It is clear that the problem cannot be solved until an additional independent equation is specified. What is required is the rheological equation of the fluid, an equation relating the stress component τ_{yx} to the rate of strain of the fluid at the point under consideration.

The simplest rheological equation for a viscous fluid assumes a linear relationship between τ_{yx} and the gradient of the velocity in the y-direction and is written

$$\tau_{yx} = +\mu\left(\frac{\partial v_x}{\partial y}\right) \tag{2-10}$$

where μ is the coefficient of viscosity of the fluid. Equation (2-10) is Newton's law of viscosity.[3]

The problem as formulated now has three independent equations. To solve, integrate equation (2-6) to obtain

$$\tau_{xy} = \tau_{yx} = C_1 \tag{2-11}$$

where C_1 is the constant of integration. Introduce this result into (2-10) and integrate to obtain

$$v_x = +\frac{C_1 y}{\mu} + C_2 \tag{2-12}$$

and then introduce equations (2-11) and (2-12) into (2-9) and integrate to obtain

$$T = -\frac{C_1^{\,2} y^2}{2\mu k} + C_3 y + C_4 \tag{2-13}$$

[3] The positive sign is a convention indicating that τ_{yx} acts on the upper side of a surface perpendicular to the y-axis. If equation (2-10) is written with a negative sign, then τ_{yx} represents the stress acting on the under side of the surface.

The arbitrary constants of integration evaluated from the boundary conditions, equations (2-8), are

$$C_1 = +\mu\left(\frac{V_x}{H}\right) \qquad C_3 = \frac{\mu H}{2k}\left(\frac{V_x}{H}\right)^2$$

$$C_2 = 0 \qquad C_4 = T_w$$

and equations (2-11), (2-12), and (2-13) reduce to

$$\tau_{yx} = \mu\left(\frac{V_x}{H}\right) \tag{2-14}$$

$$v_x = V_x\left(\frac{y}{H}\right) \tag{2-15}$$

$$(T - T_w) = \frac{\mu}{2k}\left(\frac{V_x}{H}\right)^2[Hy - y^2] \tag{2-16}$$

It is easily verified that these functions satisfy the differential equations and the boundary conditions of the problem.

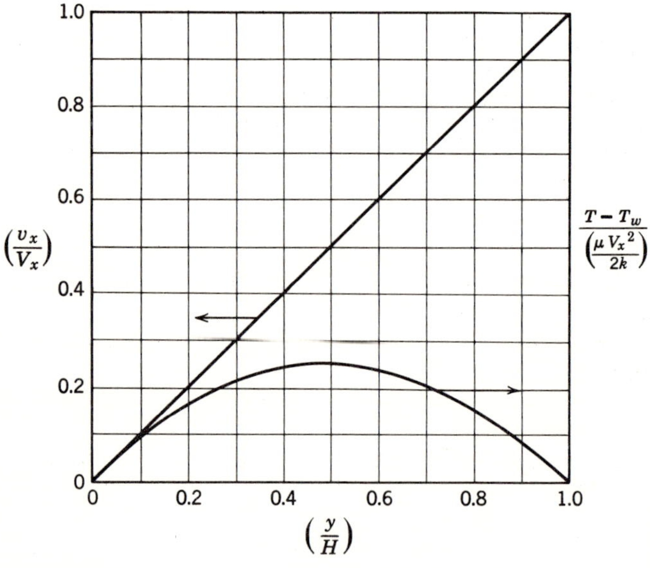

Figure 2-3. Dimensionless velocity and temperature profiles for a Newtonian fluid between parallel plates.

Equation (2-14) shows that τ_{yx} is independent of the position variable y but is directly proportional to the viscosity of the fluid and to the velocity gradient. Equation (2-15) shows that v_x is directly proportional to y. A plot of (v_x/V_x), the relative velocity, versus (y/H) is a straight line having a slope of unity, as shown in Figure 2-3.

Equation (2-16) indicates that the temperature profile is parabolic in shape. If (2-16) is written

$$(T - T_w)\left(\frac{2k}{\mu V_x^2}\right) = \left(\frac{y}{H}\right)\left[1 - \left(\frac{y}{H}\right)\right]$$

then the left side is a function only of the dimensionless position variable (y/H), as shown in Figure 2-3.

2-2 Mechanics of Newtonian Fluids

In Example 2-2, the rheological equation for the Newtonian fluid

$$\tau_{yx} = \mu\left(\frac{\partial v_x}{\partial y}\right)$$

was obtained by postulating a linear relationship between stress and rate of strain. This equation is a special case, applying to plane parallel shear, of the general rheological equation of the Newtonian fluid that is derived in Appendix A-6. It can be written

$$\tau_{ij} = \mu\Delta_{ij} + (\kappa - \tfrac{2}{3}\mu)(\nabla \cdot \boldsymbol{v})\delta_{ij} \qquad i, j = 1, 2, 3 \tag{2-17}$$

where κ is the coefficient of volume viscosity and μ is the coefficient of shear viscosity. The components Δ_{ij} of the rate of strain tensor $\boldsymbol{\Delta}$ and δ_{ij} of the unit tensor $\boldsymbol{\delta}$ are defined by the equations

$$\Delta_{ij} = \left[\left(\frac{\partial v_i}{\partial x_j}\right) + \left(\frac{\partial v_j}{\partial x_i}\right)\right]$$

$$\delta_{ij} = 0 \quad \text{if} \quad i \neq j$$

$$\delta_{ij} = 1 \quad \text{if} \quad i = j$$

where the following relationships exist between the x, y, z and the 1, 2, 3 notational systems:

$$x_1 = x \qquad v_1 = v_x$$

$$x_2 = y \qquad v_2 = v_y$$

$$x_3 = z \qquad v_3 = v_z$$

For the special case where $i = 2$ and $j = 1$, equation (2-17) becomes

$$\tau_{21} = \mu\Delta_{21} = \mu\left[\left(\frac{\partial v_2}{\partial x_1}\right) + \left(\frac{\partial v_1}{\partial x_2}\right)\right]$$

or in x, y, z notation

$$\tau_{yx} = \mu\left[\left(\frac{\partial v_y}{\partial x}\right) + \left(\frac{\partial v_x}{\partial y}\right)\right] \tag{2-18}$$

Table 2-3*

Components of the Stress Tensor $\boldsymbol{\tau}$

Rectangular coordinates:

$$\tau_{xx} = +\mu\left[2\frac{\partial v_x}{\partial x} - \tfrac{2}{3}(\nabla \cdot \boldsymbol{v})\right] + \kappa(\nabla \cdot \boldsymbol{v})$$

$$\tau_{yy} = +\mu\left[2\frac{\partial v_y}{\partial y} - \tfrac{2}{3}(\nabla \cdot \boldsymbol{v})\right] + \kappa(\nabla \cdot \boldsymbol{v})$$

$$\tau_{zz} = +\mu\left[2\frac{\partial v_z}{\partial z} - \tfrac{2}{3}(\nabla \cdot \boldsymbol{v})\right] + \kappa(\nabla \cdot \boldsymbol{v})$$

$$\tau_{xy} = \tau_{yx} = +\mu\left[\frac{\partial v_x}{\partial y} + \frac{\partial v_y}{\partial x}\right]$$

$$\tau_{yz} = \tau_{zy} = +\mu\left[\frac{\partial v_y}{\partial z} + \frac{\partial v_z}{\partial y}\right]$$

$$\tau_{zx} = \tau_{xz} = +\mu\left[\frac{\partial v_z}{\partial x} + \frac{\partial v_x}{\partial z}\right]$$

$$(\nabla \cdot \boldsymbol{v}) = \frac{\partial v_x}{\partial x} + \frac{\partial v_y}{\partial y} + \frac{\partial v_z}{\partial z}$$

Cylindrical coordinates:

$$\tau_{rr} = +\mu\left[2\frac{\partial v_r}{\partial r} - \tfrac{2}{3}(\nabla \cdot \boldsymbol{v})\right] + \kappa(\nabla \cdot \boldsymbol{v})$$

$$\tau_{\theta\theta} = +\mu\left[2\left(\frac{1}{r}\frac{\partial v_\theta}{\partial \theta} + \frac{v_r}{r}\right) - \tfrac{2}{3}(\nabla \cdot \boldsymbol{v})\right] + \kappa(\nabla \cdot \boldsymbol{v})$$

$$\tau_{zz} = +\mu\left[2\frac{\partial v_z}{\partial z} - \tfrac{2}{3}(\nabla \cdot \boldsymbol{v})\right] + \kappa(\nabla \cdot \boldsymbol{v})$$

$$\tau_{r\theta} = \tau_{\theta r} = +\mu\left[r\frac{\partial}{\partial r}\left(\frac{v_\theta}{r}\right) + \frac{1}{r}\frac{\partial v_r}{\partial \theta}\right]$$

$$\tau_{\theta z} = \tau_{z\theta} = +\mu\left[\frac{\partial v_\theta}{\partial z} + \frac{1}{r}\frac{\partial v_z}{\partial \theta}\right]$$

$$\tau_{zr} = \tau_{rz} = +\mu\left[\frac{\partial v_z}{\partial r} + \frac{\partial v_r}{\partial z}\right]$$

$$(\nabla \cdot \boldsymbol{v}) = \frac{1}{r}\frac{\partial}{\partial r}(rv_r) + \frac{1}{r}\frac{\partial v_\theta}{\partial \theta} + \frac{\partial v_z}{\partial z}$$

* Reprinted with permission from Bird, Stewart, and Lightfoot, *Transport Phenomena.*

Table 2-3 (continued)

Spherical coordinates:

$$\tau_{rr} = +\mu\left[2\frac{\partial v_r}{\partial r} - \tfrac{2}{3}(\nabla\cdot v)\right] + \kappa(\nabla\cdot v)$$

$$\tau_{\theta\theta} = +\mu\left[2\left(\frac{1}{r}\frac{\partial v_\theta}{\partial\theta} + \frac{v_r}{r}\right) - \tfrac{2}{3}(\nabla\cdot v)\right] + \kappa(\nabla\cdot v)$$

$$\tau_{\phi\phi} = +\mu\left[2\left(\frac{1}{r\sin\theta}\frac{\partial v_\phi}{\partial\phi} + \frac{v_r}{r} + \frac{v_\theta\cot\theta}{r}\right) - \tfrac{2}{3}(\nabla\cdot v)\right] + \kappa(\nabla\cdot v)$$

$$\tau_{r\theta} = \tau_{\theta r} = +\mu\left[r\frac{\partial}{\partial r}\left(\frac{v_\theta}{r}\right) + \frac{1}{r}\frac{\partial v_r}{\partial\theta}\right]$$

$$\tau_{\theta\phi} = \tau_{\phi\theta} = +\mu\left[\frac{\sin\theta}{r}\frac{\partial}{\partial\theta}\left(\frac{v_\phi}{\sin\theta}\right) + \frac{1}{r\sin\theta}\frac{\partial v_\theta}{\partial\phi}\right]$$

$$\tau_{\phi r} = \tau_{r\phi} = +\mu\left[\frac{1}{r\sin\theta}\frac{\partial v_r}{\partial\phi} + r\frac{\partial}{\partial r}\left(\frac{v_\phi}{r}\right)\right]$$

$$(\nabla\cdot v) = \frac{1}{r^2}\frac{\partial}{\partial r}(r^2 v_r) + \frac{1}{r\sin\theta}\frac{\partial}{\partial\theta}(v_\theta\sin\theta) + \frac{1}{r\sin\theta}\frac{\partial v_\phi}{\partial\phi}$$

Note that if v_y should be zero, as in plane parallel shear, equation (2-18) reduces to

$$\tau_{yx} = \mu\left(\frac{\partial v_x}{\partial y}\right)$$

The mathematical formulation of a flow problem involves the rheological equation of the fluid, the equations of continuity, momentum, energy, and the equation of state of the fluid. The solution consists of functions that satisfy these equations and, also, the specific boundary conditions of the problem.

To simplify the formulation of Newtonian flow problems, Table 2-3 gives the components of equation (2-17) in rectangular, cylindrical, and spherical coordinates. The formulation and solution of a simple Newtonian flow problem are illustrated by the following example:

Example 2-2.

This example considers steady laminar flow of an incompressible Newtonian fluid through a long tube of circular cross section having radius R. The tube wall is maintained at the constant temperature T_w. The problem is to calculate the velocity and temperature profiles across the tube at a point far enough from the entrance that neither temperature nor velocity is a function of the axial position coordinate z. In order to simplify the problem, it is assumed that the viscosity μ is constant.

In this problem, all derivatives of temperature, velocity, and stress components with respect to θ, z, and t are zero, the velocity components v_θ and v_r are zero, and because the fluid is incompressible $(\nabla \cdot v) = 0$. The equations of momentum and energy, obtained from Table 2-2, reduce to the following:

$$\left(\frac{\partial P}{\partial z}\right) = \frac{1}{r}\frac{\partial}{\partial r}[r\tau_{rz}] \tag{2-19}$$

$$\frac{1}{r}\frac{\partial}{\partial r}[rq_r] = \tau_{rz}\left(\frac{\partial v_z}{\partial r}\right) \tag{2-20}$$

Upon introducing expressions for τ_{rz} and q_r, obtained from Tables 2-1 and 2-3, the above equations become

$$\frac{\partial P}{\partial z} = \frac{\mu}{r}\frac{\partial}{\partial r}\left[r\left(\frac{\partial v_z}{\partial r}\right)\right] \tag{2-21}$$

$$-\frac{k}{r}\frac{\partial}{\partial r}\left[r\left(\frac{\partial T}{\partial r}\right)\right] = \mu\left(\frac{\partial v_z}{\partial r}\right)^2 \tag{2-22}$$

The boundary conditions for the problem are

(1) $v_z(R) = 0$ (3) $v_z'(0) = 0$ where $v_z' = \left(\frac{\partial v_z}{\partial r}\right)$

(2) $T(R) = T_w$ (4) $T'(0) = 0$ where $T' = \left(\frac{\partial T}{\partial r}\right)$

Because its left side does not depend upon r, equation (2-21) can be integrated directly to yield

$$\left(\frac{\partial v_z}{\partial r}\right) = \frac{r}{2\mu}\left(\frac{\partial P}{\partial z}\right) + \frac{C_1}{r} \tag{2-23}$$

and, because of the third boundary condition, the constant of integration C_1 must be zero.

The velocity profile, obtained by integrating equation (2-23), is

$$v_z = \frac{r^2}{4\mu}\left(\frac{\partial P}{\partial z}\right) + C_2 \tag{2-24}$$

which becomes, upon evaluating C_2 from the first boundary condition,

$$v_z = -\frac{R^2}{4\mu}\left(\frac{\partial P}{\partial z}\right)\left[1 - \left(\frac{r}{R}\right)^2\right] \tag{2-25}$$

The negative sign indicates that the fluid flows in the direction of decreasing pressure.

The temperature profile is obtained by introducing equation (2-23) into (2-22) and integrating. One obtains after a single integration

$$\left(\frac{\partial T}{\partial r}\right) = -\frac{1}{4\mu k}\left(\frac{\partial P}{\partial z}\right)^2\frac{r^3}{4} + \frac{C_3}{r} \tag{2-26}$$

and, because of the fourth boundary condition, the constant of integration C_3 is zero. The integration of equation (2-26) gives

$$T = -\frac{1}{4\mu k}\left(\frac{\partial P}{\partial z}\right)^2 \frac{r^4}{16} + C_4$$

which becomes, upon evaluating C_4 from boundary condition 3,

$$T - T_w = \frac{R^4}{64\mu k}\left(\frac{\partial P}{\partial z}\right)^2\left[1 - \left(\frac{r}{R}\right)^4\right] \tag{2-27}$$

The velocity of the fluid at the tube center, obtained by placing $r = 0$ in equation (2-25), is given by the equation

$$v_0 = -\frac{R^2}{4\mu}\left(\frac{\partial P}{\partial z}\right)$$

and the temperature of the fluid at the center of the tube, obtained from equation (2-27), is given by

$$T_0 - T_w = \frac{R^4}{64\mu k}\left(\frac{\partial P}{\partial z}\right)^2 = \left(\frac{\mu v_0^2}{4k}\right)$$

If, for example, the fluid velocity at the center of the tube should be 10 cm/sec and if the fluid should have a viscosity of 10^4 poise and a thermal conductivity of 10^{-3} cal/cm-sec-deg C, then the fluid temperature at the tube center would be 6 deg C higher than at the tube wall.

Equations (2-25) and (2-27) can be put in dimensionless form by dividing by v_0 and T_0, respectively. One obtains

$$\frac{v}{v_0} = 1 - \left(\frac{r}{R}\right)^2$$

$$\frac{T - T_w}{T_0 - T_w} = 1 - \left(\frac{r}{R}\right)^4$$

Figure 2-4 shows a plot of these dimensionless quantities as a function of the dimensionless position variable (r/R).

2-3 Rheological Equations

The response of a material to an applied stress can, in principle, be predicted if the rheological equation of the material is known. For example, the rheological equation discussed in the previous section, equation (2-17), when combined with the equations of continuity, momentum, and energy, enables one to formulate flow problems for Newtonian fluids.

The following discussion is concerned with rheological equations of some other classes of materials. To simplify it, we shall limit it to the special case where the material is subjected to plane parallel (simple) shear. A general treatment applicable to inelastic incompressible non-Newtonian fluids is presented in Section 2-5.

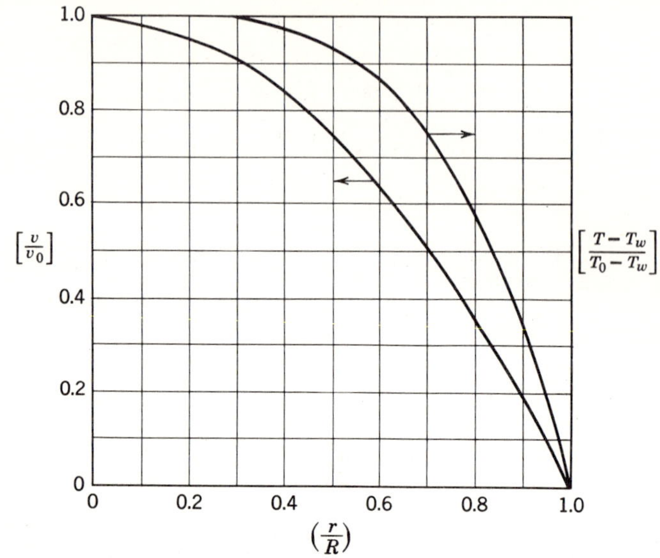

Figure 2-4. Dimensionless velocity and temperature profiles for Newtonian flow through a long tube (Example 2-2).

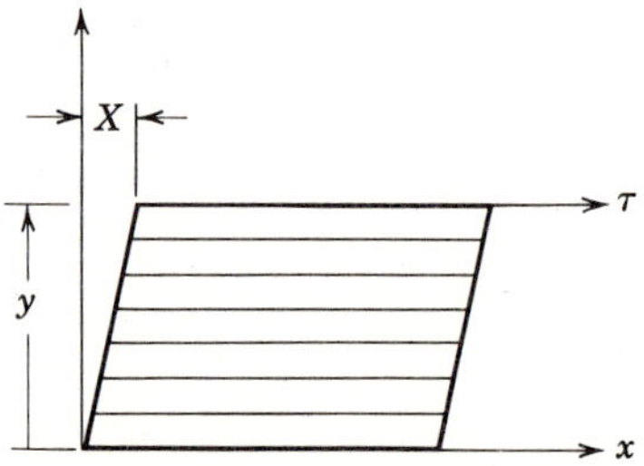

Figure 2-5. Simple shear. Elements of the body are displaced in the x-direction only.

In simple shear, all elements of a body being deformed are displaced in the same direction. This is illustrated by Figure 2-5 where the coordinate axes are so oriented that all displacements occur in the x-direction.

We consider, first, the Newtonian fluid in simple shear. With the coordinate axes oriented so that the only nonzero component of velocity is v_x, its rheological equation, (2-17), reduces to

$$\tau_{yx} = \mu\left(\frac{\partial v_x}{\partial y}\right) \tag{2-28}$$

and noting that v_x is the time derivative of the displacement X in the x-direction, equation (2-28) becomes

$$\tau_{yx} = \mu \frac{\partial}{\partial y}\left(\frac{\partial X}{\partial t}\right) = \mu \frac{\partial}{\partial t}\left(\frac{\partial X}{\partial y}\right) \tag{2-29}$$

Letting γ represent the quantity $(\partial X/\partial y)$, equation (2-29) can be written

$$\tau = \mu \frac{\partial \gamma}{\partial t} = \mu\dot{\gamma} \tag{2-30}$$

where the subscript on τ has been dropped in order to simplify the notation. This does not introduce any ambiguity because only this single stress component is considered in the following discussion. The quantity γ is a measure of strain at a point in a material subjected to simple shear, and its time derivative $\dot{\gamma}$ represents the rate of strain at that point.

Example 2-3.

Consider a Newtonian fluid contained between parallel plates separated by the distance H, as shown in Figure 2-2. The upper plate is free to move, but the lower plate is fixed in its position. At time zero, a constant stress S applied to the upper plate sets it in motion. The fluid, which was initially stationary, eventually reaches the steady state velocity profile shown in the figure. In this example, an analysis of the transient or start-up period is made for isothermal conditions.

A rectangular coordinate system oriented so that v_x is the only nonzero velocity component is used. The differential equation for the motion is obtained by combining the momentum equation (from Table 2-2)

$$\rho \frac{\partial v_x}{\partial t} = \frac{\partial \tau_{yx}}{\partial y}$$

with the rheological equation for the Newtonian fluid

$$\tau_{yx} = \mu\left(\frac{\partial v_x}{\partial y}\right)$$

to obtain

$$\frac{\partial v_x}{\partial t} = \nu \frac{\partial^2 v_x}{\partial y^2}$$

where ν, the kinematic viscosity of the fluid is defined as (μ/ρ).

The boundary conditions for the problem are as follows:

BC 1: $v_x(0, t) = 0$

BC 2: $v_x(y, 0) = 0$

BC 3: $v_x'(H, t) = S/\mu$ where $v_x' = \left(\dfrac{\partial v_x}{\partial y}\right)$

The first boundary condition is a mathematical statement that the lower plate is stationary, the second that the fluid is initially at rest, and the third that the constant stress S is applied at the upper plate.

It is easily verified that the function

$$v_x = \frac{Sy}{\mu} - \sum_{n=1,3,\cdots}^{\infty} C_n\, e^{-\nu n^2\pi^2 t/4H^2} \sin\left(\frac{n\pi y}{2H}\right)$$

is a solution of the above differential equation and also satisfies the first and third boundary conditions. It can also be shown that the functions

$$\left\{\sin\left(\frac{n\pi y}{2H}\right)\right\} \qquad n = 1, 3, \cdots$$

are orthogonal in the interval $(0, H)$. Hence the following formula for the C_n, which enables the second boundary condition to be satisfied, can be derived:

$$C_n = \frac{SH}{\mu}\frac{8}{n^2\pi^2} \sin\left(\frac{n\pi}{2}\right) \qquad n = 1,3, \cdots$$

and the equation for v_x becomes

$$v_x = \frac{HS}{\mu}\left[\left(\frac{y}{H}\right) - \frac{8}{\pi^2}\sum_{n=1,3,\cdots}^{\infty} \frac{\sin(n\pi/2)}{n^2}\, e^{-\nu n^2\pi^2 t/4H^2} \sin\left(\frac{n\pi y}{2H}\right)\right]$$

To determine the shear strain γ as a function of time and position, we note that displacement $X(y, t)$ is given by the integral

$$X(y, t) = X(y, 0) + \int_0^t v_x\, dt$$

where $X(y, 0)$ is the arbitrary initial or reference displacement. Placing $X(y, 0)$ equal to zero and differentiating the above expression with respect to y gives the shear strain

$$\gamma = \frac{\partial X}{\partial y} = \frac{\partial}{\partial y}\int_0^t v_x\, dt$$

Noting that the above integral contains y as a parameter (through v_x), it becomes

$$\gamma = \int_0^t \left(\frac{\partial v_x}{\partial y}\right) dt$$

Upon evaluating the above integral, one obtains

$$\gamma = \frac{St}{\mu}\left[1 + \frac{4}{\pi\Phi}\sum_{n=1,3,\cdots}^{\infty} \frac{\sin(n\pi/2)}{n^3}\, e^{-n^2\Phi} \cos\left(\frac{n\pi y}{2H}\right)\right]$$

where

$$\Phi = \frac{\nu\pi^2 t}{4H^2}$$

When $y = H$, and also for large values of t, the strain equation reduces to

$$\gamma = \frac{St}{\mu}$$

Hence the quantity (St/μ) represents the strain at the *upper plate* for all values of t, and the strain at all positions in the fluid when t becomes very large.

Examination of the rheological equation for the Newtonian fluid

$$\tau_{yx} = \mu\left(\frac{\partial v_x}{\partial y}\right) = \mu\left(\frac{\partial \gamma}{\partial t}\right)$$

shows that it can be integrated directly to give the result

$$\gamma = \frac{St}{\mu}$$

Hence the strain-time behavior at the upper plate can be determined from the rheological equation for the material. Only if one is interested in the strain at a point removed from the upper plate is it necessary to introduce the equation of motion for the fluid into the analysis.

As a numerical example, consider the strain at the point $y = H/2$. If the dimensionless group Φ has reached a value of 2, then the strain calculated with the above equation differs from that calculated directly from the rheological equation by the factor 1.06. If the fluid has a viscosity of 10 poise, a density of 1 gram/cm^3, and if the plate separation is 0.1 cm, then Φ has a value of 2 when $t = 2.5$ milliseconds. In other words, the shear strain at the mid-plate is within 6% of its equilibrium value 2.5 milliseconds after the stress is applied at the upper plate.

Another class of materials, *Hookean solids*, has the rheological equation

$$\tau = G\gamma \tag{2-31}$$

where G, a material constant, is the modulus of rigidity (shear modulus). Note that in Hookean solids there is a linear relationship between stress and strain while in Newtonian fluids there is a linear relationship between stress and rate of strain.

The Hookean solid and the Newtonian fluid represent two extremes in the rheological classification of materials. Many intermediate classifications exist because most materials are neither purely elastic nor purely viscous but exhibit both elastic and viscous effects. Such materials are said to be *viscoelastic*. Rheological equations for viscoelastic materials contain a minimum of two parameters, one to characterize elastic behavior and the other, viscous behavior.

The rheological equation for the *Maxwell fluid* is derived under the assumption that in simple shear the total strain γ, at a point in the fluid where the stress is τ, is the sum of an elastic strain γ_e and a flow deformation γ_f. Therefore

$$\gamma = \gamma_e + \gamma_f \tag{2-32}$$

which, upon differentiation with respect to time, becomes

$$\dot{\gamma} = \dot{\gamma}_e + \dot{\gamma}_f \tag{2-33}$$

Furthermore, the elastic response of the Maxwell fluid is Hookean and the viscous response is Newtonian. Upon introducing the rheological equations (2-30) and (2-31) into the above equation and rearranging, one obtains the rheological equation for the Maxwell fluid:

$$\tau = \mu\dot{\gamma} - \left(\frac{\mu}{G}\right)\dot{\tau} \tag{2-34}$$

The coefficient of the $\dot{\tau}$ term in equation (2-34) has dimensions of time. It is shown below that it represents the time required for a stress in a Maxwell fluid to decay or relax to 37% of its initial value if the fluid is held at constant strain. Hence it is often called the relaxation time.

For constant strain, the rheological equation (2-34) becomes

$$\dot{\tau} + \left(\frac{1}{\lambda}\right)\tau = 0$$

where λ represents the relaxation time (μ/G). The general solution of the above equation is written

$$\tau = C_1 e^{-t/\lambda}$$

where the constant of integration C_1 represents the stress S at time zero. Therefore

$$\frac{\tau}{S} = e^{-t/\lambda}$$

and we see that for $t = \lambda$

$$\frac{\tau}{S} = \frac{1}{e} = 0.37$$

Hence λ represents the time for the stress to decay by a factor of $1/e$.

The rheological equation for the *Voigt solid* is derived under the assumption that in simple shear, the total stress τ at a point in the material where the strain is γ is the sum of the stresses arising from the elasticity of the fluid τ_e and from the viscosity of the fluid τ_v. Therefore

$$\tau = \tau_e + \tau_v \tag{2-35}$$

and because the elastic response is Hookean and the viscous response Newtonian, the rheological equation for the Voigt solid is

$$\tau = \mu\dot{\gamma} + G\gamma \tag{2-36}$$

The rheological equation of the *Bingham plastic*, derived under the assumption that the material is perfectly rigid below a certain yield stress and a viscous fluid above the yield stress, is

$$\gamma = 0 \qquad |\tau| < S_0 \tag{2-37}$$

$$(\tau \pm S_0) = \zeta\dot{\gamma} \qquad |\tau| > S_0 \tag{2-38}$$

where ζ is a parameter (the plastic viscosity) that characterizes the flow behavior of the material and S_0 is the yield stress. The minus sign is used when τ is positive and the positive sign when τ is negative.

The rheological equation of the Newtonian fluid is a special case of the more general rheological equation

$$\tau = \eta\dot{\gamma} \tag{2-39}$$

where the coefficient η can be a function of shear rate (or of shear stress if it is selected as the independent variable). If η is constant, then equations (2-39) and (2-30) are identical and equation (2-39) becomes the rheological equation of a Newtonian fluid. If η is not independent of the state of shear, then the fluid is said to be non-Newtonian. Non-Newtonian fluids are classified and various empirical equations relating η to $\dot{\gamma}$ presented in Section 2-4.

Example 2-4.

In this example, intended to illustrate the difference in the nature of the materials discussed above, the material is considered to be contained between parallel plates, as in Example 2-3. A particular sequence of stresses is then applied to the upper plate and the resulting strain, at the upper plate, is calculated.

The stress-time history at the upper plate

I	$\tau = 0$	$(0 < t < t_1)$
II	$\tau = S$	$(t_1 < t < t_2)$
III	$\tau = 0$	$(t_2 < t < t_3)$
IV	$\tau = 2S$	$(t_3 < t < t_4)$
V	$\tau = 0$	$(t > t_4)$

is shown graphically in Figure 2-6*a*. The periods I through V are of equal duration.

Consider first the response, shown in Figure 2-6*b* and 2-6*c*, of the Newtonian and the non-Newtonian fluid. During period I, neither material is strained because the applied stress is zero. In period II, where the applied stress is S, the rheological equation for the Newtonian fluid shows that its rate of strain is (S/μ). In this period, the rate of strain of the non-Newtonian fluid is (S/η_{II}) where η_{II} is the non-Newtonian viscosity corresponding to the stress S. Upon removal of the stress in period III, both fluids retain their maximum strain, $(S\Delta t/\mu)$ for the Newtonian and $(S\Delta t/\eta_{II})$ for the non-Newtonian fluid. In period IV, where the applied stress is twice that in period II, the rate of strain of the Newtonian fluid is doubled over that in period II, but the rate of strain of the non-Newtonian fluid $(2S/\eta_{IV})$ is not twice its rate in II because η_{II} and η_{IV}, being functions of stress, are different. Upon removal of the stress at the start of period V, the total flow deformation for the Newtonian fluid is $(3S\Delta t/\mu)$ and for the non-Newtonian fluid is

$$S\,\Delta t\left[\frac{1}{\eta_{II}} + \frac{2}{\eta_{IV}}\right]$$

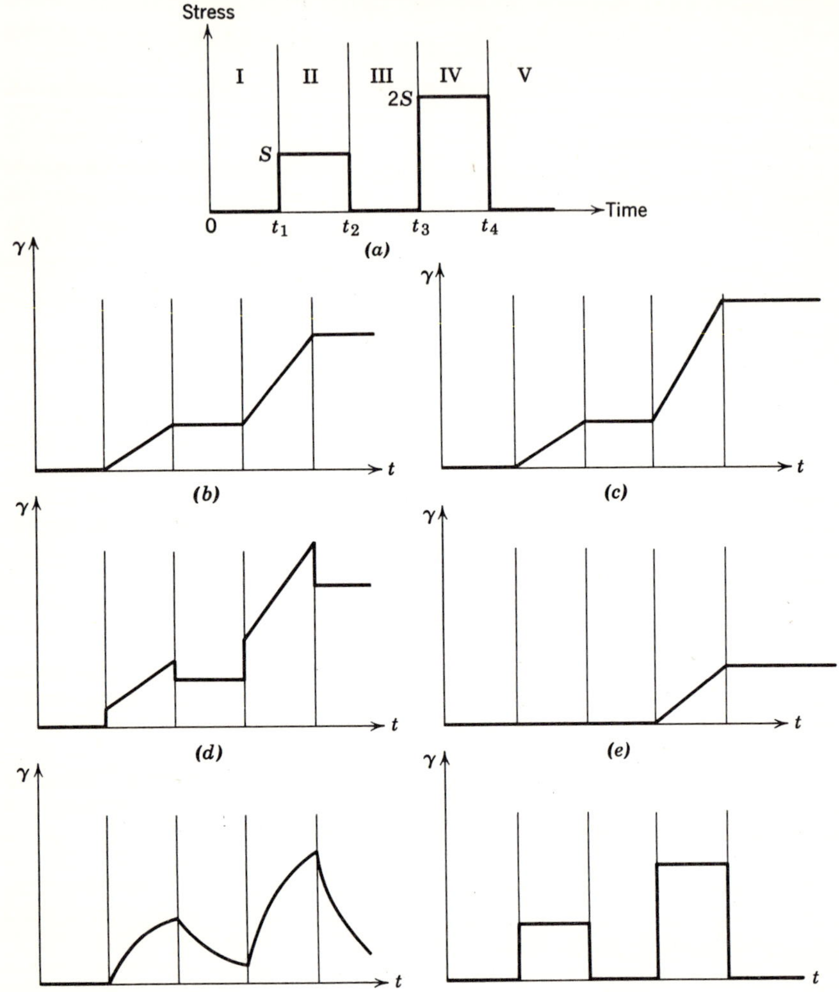

Figure 2-6. Strain history in various materials subjected to stress history shown in (*a*) Diagram, (*b*) Newtonian fluid, (*c*) Non-Newtonian fluid, (*d*) Maxwell fluid, (*e*) Bingham plastic, (*f*) Voigt solid, (*g*) Hookian solid.

The response of the Maxwell fluid is shown in Figure 2-6*d*. The strain in period II is obtained by integrating equation (2-34) with respect to time. One obtains

$$\gamma = \frac{1}{\mu}\int_{t_1}^{t} \tau \, dt + \frac{\tau}{G} \qquad (t_1 < t < t_2) \tag{2-40}$$

which becomes

$$\gamma = \frac{S}{\mu}(t - t_1) + \frac{S}{G} \qquad (t_1 < t < t_2) \tag{2-41}$$

upon introducing S for τ. Equation (2-41) shows that the material undergoes an initial instantaneous strain (S/G) followed by flow, which is characterized by the constant rate of strain (S/μ).

Upon removal of the stress at time t_2, the total flow deformation ($S\Delta t/\mu$) is retained, but the elastic deformation (S/G) is recovered instantaneously, as shown in the figure. The strain as a function of time in period IV is given by the equation

$$\gamma = \gamma_i + \frac{1}{\mu}\int_{t_3}^{t} \tau\, dt + \frac{\tau}{G} \qquad (t_3 < t < t_4) \tag{2-42}$$

where γ_i the initial strain at time t_3 is ($S\Delta t/\mu$). Equation (2-42) is analogous to (2-41). Introducing ($2S$) for the stress into it, one obtains

$$\gamma = \frac{S\Delta t}{\mu} + \frac{2S}{G} + \frac{2S}{\mu}(t - t_3) \qquad (t_3 < t < t_4) \tag{2-43}$$

which shows that the fluid in period IV is characterized by a rate of strain ($2S/\mu$). Upon removal of the stress at time t_4, the total flow deformation ($3S\Delta t/\mu$) is retained, but the elastic deformation ($2S/G$) is recovered.

The response of a Bingham plastic for which

$$(S < S_0 < 2S)$$

is shown in Figure 2-6*e*. There is no strain in period II because the applied stress is less than the yield stress of the material. In period IV, the applied stress exceeds the yield stress, and the resulting flow is characterized by the rate of strain ($2S - S_0/\zeta$).

The response of the Voigt solid is shown in Figure 2-6*f*. The strain time curve for period II is obtained by integrating its rheological equation. Assuming τ to be constant, the general solution of equation (2-36) is

$$\gamma = Ce^{-Gt/\mu} + \frac{\tau}{G} \tag{2-44}$$

where C is the arbitrary constant of integration. For period II, it is evaluated from the condition $\gamma(t_1) = 0$ so that equation (2-44) becomes

$$\gamma = \frac{S}{G}[1 - e^{-G(t-t_1)/\mu}] \tag{2-45}$$

upon replacing τ with S. Note that as $t \to \infty$, the strain approaches its equilibrium value (S/G).

At the end of period II, the total strain γ_2 is

$$\gamma_2 = \frac{S}{G}[1 - e^{-G\Delta t/\mu}] \tag{2-46}$$

Upon removal of the stress at time t_2, this strain is eventually recovered but at a rate that decreases exponentially with time. The equation of the strain recovery curve is obtained from equation (2-45) by evaluating C from the condition $\gamma(t_2) = \gamma_2$. One obtains

$$\gamma = \gamma_2 e^{-G(t-t_2)/\mu} \tag{2-47}$$

A similar analysis will yield the equations of the strain time curves for periods IV and V.

The response of the Hookean solid is shown in Figure 2-6*g*. Upon application of the stress S at time t_1, the solid is instantaneously strained an amount (S/G). This strain is recovered instantaneously at time t_2 when the stress is removed. At time t_3, when the stress $2S$ is applied, the solid is instantaneously strained an amount $(2S/G)$.

Some of the above materials have been referred to as fluids and others as solids. Referring to Figure 2-6, it can be seen that a stress causes the strain in a *fluid* to increase continuously, irrecoverably, and indefinitely. With *solids*, a stress causes an elastic deformation so that upon removal of the stress, the strain is recovered, in some materials instantaneously and in others at a finite rate. The Bingham plastic falls into neither category. Stresses smaller than the yield stress cause no deformation of the material and thus it is not a fluid by the above criterion. On the other hand, stresses greater than the yield stress cause flow; hence it cannot be called a solid.

It should be noted that the apparent behavior of a material depends upon the experimental conditions under which the material is tested. For example, consider a Voigt solid tested by applying a constant stress S and measuring the strain as a function of time. Equation (2-45) describes the observed behavior. If the timescale of the experimental measurement is very large, equation (2-45) reduces to

$$\gamma = \frac{S}{G} \tag{2-48}$$

which is the rheological equation of the Hookean solid. Hence the Voigt solid appears to be a Hookean solid if the measurements are made over a long period of time.

The rate of strain, found by differentiating equation (2-45), is

$$\dot{\gamma} = \frac{S}{\mu} e^{-G(t-t_1)/\mu} \tag{2-49}$$

If the timescale of the experimental measurements is very *small*, equation (2-49) reduces to

$$\dot{\gamma} = \frac{S}{\mu} \tag{2-50}$$

which is the rheological equation of the Newtonian fluid. Hence the Voigt solid appears to be a Newtonian fluid if the measurements are made over a short period of time.

The behavior in simple shear (stress-strain-time at a point) of the materials discussed above can be deduced, qualitatively, from mechanical and electrical analogs of the rheological equations. Analogs or models of viscoelastic materials contain a minimum of two elements or parameters, one to characterize the viscous and the other the elastic behavior of the material. Models of more complex viscoelastic materials are made by combining additional model parameters or elements. Analogs for the simple materials discussed above are developed here. Analogs for more complex viscoelastic behavior are developed in Section 2-7.

Consider a spring that elongates instantaneously, elastically, and in direct proportion to the applied force F. Its deformation is described by the equation

$$F = k_1 X \tag{2-51}$$

where k_1 is the spring constant. If one compares this equation with the rheological equation for the Hookean solid

$$\tau = G\gamma \tag{2-52}$$

it is clear that the spring is a mechanical analog of the Hookean solid in simple shear, provided one identifies the force F with the stress τ and the displacement X with the strain γ.

Similarly, if the equation for an electrical capacitor

$$Q = CV \tag{2-53}$$

is written in the form

$$V = k_2 Q \tag{2-54}$$

where k_2 is a constant, it is clear that the capacitor is an electrical analog of the Hookean solid in simple shear.

The mechanical analog of the Newtonian fluid in simple shear is a dashpot. A dashpot consists of a piston contained within a cylinder but separated from it by a layer of a viscous liquid. The rate $\dot{X}$ at which the piston can be pulled through the cylinder is directly proportional to the applied force F, so that

$$F = k_3 \dot{X} \tag{2-55}$$

where k_3 is a constant. Comparison of equation (2-55) with (2-30) shows that the dashpot is the mechanical analog of the Newtonian fluid in simple shear if one identifies the rate of strain $\dot{\gamma}$ with the rate of displacement of the cylinder $\dot{X}$.

If the equation for an electrical resistor

$$V = IR \tag{2-56}$$

is written in the form

$$V = k_4\dot{Q} \tag{2-57}$$

and then compared with equation (2-30), it can be seen that the resistor is the electrical analog of the Newtonian fluid.

Mechanical and electrical analogs of the rheological equations for viscoelastic materials in simple shear can be made by combining capacitors

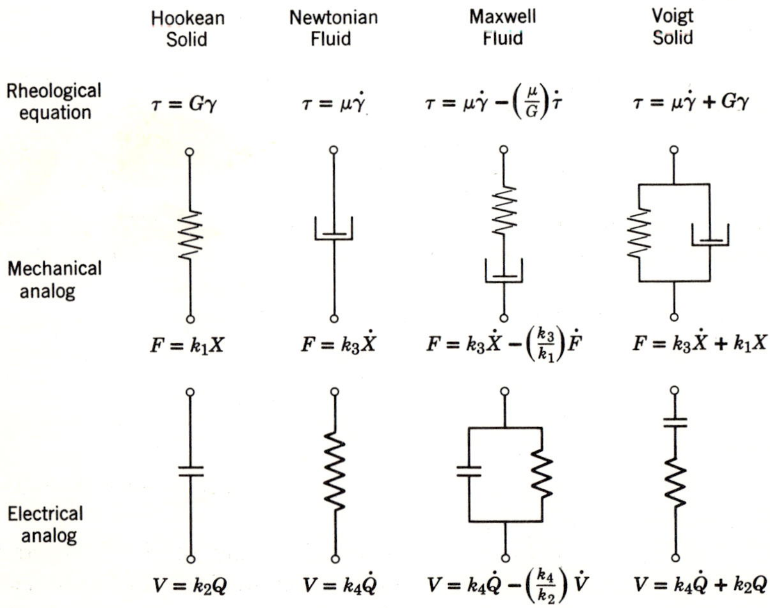

Figure 2-7. Mechanical and electrical analogs of rheological equations.

and resistors, and springs and dashpots. For example, consider a series connection of a capacitor and resistor. The total potential drop V, obtained by adding the potential drop across each element, is

$$V = V_r + V_c \tag{2-58}$$

and upon introducing equations (2-57) and (2-54), one obtains

$$V = k_4\dot{Q} + k_2 Q \tag{2-59}$$

which is analogous to the rheological equation for the Voigt solid in simple shear

$$\tau = \mu\dot{\gamma} + G\gamma \tag{2-60}$$

It is easily shown that the mechanical analog of the Voigt solid is the parallel coupling of the spring and dashpot. Similarly, the mechanical analog of the Maxwell fluid is the series coupling of the spring and dashpot, while the electrical analog is a parallel coupled resistor and capacitor. These results are summarized in Figure 2-7.

2-4 Non-Newtonian Fluids

In this section, non-Newtonian fluids are classified and various empirical and theoretical equations relating the non-Newtonian viscosity η to shear rate (or shear stress) for isothermal simple shear are discussed. The effect of temperature on η is discussed in Section 2-6. We distinguish, first, between those fluids for which η depends on the previous shear history of the sample as well as on the state of shear and those for which η depends only on the state of shear.

The first class of fluids is said to have a time dependent viscosity and is classified on the basis of whether the time effects are reversible or irreversible. If the time effects are reversible, the fluids are either *thixotropic* or *rheopectic*. If a thixotropic fluid that has been at rest for a long period of time is sheared at a constant rate, its viscosity η will decrease with time, eventually reaching an equilibrium value corresponding to the particular state of shear. With rheopectic fluids, η increases to reach eventually an equilibrium value.

Those fluids for which η depends only on the instantaneous state of shear are either *pseudoplastic* or *dilatant*. With pseudoplastic fluids, η decreases with increasing τ or $\dot{\gamma}$, and with dilatant fluids, η increases with increasing τ or $\dot{\gamma}$. The fluids dealt with in polymer processing, primarily molten polymers, are usually pseudoplastic.

Table 2-4 presents a summary of viscosity relations applicable to non-Newtonian fluids in isothermal simple shear. The first relation, that of Newton, is the special case where η is independent of the state of shear. τ° and $\dot{\gamma}^\circ$ in Table 2-4 represent values of shear stress and shear rate in an arbitrary standard state and η° is the non-Newtonian viscosity in this state. μ_0 represents the limiting Newtonian viscosity as the shear rate (or shear stress) approaches zero, while μ_∞ represents the limiting Newtonian viscosity as the shear rate (or shear stress) becomes infinitely large. The quantities m, n, A, B, α, and β are parameters that depend upon the nature of the fluid. The functions $f(\alpha\dot{\gamma})$ and $g(\beta\dot{\gamma})$ are shown graphically in Figure 2-8.

The second relation, sometimes referred to as the Ostwald-deWaele equation[4] but usually called the power law, is an empirical equation having two adjustable parameters. It is the most widely used viscosity relation for non-Newtonian fluids and has found many important engineering applications. It has been discussed in some detail by Reiner.[5]

Table 2-4

Summary of Viscosity Relations for Non-Newtonian Fluids in Simple Shear

(1) Newton
$$\eta = \mu$$

(2) Power law
$$\eta = \eta^\circ \left|\frac{\dot{\gamma}}{\dot{\gamma}^0}\right|^{n-1} = \eta^\circ \left|\frac{\tau}{\tau^\circ}\right|^{(n-1)/n}$$

(3) Reiner-Philippoff
$$\eta = \mu_\infty + \left[\frac{\mu_0 - \mu_\infty}{1 + (\tau/A)^2}\right]$$

(4) Ellis
$$\eta = \frac{\mu_0}{1 + \dfrac{\mu_0}{\eta^\circ}\left|\dfrac{\tau}{\tau^\circ}\right|^{m-1}}$$

(5) Powell-Eyring
$$\eta = \mu_0\left[\frac{(\tau/B)}{\sinh(\tau/B)}\right]$$
$$\eta = \mu_0\left[\frac{\operatorname{arcsinh}(\mu_0\dot{\gamma}/B)}{(\mu_0\dot{\gamma}/B)}\right]$$

(6) Bueche
$$\eta = \mu_0 f(\alpha\dot{\gamma})$$

(7) Rouse
$$\eta = \mu_0 g(\beta\dot{\gamma})$$

In the power law $\dot{\gamma}^\circ$ and τ° represent values of shear rate and shear stress in an arbitrarily chosen standard state and η° is the non-Newtonian viscosity of the fluid in this state. For convenience, $\dot{\gamma}^\circ$ is usually chosen as 1 $\sec^{-1}$ or τ° as 1 dyne/cm^2. The parameter n is the flow index for the fluid. When n is less than unity, the viscosity of the fluid decreases with increasing shear rate (or shear stress). Hence the power law describes the behavior of a pseudoplastic fluid when $n < 1$ and a dilatant fluid when $n > 1$.

Absolute values are used in the power law because $\dot{\gamma}$ (or τ) can be either positive or negative, but η must always be positive. Note that when the flow index n is unity, the power law reduces to Newton's law. Hence the extent of deviation of n from unity is an indication of the non-Newtonian nature of a fluid. It should also be noted that the power law requires the

[4] Bird, Stewart, and Lightfoot, *op. cit.*, p. 11.

[5] M. Reiner, *Deformation, Strain, and Flow*, Interscience Publishers, 1960, p. 243.

viscosity of pseudoplastic fluids to become infinitely large as $\dot{\gamma}$ (or τ) approaches zero, and for η to approach zero as $\dot{\gamma}$ (or τ) becomes infinitely large. The converse is true for dilatant fluids.

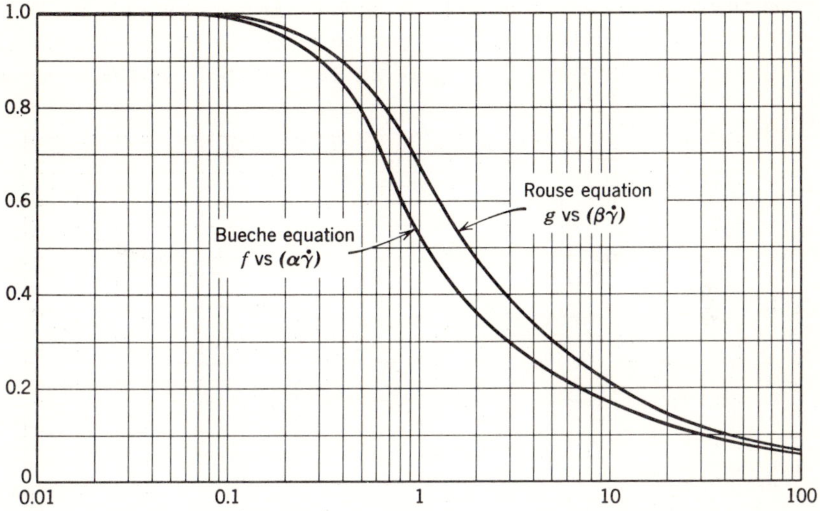

Figure 2-8. "Master curves" for the functions f and g of the Bueche and Rouse equations.

Because of the general relation, equation (2-39), the relationship between shear stress and shear rate for a power law fluid is

$$\pm\tau = \tau^\circ \left| \frac{\dot{\gamma}}{\dot{\gamma}^\circ} \right|^n \tag{2-61}$$

where the sign of τ must be the same as $\dot{\gamma}$. If $\dot{\gamma}^\circ = 1$ and if $\dot{\gamma}$ is positive, then the logarithm of equation (2-61) is

$$\log \tau = \log \tau^\circ + n \log \dot{\gamma}$$

showing that a plot of log (τ) versus log ($\dot{\gamma}$), which is called the logarithmic flow curve, will be a straight line for those fluids that obey the power law.

Logarithmic flow curves of molten polymers often appear to be straight if restricted to one or two decades of shear rate, but when extended over several decades, they are usually not straight. The logarithmic flow curve[6] for molten polyethylene ("Alathon-14" at 126° C) shown in Figure 2-9 is similar to those of many molten polymers. At very low shear rates, the

[6] T. G. Smith, "Development of a Rheometer for Molten Thermoplastics," doctoral thesis, Washington University, St. Louis, 1960.

curve becomes straight and has a slope of unity, indicating Newtonian behavior in this region. At shear rates above about 0.01 sec^{-1}, the slope of the curve is less than unity, indicating pseudoplastic behavior. Furthermore, the slope of the curve continually decreases, indicating a greater deviation from Newtonian behavior at high shear rates than at low.

Although it is clearly impossible to describe the entire curve shown in Figure 2-9 with the power law, portions can be approximated with it.

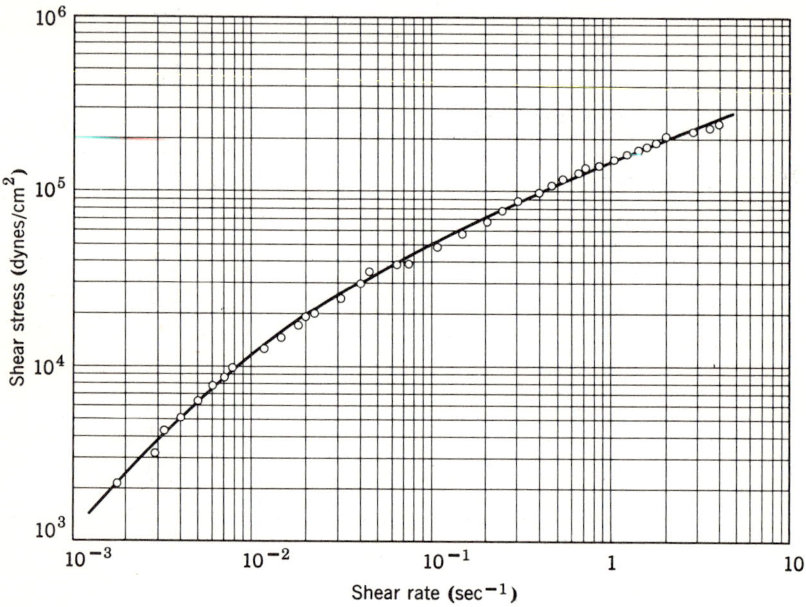

Figure 2-9. Logarithmic flow curve for "Alathon-14" polyethylene resin at 126°C (Ref. 6).

For example, in the region between 0.1 and 1.0 sec^{-1}, the straight line that best fits the experimental points has a slope of about 0.59. Choosing $\dot{\gamma}^\circ$ as 1 sec^{-1}, the value of τ° is 1.5×10^5 dynes/cm^2. Hence η° is 1.5×10^5 poise, and the viscosity equation becomes

$$\eta = \eta^\circ \left| \frac{\dot{\gamma}}{\dot{\gamma}^\circ} \right|^{n-1} = \frac{1.5 \times 10^5}{(\dot{\gamma})^{0.41}}$$

and this can be used in the range from 0.1 to 1.0 sec^{-1} with less than about 5% error.

The Reiner-Philippoff equation,[7] which has three parameters, μ_0, μ_∞,

[7] Reiner, *op. cit.*, p. 259.

and A, requires η to decrease with increasing τ. Hence it is applicable only to pseudoplastic fluids. Limiting values of the viscosity and the slope of the $\eta - \tau$ curve are as follows:

$$\lim_{\tau \to 0} \eta = \mu_0 \qquad \lim_{\tau \to 0} \left(\frac{d\eta}{d\tau}\right) = 0$$

$$\lim_{\tau \to \infty} \eta = \mu_\infty \qquad \lim_{\tau \to \infty} \left(\frac{d\eta}{d\tau}\right) = 0$$

The Ellis equation,[8] which has three parameters, m, η°, and μ_0, describes the behavior of a pseudoplastic fluid when $m > 1$ and a dilatant fluid when $m < 1$. If $m = 1$, the equation reduces to Newton's law. Limiting values of the viscosity and the slope of the $\eta - \tau$ curve are as follows:

$$\lim_{\tau \to 0} \eta = \mu_0 \qquad \lim_{\tau \to 0} \left(\frac{d\eta}{d\tau}\right) = 0$$

$$\lim_{\tau \to \infty} \eta = 0 \qquad \lim_{\tau \to \infty} \left(\frac{d\eta}{d\tau}\right) = 0$$

The Powell-Eyring equation[9] has two parameters, μ_0 and B, and requires η to decrease with increasing τ. Limiting values of the viscosity and the slope of the $\eta - \tau$ curve are as follows:

$$\lim_{\tau \to 0} \eta = \mu_0 \qquad \lim_{\tau \to 0} \left(\frac{d\eta}{d\tau}\right) = 0$$

$$\lim_{\tau \to \infty} \eta = 0 \qquad \lim_{\tau \to \infty} \left(\frac{d\eta}{d\tau}\right) = 0$$

The Bueche[10] and Rouse[11] equations are based on theoretical analyses of polymer flow that yield the universal functions f and g. These functions depend uniquely on the quantities $(\alpha\dot{\gamma})$ and $(\beta\dot{\gamma})$, respectively, where α and β are parameters representing a characteristic relaxation time. The Bueche and Rouse "master curves," which have been evaluated from the theoretical expressional for f and g, are shown in Figure 2-8 and apply to all fluids obeying these viscosity relations.

If the value of (η/μ_0) is determined experimentally at a given shear rate, then, by using the master curves, the value of $\alpha\dot{\gamma}$ or $\beta\dot{\gamma}$ can be determined and hence the value of the parameters α and β. Figure 2-10 shows the data of Figure 2-8 plotted as (η/μ_0) versus log $\dot{\gamma}$. Curves representing

[8] Bird, Stewart, and Lightfoot, *op. cit.*, p. 14.

[9] R. E. Powell and H. Eyring, *Nature*, **154,** 427 (1944).

[10] F. Bueche, *J. Chem. Phys.*, **22,** 1570 (1954).

[11] P. E. Rouse, *J. Chem. Phys.*, **21,** 1272 (1953).

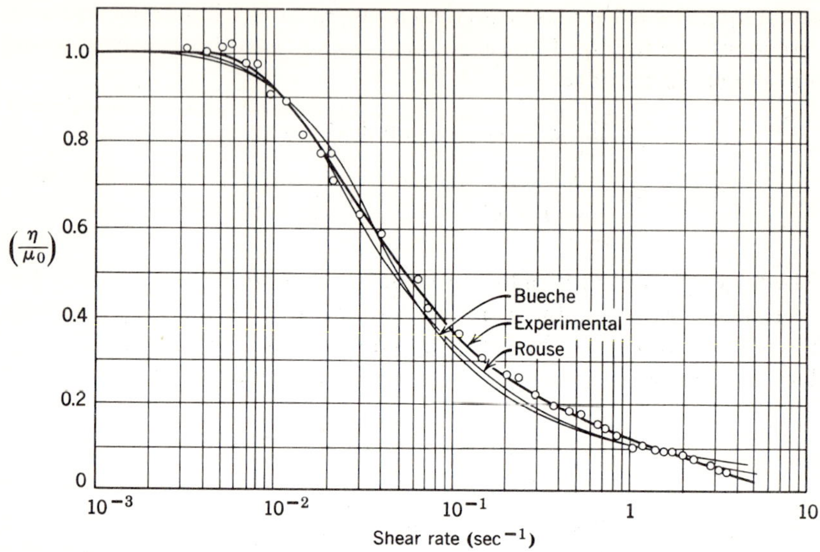

Figure 2-10. Shear dependence of "Alathon-14" polyethylene resin compared with the theoretical curves of Bueche and Rouse (Ref. 6).

both the Bueche and Rouse equations are also plotted and show reasonably good agreement with the experimental data over the entire range of shear rates.

2-5 Mechanics of Incompressible Non-Newtonian Fluids

The general rheological equation for *incompressible* Newtonian fluids can be written

$$\boldsymbol{\tau} = \mu \boldsymbol{\Delta} \tag{2-62}$$

where $\boldsymbol{\Delta}$, a rate of strain tensor, has elements defined by the equation

$$\Delta_{ij} = \left(\frac{\partial v_i}{\partial x_j}\right) + \left(\frac{\partial v_j}{\partial x_i}\right) \tag{2-63}$$

which is a special case of equation (2-17), obtained by introducing the condition of incompressibility

$$(\boldsymbol{\nabla} \cdot \boldsymbol{v}) = 0$$

into it. The coefficient μ in equation (2-62) is independent of the state of shear.

The analogous equation [12]

$$\boldsymbol{\tau} = \eta \boldsymbol{\Delta} \tag{2-64}$$

[12] See, for example, R. B. Bird, *Physics of Fluids*, **3**, 539 (1960); J. G. Oldroyd, "Non-Newtonian Flow of Liquids and Solids," Ch. 16 of *Rheology*, Vol. I, edited by F. R. Eirich, Academic Press, 1956.

has been postulated for incompressible non-Newtonian fluids. In equation (2-64), the viscosity η is considered to be a function of the state of shear, as characterized by either the shear stress or the shear rate.

Because η is a scalar, it can be a function only of the scalar invariants (scalar quantities consisting of components of $\boldsymbol{\Delta}$ that remain unchanged upon a rotation of the coordinate axes) of $\boldsymbol{\Delta}$. The three scalar invariants of the rate of strain tensor $\boldsymbol{\Delta}$ are defined by the equations

$$I_1 = \sum_i \Delta_{ii} = 2(\boldsymbol{\nabla} \cdot \boldsymbol{v}) \tag{2-65}$$

$$I_2 = \sum_i \sum_j (\Delta_{ij})^2 = (\boldsymbol{\Delta} : \boldsymbol{\Delta}) \tag{2-66}$$

$$I_3 = \sum_i \sum_j \sum_k \epsilon_{ijk}\Delta_{1i}\Delta_{2j}\Delta_{3k} = \det \boldsymbol{\Delta} \tag{2-67}$$

For incompressible fluids, the first invariant vanishes. Hence the viscosity of incompressible non-Newtonian fluids depends only upon the second and third invariants. In rectilinear flow and in two-dimensional flow, the third invariant is zero and is considered not very important in other types of flow. Hence the viscosity η is taken to be a function only of the second invariant of $\boldsymbol{\Delta}$.

By expanding $\boldsymbol{\tau}$ in a power series of $\boldsymbol{\Delta}$, Reiner[13] has shown that the most general rheological equation for incompressible non-Newtonian fluids has the form

$$\boldsymbol{\tau} = \eta\boldsymbol{\Delta} + \tfrac{1}{2}\eta_c(\boldsymbol{\Delta} \cdot \boldsymbol{\Delta})$$

where η_c, the coefficient of cross viscosity, is a scalar function of invariants I_1 and I_2.

The generalized form of the power law can be written

$$\eta = \eta^\circ \left[\frac{(\boldsymbol{\Delta} : \boldsymbol{\Delta})}{2}\right]^{(n-1)/2} \tag{2-68}$$

where the standard state is taken as a shear rate of 1 sec^{-1}. The scalar quantity $(\boldsymbol{\Delta} : \boldsymbol{\Delta})$ is given in Table 2-5 in rectangular, cylindrical, and spherical coordinates.

For simple shear, using rectangular coordinates, the only nonzero velocity component is v_x and the only nonzero derivative of v_x is $(\partial v_x/\partial y)$ so that the expression for $(\boldsymbol{\Delta} : \boldsymbol{\Delta})$ reduces to

$$(\boldsymbol{\Delta} : \boldsymbol{\Delta}) = 2\left(\frac{\partial v_x}{\partial y}\right)^2$$

and equation (2-68) becomes

$$\eta = \eta^\circ \left(\frac{\partial v_x}{\partial y}\right)^{n-1} = \eta^\circ \dot{\gamma}$$

[13] M. Reiner, *Am. J. Math.*, **67**, 350 (1945).

which shows that equation (2) of Table 2-4 is a special case of the general relationship, equation (2-68).

Table 2-5*

The Function ($\Delta : \Delta$) Expressed in Rectangular, Cylindrical, and Spherical Coordinates

Rectangular

$$\tfrac{1}{2}(\Delta : \Delta) = 2\left[\left(\frac{\partial v_x}{\partial x}\right)^2 + \left(\frac{\partial v_y}{\partial y}\right)^2 + \left(\frac{\partial v_z}{\partial z}\right)^2\right] + \left[\frac{\partial v_y}{\partial x} + \frac{\partial v_x}{\partial y}\right]^2 + \left[\frac{\partial v_z}{\partial y} + \frac{\partial v_y}{\partial z}\right]^2 + \left[\frac{\partial v_x}{\partial z} + \frac{\partial v_z}{\partial x}\right]^2$$

Cylindrical

$$\tfrac{1}{2}(\Delta : \Delta) = 2\left[\left(\frac{\partial v_r}{\partial r}\right)^2 + \left(\frac{1}{r}\frac{\partial v_\theta}{\partial \theta} + \frac{v_r}{r}\right)^2 + \left(\frac{\partial v_z}{\partial z}\right)^2\right] + \left[r\frac{\partial}{\partial r}\left(\frac{v_\theta}{r}\right) + \frac{1}{r}\frac{\partial v_r}{\partial \theta}\right]^2 + \left[\frac{1}{r}\frac{\partial v_z}{\partial \theta} + \frac{\partial v_\theta}{\partial z}\right]^2 + \left[\frac{\partial v_r}{\partial z} + \frac{\partial v_z}{\partial r}\right]^2$$

Spherical

$$\tfrac{1}{2}(\Delta : \Delta) = 2\left[\left(\frac{\partial v_r}{\partial r}\right)^2 + \left(\frac{1}{r}\frac{\partial v_\theta}{\partial \theta} + \frac{v_r}{r}\right)^2 + \left(\frac{1}{r\sin\theta}\frac{\partial v_\phi}{\partial \phi} + \frac{v_r}{r} + \frac{v_\theta\cot\theta}{r}\right)^2\right] + \left[r\frac{\partial}{\partial r}\left(\frac{v_\theta}{r}\right) + \frac{1}{r}\frac{\partial v_r}{\partial \theta}\right]^2 + \left[\frac{\sin\theta}{r}\frac{\partial}{\partial \theta}\left(\frac{v_\phi}{\sin\theta}\right) + \frac{1}{r\sin\theta}\frac{\partial v_\theta}{\partial \phi}\right]^2 + \left[\frac{1}{r\sin\theta}\frac{\partial v_r}{\partial \phi} + r\frac{\partial}{\partial r}\left(\frac{v_\phi}{r}\right)\right]^2$$

* Abstracted with permission from Table 3.4-8 of Bird, Stewart, and Lightfoot, *Transport Phenomena*.

The other equations of Table 2-4 can be generalized by replacing $\dot{\gamma}$ with $[\tfrac{1}{2}(\Delta : \Delta)]^{1/2}$ or τ with $[\tfrac{1}{2}(\boldsymbol{\tau} : \boldsymbol{\tau})]^{1/2}$ where $(\boldsymbol{\tau} : \boldsymbol{\tau})$ is defined by the equation

$$(\boldsymbol{\tau} : \boldsymbol{\tau}) = \sum_i \sum_j \tau_{ij}{}^2 \tag{2-69}$$

For example, the generalization of the Reiner-Philippoff equation is

$$\eta = \mu_\infty + \frac{\mu_0 - \mu_\infty}{1 + [(\boldsymbol{\tau} : \boldsymbol{\tau})/2A^2]} \tag{2-70}$$

Example 2-5.

The problem considered here is the same as that in Example 2-2, except that the fluid is now assumed to be a non-Newtonian fluid obeying the power law. Equations (2-21) and (2-22), for non-Newtonian flow, are written

$$\left(\frac{\partial P}{\partial z}\right) = \frac{1}{r}\frac{\partial}{\partial r}\left[r\eta\left(\frac{\partial v_z}{\partial r}\right)\right] \tag{2-71}$$

$$-\frac{k}{r}\frac{\partial}{\partial r}\left[r\frac{\partial T}{\partial r}\right] = \eta\left(\frac{\partial v_z}{\partial r}\right)^2 \tag{2-72}$$

where η is defined by equation (2-68).

In cylindrical coordinates, v_z is the only nonzero velocity component and, because derivatives of it with respect to z and θ are zero, the scalar invariant $(\boldsymbol{\Delta} : \boldsymbol{\Delta})$ reduces to

$$(\boldsymbol{\Delta} : \boldsymbol{\Delta}) = 2\left(\frac{\partial v_z}{\partial r}\right)^2 = 2\dot{\gamma}^2$$

so that η is given by the expression

$$\eta = \eta^\circ\, |\dot{\gamma}|^{n-1} \tag{2-73}$$

where the standard shear rate $\dot{\gamma}^\circ$ is taken as 1 sec^{-1}.

If the z-axis is oriented so that the pressure increases with increasing z, the fluid will flow in the negative z-direction and the shear rate $\dot{\gamma}$ will be positive for all values of r. Hence the absolute value sign in equation (2-73) can be dropped. Upon introducing (2-73) into (2-71), one obtains

$$\left(\frac{\partial P}{\partial z}\right) = \frac{\eta^\circ}{r}\frac{\partial}{\partial r}[r\,\dot{\gamma}^n] \tag{2-74}$$

which, upon integration, becomes

$$\dot{\gamma} = \left(\frac{\partial v_z}{\partial r}\right) = \left[\frac{1}{2\eta^\circ}\left(\frac{\partial P}{\partial z}\right)r - \frac{C_1}{r}\right]^{1/n} \tag{2-75}$$

and because the shear rate at the center of the tube must be zero, the constant of integration is zero. Upon integrating (2-75), one obtains

$$v_z = \left(\frac{n}{n+1}\right)\left[\left(\frac{1}{2\eta^\circ}\right)\left(\frac{\partial P}{\partial z}\right)\right]^{1/n} r^{(n+1)/n} + C_2 \tag{2-76}$$

where the constant of integration C_2 is to be evaluated from the condition

$$v_z(R) = 0 \tag{2-77}$$

Upon evaluating C_2, equation (2-76) becomes

$$v_z = -\left(\frac{n}{n+1}\right)\left[\frac{1}{2\eta^\circ}\left(\frac{\partial P}{\partial z}\right)\right]^{1/n} R^{(n+1)/n}\left[1 - \left(\frac{R}{r}\right)^{(n+1)/n}\right] \tag{2-78}$$

The velocity at the center of the tube v_0, obtained by placing $r = 0$ in the above equation, is given by the equation

$$v_0 = -\left[\frac{nR}{n+1}\right]^{(n+1)/n}\left[\frac{1}{\eta^\circ}\left(\frac{\partial P}{\partial z}\right)\right]^{1/n} \tag{2-79}$$

so that the equation for the velocity profile can be written

$$v = v_0\left[1 - \left(\frac{r}{R}\right)^{(n+1)/n}\right] \tag{2-80}$$

Note that for the special case where $n = 1$, equation (2-80) reduces to the result obtained for Newtonian flow in Example 2-2.

To obtain the temperature profile, introduce (2-73) and (2-75) into equation (2-72). Upon integrating once, one obtains

$$\frac{\partial T}{\partial r} = -\left(\frac{\eta^\circ}{k}\right)\left(\frac{n}{3n+1}\right)\left[\frac{1}{2\eta^\circ}\left(\frac{\partial P}{\partial z}\right)\right]^{(n+1)/n} r^{(2n+1)/n} + \frac{C_3}{r} \tag{2-81}$$

where the constant C_3 must be zero because the temperature gradient is zero at the center of the tube. Upon integrating (2-81), one obtains

$$T = -\frac{\eta^\circ}{k}\left(\frac{n}{3n+1}\right)^2\left[\frac{1}{2\eta^\circ}\left(\frac{\partial P}{\partial z}\right)\right]^{(n+1)/n} r^{(3n+1)/n} + C_4 \tag{2-82}$$

where the constant of integration C_4 is evaluated from the condition $T(R) = T_w$, where T_w is the temperature of the tube wall. Hence

$$T - T_w = \frac{\eta^\circ}{k}\left(\frac{n}{3n+1}\right)^2\left[\frac{1}{2\eta^\circ}\left(\frac{\partial P}{\partial z}\right)\right]^{(n+1)/n} R^{(3n+1)/n}\left[1 - \left(\frac{r}{R}\right)^{(3n+1)/n}\right] \tag{2-83}$$

The difference in the fluid temperature between the center of the tube and the wall, obtained by placing $r = 0$ in the above equation, is

$$T_0 - T_w = \frac{\eta^\circ}{k}\left(\frac{n}{3n+1}\right)^2\left[\frac{1}{2\eta^\circ}\left(\frac{\partial P}{\partial z}\right)\right]^{(n+1)/n} R^{(3n+1)/n} \tag{2-84}$$

so that equation (2-83) can be written in the equivalent form

$$\frac{T - T_w}{T_0 - T_w} = 1 - \left(\frac{r}{R}\right)^{(3n+1)/n} \tag{2-85}$$

Note that for the special case where $n = 1$, equation (2-85) reduces to the result obtained for Newtonian flow in Example 2-2.

2-6 Temperature Dependence of Viscosity

It is well known that temperature has an important effect on the rheological properties of matter. This is particularly true in the case of high polymers. This section discusses the effect of temperature on viscosity. The objective is to present engineering procedures for treating the temperature dependence of the viscosity of non-Newtonian liquids, a problem of major importance in polymer processing.

The following discussion is limited to liquids. In general, liquid viscosities decrease and gas viscosities increase with increasing temperature, indicating that there is an essential difference in the flow mechanisms of liquids and gases. At low densities, gas viscosities can be analyzed on the basis of momentum transfer between adjacent layers of molecules moving at different velocities. In the liquid state, the situation is much more complicated, and no attempt is made here to give a molecular interpretation of the flow of polymers.

The flow behavior of a Newtonian liquid depends upon a single quantity, the viscosity, which is a function of temperature only. The most commonly used expression relating viscosity to temperature is the Arrhenius equation

$$\mu = Ae^{E/RT} \tag{2-86}$$

where R is the gas constant, E the energy of activation for flow, and A a coefficient depending upon the nature of the liquid.

Equation (2-86) can be tested by plotting log μ versus $(1/T)$. For most liquids, a reasonably good straight line can be drawn over a temperature range of 100 deg F. Beyond this, curvature usually becomes pronounced, indicating that variation in the quantities A and E is becoming significant. Another equation sometimes useful in correlating viscosity-temperature data is

$$\mu = ae^{-bT} \tag{2-87}$$

where a and b are empirical constants. It has about the same range of usefulness as the other.

At very low shear rates most non-Newtonian fluids exhibit Newtonian behavior. This effect is illustrated in Figure 2-9, where the logarithmic flow curve becomes straight and has a slope equal to unity below the limiting value of the shear rate. In this region the flow behavior can be characterized by a single constant, the limiting viscosity.

The flow measurements of Philippoff and Gaskins[14] on two polyethylene samples and a plasticized polyvinylbutyral sample extended into this region of Newtonian behavior. In Figure 2-11 their data are plotted as log μ_0 versus $(1/T)$, as suggested by equation (2-86). Note that the points fall on straight lines. Figure 2-12 shows the same data plotted as log μ_0 versus T, as suggested by equation (2-87). Note that there is some curvature to the line drawn through the points for the polyethylene, but that over a slightly narrower temperature range the curves can be approximated by a straight line with sufficient accuracy for most engineering purposes.

Flow data for many commercially important thermoplastics have been compiled by Westover.[15] He presents plots of viscosity η versus temperature

[14] W. Philippoff and F. H. Gaskins, *J. Poly. Sci.*, **21**, 205 (1956).

[15] R. F. Westover, Section III of *Processing of Thermoplastic Materials*, edited by E. C. Bernhardt, Reinhold Publishing Corp., 1959.

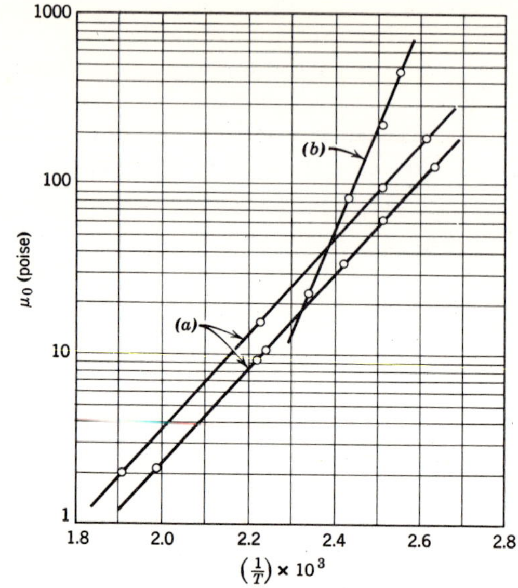

Figure 2-11. Temperature dependence of the limiting Newtonian viscosity μ_0 of (*a*) polyethylene and (*b*) polyvinyl butyral (data from Ref. 14).

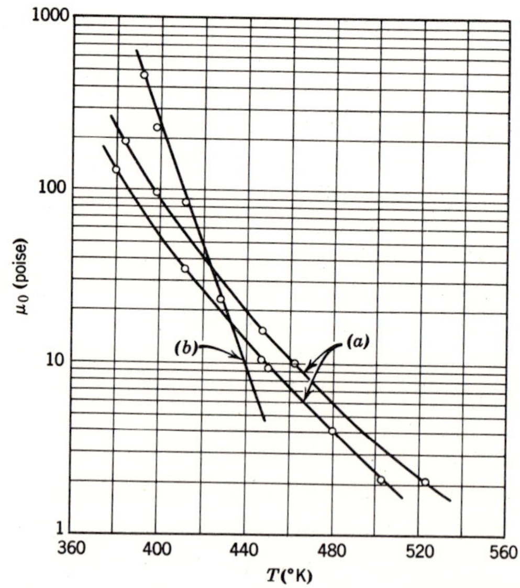

Figure 2-12. Viscosity data of Figure 2-11 replotted with absolute temperature as the abscissa.

at constant shear rate. Over a temperature range of about 100 deg F, most of the curves are nearly straight, indicating that a relationship of the form of (2-87) can be used to correlate the data. Table 2-6 shows values of $(1/b_{\dot{\gamma}})$, which have been derived from these data.

Table 2-6*

Temperature Dependence of the Viscosity of Some Common Thermoplastics at Constant Rate of Shear

Material	Shear Rate (sec^{-1})	$\frac{1}{b_{\dot{\gamma}}}$ (deg C)	Trade Name
Polymethyl methacrylate	100	24	"Lucite 140"
Polymethyl methacrylate	27	18	"Plexiglas V100"
Cellulose acetate	100	32	"Tenite Acetate 036-H2"
Type 6 Nylon	100	60	"Plaskon Nylon 8206"
Type 66 Nylon	100	56	"Zytel 101 NC10"
Polyethylene	100	85	"Bakelite DYNH"
Polyethylene	100	70	"Alathon-10"
Polystyrene	100	73	"Styron 475"
Polystyrene	100	46	"Lustrex HF 77-2020"
Polyvinylchloride	40	51	"Geon 8750"
Plasticized polyvinylchloride	100	40	"Opalon 71329"

* Based on data from Ref. 15.

The quantity $(1/b_{\dot{\gamma}})$ represents the number of degrees that the temperature of the polymer must be raised at constant shear rate in order to decrease the viscosity by the factor $(1/e)$. To show this, write (2-87) in the form

$$\frac{\mu_1}{\mu_2} = \frac{a \exp\left[-b_{\dot{\gamma}} T_1\right]}{a \exp\left[-b_{\dot{\gamma}} T_2\right]}$$

which can be put in the equivalent form

$$\mu_1 = \mu_2 \exp\left[b_{\dot{\gamma}}(T_2 - T_1)\right] = \mu_2 \exp\left(b_{\dot{\gamma}}\, \Delta T\right)$$

Note that when $\Delta T = (1/b_{\dot{\gamma}})$

$$\mu_2 = \left(\frac{1}{e}\right)\mu_1$$

Hence the meaning of $(1/b_{\dot{\gamma}})$ is clear.

Note that polymethyl methacrylate has the most temperature sensitive viscosity while polyethylene has the least. This is an important consideration in determining adiabatic heating effects in extruders and in determining uniformities in the extrusion of films and sheets.

The non-Newtonian viscosity η is a function of the state of shear (as characterized by either $\dot{\gamma}$ or τ) as well as a function of temperature. This is expressed by the functional equations

$$\eta = \eta(\tau, T) \tag{2-88}$$

$$\eta = \eta(\dot{\gamma}, T) \tag{2-89}$$

Therefore when the variation of η with temperature is investigated, it is necessary to specify whether the variation is to occur at constant shear rate or at constant shear stress. In general, $(\partial\eta/\partial T)_\tau$ will not be equal to $(\partial\eta/\partial T)_{\dot{\gamma}}$. Each type of variation gives rise to a different energy of activation. Bestul and Belcher[16] have shown theoretically that, for pseudoplastic fluids,

$$\left(\frac{\partial\eta}{\partial T}\right)_\tau > \left(\frac{\partial\eta}{\partial T}\right)_{\dot{\gamma}} \tag{2-90}$$

The total differential of η, considering η to be a function of τ and T, is

$$d\eta = \left(\frac{\partial\eta}{\partial T}\right)_\tau dT + \left(\frac{\partial\eta}{\partial\tau}\right)_T d\tau \tag{2-91}$$

For a process occurring at constant shear rate, equation (2-91) can be written

$$\left(\frac{\partial\eta}{\partial T}\right)_{\dot{\gamma}} = \left(\frac{\partial\eta}{\partial\tau}\right)_T \left(\frac{\partial\tau}{\partial T}\right)_{\dot{\gamma}} + \left(\frac{\partial\eta}{\partial T}\right)_\tau \tag{2-92}$$

which can be rearranged to the following:

$$\frac{(\partial\eta/\partial T)_\tau}{(\partial\eta/\partial T)_{\dot{\gamma}}} = 1 - \left(\frac{\partial\eta}{\partial\tau}\right)_T \left(\frac{\partial\tau}{\partial\eta}\right)_{\dot{\gamma}} \tag{2-93}$$

Differentiating equation (2-39) with respect to η at constant $\dot{\gamma}$ gives

$$\left(\frac{\partial\tau}{\partial\eta}\right)_{\dot{\gamma}} = \dot{\gamma} \tag{2-94}$$

which reduces (2-93) to

$$\frac{(\partial\eta/\partial T)_\tau}{(\partial\eta/\partial T)_{\dot{\gamma}}} = 1 - \dot{\gamma}\left(\frac{\partial\eta}{\partial\tau}\right)_T \tag{2-95}$$

By definition, pseudoplastic fluids are those for which η decreases as τ increases. Therefore $(\partial\eta/\partial\tau)_T$ is negative and the right side of (2-95) must be greater than unity. Hence equation (2-90) is proved.

---

[16] A. B. Bestul and H. V. Belcher, *J. Appl. Phys.*, **24**, 696 (1953).

Assuming that the temperature dependence of η can be expressed through an equation of the form of the Arrhenius equation, (2-86), we have

$$\eta = Ae^{E_\tau/RT} \qquad \eta = Ae^{E_{\dot\gamma}/RT} \tag{2-96}$$

The first equation applies to changes occurring at constant shear stress and the second to changes occurring at constant shear rate. Values of E_τ and $E_{\dot\gamma}$ have been determined by Philippoff and Gaskins for a polyethylene in the range of 108° C to 230° C. They are shown in Table 2-7.

Table 2-7*

Activation Energies for Polyethylene
Temperature Range: 108° C–230° C

$\dot\gamma$ (sec^{-1})	$E_{\dot\gamma}$ (kcal/mole)	τ (dynes/cm^2)	E_τ (kcal/mole)
0	12.8	0	12.8
10^{-1}	11.4	10^4	15.0
10^0	10.3	10^5	17.8
10^1	8.5	10^6	19.0
10^2	7.2		
10^3	6.1		

* From Ref. 14 with permission of Interscience Publishers.

Note that at very low shear stresses, the activation energies are the same. This result could have been predicted from equation (2-95) by introducing the partial derivatives of (2-96)

$$\left(\frac{\partial \eta}{\partial T}\right)_\tau = \frac{-\eta E_\tau}{RT^2} \qquad \left(\frac{\partial \eta}{\partial T}\right)_{\dot\gamma} = -\frac{\eta E_{\dot\gamma}}{RT^2} \tag{2-97}$$

into it to obtain

$$\frac{E_\tau}{E_{\dot\gamma}} = 1 - \dot\gamma\left(\frac{\partial \eta}{\partial \tau}\right)_T \tag{2-98}$$

which shows that E_τ becomes equal to $E_{\dot\gamma}$ as $\dot\gamma$ approaches zero. This is a general result applying to all fluids. For Newtonian fluids, E_τ is equal to $E_{\dot\gamma}$ under all conditions because $(\partial\eta/\partial\tau)_T$ is zero.

Example 2-6.

This example illustrates how the same temperature change can produce different changes in the viscosity of a non-Newtonian fluid. Consider a non-Newtonian fluid in the annular space between two coaxial cylinders, the inner cylinder being driven and the outer cylinder being stationary. If the drive mechanism is a constant speed device and if the fluid temperature is increased, the viscosity will decrease and less torque will be required to rotate the inner cylinder. Then shear rate is constant and the viscosity change is given by

$$d\eta = \left(\frac{\partial \eta}{dT}\right)_{\dot{\gamma}} \partial T \tag{2-99}$$

which, because of equation (2-97), can be written

$$d\eta = -\left(\frac{\eta E_{\dot{\gamma}}}{RT^2}\right) dT \tag{2-100}$$

Integration of equation (2-100) gives

$$\ln\left(\frac{\eta_2}{\eta_1}\right) = -\left(\frac{E_{\dot{\gamma}}\,\Delta T}{RT_1T_2}\right) \tag{2-101}$$

where ΔT represents the temperature increase. The negative sign indicates that viscosity is reduced upon an increase in temperature.

If the drive were a constant torque device, a different result would be obtained. The shear stress would remain constant, but the shear rate would increase upon an increase of temperature. By an entirely analogous procedure, the viscosity change is found to be

$$\ln\left(\frac{\eta_2}{\eta_1}\right) = -\left(\frac{E_\tau \Delta T}{RT_1T_2}\right) \tag{2-102}$$

Since E_τ must be greater than $E_{\dot{\gamma}}$, the decrease in viscosity at constant $\dot{\gamma}$ is not as great as at constant τ.

As a specific example, consider the behavior of the polyethylene described in reference 14. If the initial conditions are

$$\dot{\gamma} = 1.0 \text{ sec}^{-1}$$
$$\eta = 10^4 \text{ poise}$$
$$\tau = 10^4 \text{ dynes/cm}^2$$

then the values of the activation energies are

$$E_{\dot{\gamma}} = 10.3 \text{ kcal/mole}$$
$$E_\tau = 15.0 \text{ kcal/mole}$$

If the temperature change is not too large, the activation energies are nearly constant; then the reduction in viscosity at constant shear rate would be 50% of the reduction at constant shear stress for the same temperature change.

If the fluid obeys the power law over the range of interest, certain simplifications can be made. First, it can be shown that E_τ and $E_{\dot{\gamma}}$ are not independent quantities but are related by the simple expression

$$E_\tau = \left(\frac{1}{n}\right) E_{\dot{\gamma}} \tag{2-103}$$

where n, the flow index, is always less than unity for pseudoplastic fluids.

A proof of (2-103) is obtained by introducing the equations

$$\left(\frac{\partial \eta}{\partial \tau}\right)_T = \left(\frac{n-1}{n}\right)\left(\frac{\eta^\circ}{\tau^\circ}\right)\left|\frac{\tau}{\tau^\circ}\right|^{-1/n}$$

$$\dot{\gamma} = \dot{\gamma}^\circ \left|\frac{\tau}{\tau^\circ}\right|^{1/n}$$

which are obtained from the power law viscosity relationship given in Table 2-4, into equation (2-95) and noting that $\tau^\circ = \eta^\circ \dot{\gamma}^\circ$.

Example 2-7.

At a temperature of 174° C and a shear rate of 1 $\sec^{-1}$, the viscosity of the polyethylene of Ref. 14 is 31,500 poise. Assuming that the power law is applicable over the range of temperature and shear rate involved, estimate the viscosity of the polyethylene at a temperature of 230° C and a shear rate of 100 $\sec^{-1}$, using the data of Table 2-7.

If we choose the standard shear rate $\dot{\gamma}^\circ$ as 1 $\sec^{-1}$, then the value of the standard viscosity η° at 230° C can be obtained in the following manner: Differentiate the expression for the viscosity, equation (2) of Table 2-4, to obtain

$$\left(\frac{\partial \ln \eta^\circ}{\partial T}\right)_{\dot{\gamma}} = \left(\frac{\partial \ln \eta}{\partial T}\right)_{\dot{\gamma}} \tag{2-104}$$

which becomes

$$\left(\frac{\partial \ln \eta^\circ}{\partial T}\right)_{\dot{\gamma}} = -\frac{E_{\dot{\gamma}}}{RT^2} \tag{2-105}$$

upon introducing equation (2-97). Upon integration, equation (2-105) becomes

$$\ln\left(\frac{\eta_2^\circ}{\eta_1^\circ}\right) = \frac{E_{\dot{\gamma}}}{R}\left(\frac{1}{T_2} - \frac{1}{T_1}\right) \tag{106}$$

Taking the value of $E_{\dot{\gamma}}$ from Table 2-7 as 10.3 kcal/mole, the calculated value of η_2° is 8720 poise.

The viscosity at 230° C and 100 $\sec^{-1}$ is calculated with the power law

$$\eta_2 = \eta_2^\circ \dot{\gamma}^{\bar{n}-1}$$

where $\bar{n}$ represents the average value of the flow index between 1 $\sec^{-1}$ and 100. From the data of Table 2-7, the value of $E_{\dot{\gamma}}$ at 1 $\sec^{-1}$ is 10.3 kcal/mole. The stress in this state is about 10^4 dynes/cm^2, so that E_τ is about 15 kcal/mole and the value of n_1 is

$$n_1 = \left(\frac{E_{\dot{\gamma}}}{E_\tau}\right) = 0.68$$

In state 2 where the shear rate is 100 $\sec^{-1}$, $E_{\dot{\gamma}} = 7.2$ but the stress is unknown. If an initial estimate of 10^5 dynes/cm^2 is made, then $E_\tau = 17.8$ and

$$n_2 = \left(\frac{E_{\dot{\gamma}}}{E_\tau}\right) = 0.4$$

so that $\bar{n}$ is about 0.55. Therefore

$$\eta_2 = 8720(100)^{-0.45} = 1100 \text{ poise}$$

The stress in state 2, 1.1×10^5 dynes/cm^2, is very nearly equal to the first estimate of 10^5 dynes/cm^2. Therefore it is not necessary to correct the value of $\bar{n}$ used in the calculation.

2-7 Introduction to Viscoelasticity

According to Alfrey and Gurnee,[17] a "crude, qualitative representation" of the molecular mechanisms responsible for the viscoelastic behavior of amorphous linear high polymers is provided by the mechanical model shown in Figure 2-13. This model consists of a Voigt element connected

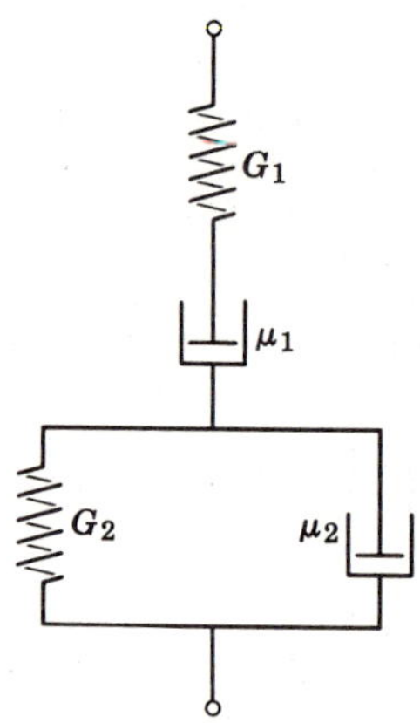

Figure 2-13. Four-element mechanical model that simulates the shear behavior of amorphous linear polymers.

in series with a Maxwell element. The total strain in the model is due to instantaneous elastic deformation, irrecoverable viscous flow, and retarded elastic deformation.

The instantaneous elastic deformation is due to the bending and stretching of primary valence bonds. The shear modulus associated with this process is G_1. The irrecoverable viscous flow is due to the slippage of polymer chains or chain segments past one another. The viscosity associated with this process is μ_1. The retarded elastic deformation arises from the transformation of a given equilibrium molecular configuration into a biased configuration in which elongated and oriented structures are favored. The driving force for this process is the shear stress. The rate at which it occurs is determined by the ratio (μ_2/G_2), called the retardation time. The equilibrium strain in the biased state determines the modulus G_2.

[17] T. Alfrey and E. F. Gurnee, "Dynamics of Viscoelastic Behavior," ch. 11 of *Rheology*, vol. I, edited by F. R. Eirich, Academic Press, 1956.

The rheological equation for the model shown in Figure 2-13 is now derived. If γ represents the total strain, then

$$\gamma = \gamma_1 + \gamma_2 \tag{2-107}$$

where γ_1 represents the strain in the Voigt element and γ_2 the strain in the Maxwell element. Using this notation, the rheological equation for the Voigt element, (2-36), is written

$$\tau = \mu_2\left(\frac{d\gamma_1}{dt}\right) + G_2\gamma_1 \tag{2-108}$$

and the rheological equation for the Maxwell element, (2-34), becomes

$$\frac{d\gamma_2}{dt} = \frac{1}{G_1}\left(\frac{d\tau}{dt}\right) + \frac{\tau}{\mu_1} \tag{2-109}$$

The above three equations constitute a system containing, as dependent variables, γ_1, γ_2, γ, and τ. By suitably combining these equations, it is possible to eliminate any two of the four dependent variables. The rheological equation for the model is obtained by eliminating γ_1 and γ_2. It is the second order differential equation

$$\frac{d^2\tau}{dt^2} + \left[\frac{G_1}{\mu_2} + \frac{G_1}{\mu_1} + \frac{G_2}{\mu_2}\right]\frac{d\tau}{dt} + \left(\frac{G_1G_2}{\mu_1\mu_2}\right)\tau = G_1\left(\frac{d^2\gamma}{dt^2}\right) + \left(\frac{G_1G_2}{\mu_2}\right)\frac{d\gamma}{dt} \tag{2-110}$$

A creep curve is obtained when a sample is subjected to a constant stress and the deformation determined as a function of time. Upon removal of the stress there is an instantaneous strain recovery, which is followed by a retarded recovery. The equations of the creep and strain recovery curves are derived here for the four-parameter model of Figure 2-13.

For a constant applied stress S, equation (2-110) becomes

$$\frac{d^2\gamma}{dt^2} + \left(\frac{G_2}{\mu_2}\right)\frac{d\gamma}{dt} = \left(\frac{G_2}{\mu_1\mu_2}\right)S \tag{2-111}$$

If the stress is suddenly applied at time zero, the model will deform instantaneously an amount (S/G_1) and then will undergo both retarded elastic deformation and steady flow. The initial rate (at $t = 0$) of deformation is obtained by adding the rates of both of these processes to obtain $(S/\mu_1 + S/\mu_2)$. Therefore the initial conditions are formulated as

$$\gamma(0) = \frac{S}{G_1}$$

$$\dot{\gamma}(0) = \left(\frac{S}{\mu_1} + \frac{S}{\mu_2}\right)$$

and the solution of equation (2-111) satisfying these conditions is

$$\gamma = \frac{S}{G_1} + \left(\frac{S}{\mu_1}\right)t + \frac{S}{G_2}\,[1 - e^{-t/\lambda}] \tag{2-112}$$

where λ, the retardation time, is defined by the equation

$$\lambda = \left(\frac{\mu_2}{G_2}\right)$$

Creep and strain recovery curves for the four-parameter model are shown in Figure 2-14. In principle, the values of the model parameters can be deduced from the creep curve. The value of G_1 can be calculated

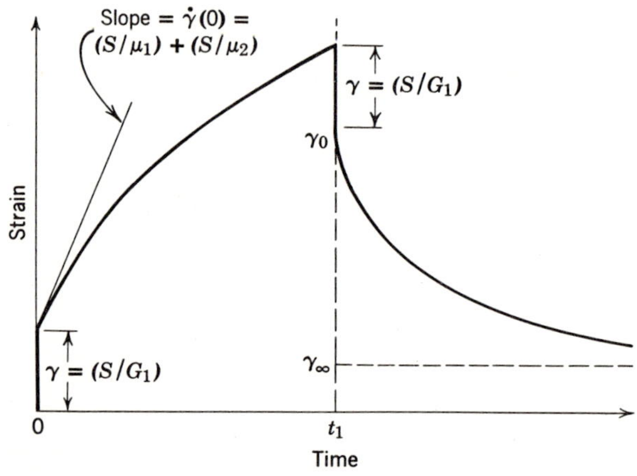

Figure 2-14. Creep and strain recovery curves for the four-element model shown in Figure 2-13.

from the initial strain $\gamma(0)$, the value of μ_1 from $\gamma(\infty)$, the value of μ_2 from $\dot{\gamma}(0)$ and G_2 from $\ddot{\gamma}(0)$. The retardation time λ, which represents the time for 63% of the retarded elastic strain to be recovered, can be calculated directly from the strain recovery curve if sufficient time is allowed for strain recovery to become virtually complete.

Upon removal of stress, $S = 0$, and equation (2-111) becomes

$$\frac{d^2\gamma}{dt^2} + \frac{1}{\lambda}\frac{d\gamma}{dt} = 0 \tag{2-113}$$

which is the differential equation of the strain recovery curve. Its general solution is written

$$\gamma = C_1 + C_2 e^{-t/\lambda} \tag{2-114}$$

where C_1 and C_2 are the constants of integration.

If γ_0 represents the strain when the stress is first removed and γ_∞ the strain after complete recovery, then

$$\gamma_0 = C_1 + C_2$$
$$\gamma_\infty = C_1$$

and the fractional approach to strain recovery f is given by the equation

$$f = \frac{\gamma_0 - \gamma}{\gamma_0 - \gamma_\infty} = 1 - e^{-t/\lambda}$$

When $t = \lambda$

$$f = 1 - \frac{1}{e} = 0.63$$

The behavior of materials subjected to stresses or strains that vary periodically with time is now considered. Consider first a Hookean solid subjected to simple shear in which the strain varies periodically with time according to the equation

$$\gamma = \gamma_0 \sin(\omega t) \tag{2-115}$$

where γ_0 is the maximum strain and $\omega = 2\pi f$ where f is the frequency of the strain expressed in cycles per second. The rheological equation of the Hookean solid $\tau = G\gamma$ shows that the stress is

$$\tau = G\gamma_0 \sin(\omega t) \tag{2-116}$$

and therefore that, in Hookean solids, stress and strain are exactly in phase.

The period of a complete cycle is $1/f$ seconds. In the first quarter of the cycle, positive strain increases and reaches a maximum at $1/4f$ seconds. During the second quarter of the cycle, the strain continually decreases and returns to zero at $1/2f$ seconds. The second half of the cycle is identical to the first, except that the strain is negative instead of positive. The following calculation shows that the energy stored during the first and third quarters of the cycle is exactly recovered in the second and fourth quarters of the cycle, so that in a Hookean solid subjected to a strain that varies sinusoidally with time, there is neither dissipation nor a net energy storage over a complete cycle.

The instantaneous rate p_i of energy absorption per unit volume is given by the equation

$$p_i = \tau\dot{\gamma}$$

and the total energy stored in the material during the first quarter of a cycle is given by the integral

$$e = \int_0^{1/4f} p_i\, dt = \int_0^{1/4f} \tau\dot{\gamma}\, dt$$

Upon introducing equations (2-115) and (2-116) into the above equation and performing the integration, one obtains $e = G\gamma_0^2/2$, but if one integrates over the half cycle or the complete cycle $e = 0$.

If a Newtonian fluid is subjected to the sinusoidal strain $\gamma_0 \sin(\omega t)$, its rheological equation $\tau = \mu\dot{\gamma}$ shows that the stress is

$$\tau = \mu\omega\gamma_0 \cos(\omega t) \tag{2-117}$$

Hence, in Newtonian fluids, stress is also a sinusoidal function of time, but it is exactly 90° out of phase with the strain. Stress and strain rates are, however, in phase.

For a Newtonian fluid, the energy expended during the first quarter cycle is

$$e = \int_0^{1/4f} \tau\dot{\gamma}\, dt = \mu\omega^2\gamma_0^2 \int_0^{1/4f} \cos^2[\omega t]\, dt = \left(\frac{\pi}{4}\right) \mu\omega\gamma_0^2$$

The same result is obtained for the second, third, and fourth quarters of the cycle. Hence the total energy expended during a complete cycle is $\pi\mu\omega\gamma_0^2$. Because viscous flow is an irreversible process, this energy appears as heat in the material.

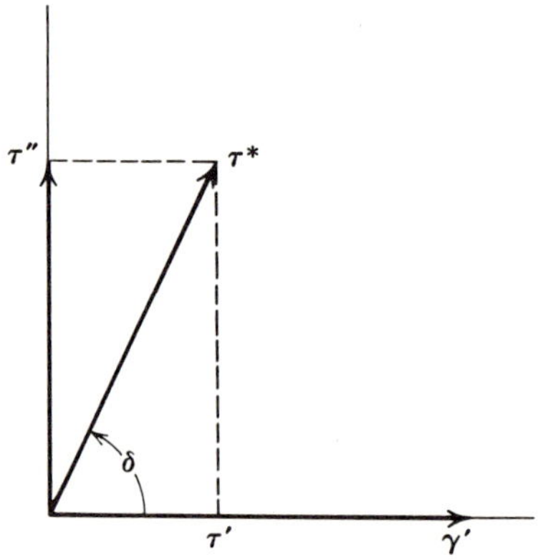

Figure 2-15. Stress and strain represented as vectors in the complex plane.

If a viscoelastic material is subjected to a strain that is a sinusoidal function of time, the stress will, in general, be neither in phase nor exactly 90° out of phase with the strain. The analysis of the behavior of such materials is facilitated by introducing certain complex quantities (for example, τ^*, whose magnitude represents the maximum value of stress) and representing these quantities as vectors in the complex plane. The vectors representing τ^* and γ^* intersect with the angle δ at the origin, where δ is the phase angle of the material. Their projections on the coordinate axes represent their real and imaginary components. The orientation of the reference axes is arbitrary and, for convenience, the real axis is made to coincide with γ^*. Then γ^* has a real component γ' only, but τ^* has both real τ' and imaginary τ'' components, as shown in Figure 2-15.

The complex shear modulus G^* can be defined by the equation

$$G^* = \frac{\tau^*}{\gamma^*} = G' + iG'' \tag{2-118}$$

which can be written

$$G^* = \frac{\tau' + i\tau''}{\gamma'} = \frac{\tau'}{\gamma'} + i\frac{\tau''}{\gamma'}$$

Upon comparing real and imaginary components of the above equations, we have

$$G' = \frac{\tau'}{\gamma'}$$
$$G'' = \frac{\tau''}{\gamma'} \tag{2-119}$$

Hence G' represents the ratio of the in-phase component of the stress to the strain. It is associated with energy storage and is called the storage modulus. Similarly, G'' represents the ratio of the out-of-phase component of the stress to strain. It is associated with energy dissipation and is called the loss modulus.

It is clear from Figure 2-15 that

$$\tan \delta = \frac{\tau''}{\tau'} \tag{2-120}$$

and from equations (2-119) and the above equation we see that

$$\tan \delta = \frac{G''}{G'} \tag{2-121}$$

The complex compliance J^* is defined by the equation

$$J^* = \frac{1}{G^*} = \frac{\gamma^*}{\tau^*} \tag{2-122}$$

from which it can be seen that

$$J' = \frac{G'}{(G')^2 + (G'')^2}$$
$$J'' = \frac{G''}{(G')^2 + (G'')^2}$$

and therefore that

$$\tan \delta = \frac{J''}{J'}$$

The significance of J' and J'' is easily seen if the reference axes of Figure 2-15 are rotated so that τ^* lies on the real axis. Then $J' = \gamma'/\tau'$ and represents the ratio of the in-phase component of strain to stress. It is associated with the storage of energy and is called the storage compliance. Similarly, $J'' = \gamma''/\tau'$ and represents the ratio of the out-of-phase component of strain to stress. It is associated with dissipation of energy and is called the loss compliance.

The complex viscosity μ^* can be defined by the equation

$$\mu^* = \frac{\tau^*}{\dot{\gamma}^*} = \mu' - i\mu''$$

where $\dot{\gamma}^*$, the complex shear rate is the time derivative of γ^*. The following relationship (which is derived below) exists between $\dot{\gamma}^*$ and γ^*:

$$\dot{\gamma}^* = i\omega\gamma^* \tag{2-123}$$

Upon introducing equation (2-123) into (2-118), we have

$$G^* = \frac{\tau^*}{\gamma^*} = i\omega\frac{\tau^*}{\dot{\gamma}^*} \tag{2-124}$$

and because of the definition of μ^* we see that the following relationship exists between the complex shear modulus and the complex viscosity:

$$G^* = i\omega\mu^* \tag{2-125}$$

It is easily shown that

$$G' = \omega\mu'' = \omega\frac{\tau'}{\dot{\gamma}''}$$

$$G'' = \omega\mu' = \omega\frac{\tau''}{\dot{\gamma}''}$$

and therefore that

$$\tan\delta = \frac{\mu'}{\mu''}$$

If γ_0 represents the magnitude of γ^*, then for any orientation of γ^* in the complex plane

$$\gamma^* = \gamma_0[\cos\theta + i\sin\theta] \tag{2-126}$$

where θ represents the argument[18] of the complex number γ^*. As shown by equation (2-115), the strain $\gamma(t)$ is the sinusoidal function $[\gamma_0 \sin(\omega t)]$. Hence the strain rate $\dot{\gamma}(t)$ is the sinusoidal function $[\omega\gamma_0\cos(\omega t)]$, which shows that the maximum value of the function $\dot{\gamma}(t)$ is $(\omega\gamma_0)$. Hence the magnitude of the complex strain rate $\dot{\gamma}^*$ is $\omega\gamma_0$ and

$$\dot{\gamma}^* = \omega\gamma_0[\cos\phi + i\sin\phi] \tag{2-127}$$

where ϕ represents the argument of $\dot{\gamma}^*$. Because strain and strain rate are 90° out of phase, the angles θ and ϕ must differ by 90°. Hence $\theta = (\phi - 90°)$, and equation (2-126) can be written

$$\gamma^* = \gamma_0[\cos(\phi - 90°) + i\sin(\phi - 90°)] = \gamma_0[\sin\phi - i\cos\phi]$$

[18] See, for example, *Mathematics of Physics and Modern Engineering*, by I. S. Sokolnikoff and R. M. Redheffer, McGraw-Hill, 1958, p. 528.

Multiplying the above equation by the unit imaginary gives

$$i\gamma^* = \gamma_0[\cos\phi + i\sin\phi] \tag{2-128}$$

and comparing (2-128) with (2-127) it can be seen that

$$\dot{\gamma}^* = i\omega\gamma^* \tag{2-129}$$

and therefore that

$$\begin{aligned} \dot{\gamma}' &= -\omega\gamma'' \\ \dot{\gamma}'' &= \omega\gamma' \end{aligned} \tag{2-130}$$

In a viscoelastic material, the instantaneous rate of energy dissipation per unit volume is given by the equation

$$p_i = \tau''(t)\dot{\gamma}''(t) \tag{2-131}$$

where $\tau''(t)$ represents the instantaneous value of the imaginary component of τ^* and $\dot{\gamma}''(t)$ the instantaneous value of the time derivative of the imaginary component of γ^*. Because of equations (2-119) and (2-130), it is possible to reduce (2-131) to the following:

$$p_i = \omega G''[\gamma'(t)]^2 \tag{2-132}$$

The total energy dissipated during one cycle is therefore given by

$$e = \int_0^{1/f} p_i\,dt = \omega G''\int_0^{1/f}[\gamma'(t)]^2\,dt$$

Upon introducing the expression

$$\gamma'(t) = \gamma_0 \sin(2\pi ft)$$

obtained from equation (2-115), into the above equation and performing the indicated integration, one obtains

$$e = \pi G''\gamma_0^2 \tag{2-133}$$

The average power dissipation $\bar{p}$ is obtained by dividing e by the period of a cycle. One obtains

$$\bar{p} = \pi f G''\gamma_0^2$$

which can also be expressed in terms of the real part of the complex viscosity by introducing equation (2-125) to obtain

$$\bar{p} = \tfrac{1}{2}\mu'\omega^2\gamma_0^2$$

In general, viscoelastic properties are temperature and frequency dependent. If the rheological equation of a material is known, it is possible to derive expressions for the frequency dependence of its viscoelastic properties. This is illustrated by the following derivation of expressions for G' and G'' of a Maxwell fluid. The mechanical model of the Maxwell fluid is shown in Figure 2-16*a*.

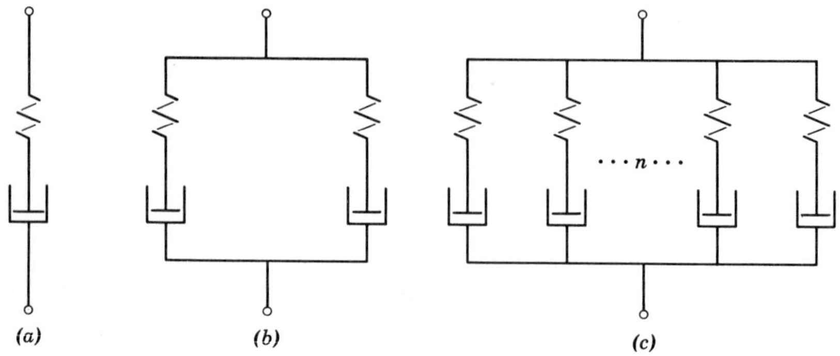

Figure 2-16. Mechanical analogs of viscoelastic liquids. (*a*) Single Maxwell element. (*b*) Two Maxwell elements connected in parallel. (*c*) n Maxwell elements connected in parallel.

Referring to equation (2-34), the rheological equation of the Maxwell fluid can be written

$$\frac{d\tau}{dt} + \frac{\tau}{\lambda} = G\dot{\gamma} \tag{2-134}$$

where the relaxation time λ is defined as the ratio μ/G. Upon introducing the periodic strain γ_0 sin (ωt) into the above equation and then solving for τ as a function of time t, one obtains

$$\tau = C_1 e^{-t/\lambda} + \frac{G\gamma_0\omega\lambda}{1 + \omega^2\lambda^2}[\omega\lambda \sin(\omega t) + \cos(\omega t)] \tag{2-135}$$

where C_1 is the constant of integration. Note that as t becomes large, the first term of equation (2-135) disappears and we are left with

$$\tau = \frac{G\gamma_0\omega\lambda}{1 + \omega^2\lambda^2}[\omega\lambda \sin(\omega t) + \cos(\omega t)] \tag{2-136}$$

where τ is now a purely periodic function.

Considering the strain to have a real component only, $\gamma'(t) = \gamma_0$ sin (ωt) and hence

$$\gamma' = \gamma_0$$

Therefore that part of the stress that is in phase with the strain is

$$\tau'(t) = \frac{G\gamma_0\omega^2\lambda^2}{1 + \omega^2\lambda^2} \sin(\omega t)$$

and therefore

$$\tau' = \frac{G\gamma_0\omega^2\lambda^2}{1 + \omega^2\lambda^2}$$

Similarly

$$\tau'' = \frac{G\gamma_0\omega\lambda}{1 + \omega^2\lambda^2}$$

and thus from equations (2-119) we see that

$$G' = \frac{G\omega^2\lambda^2}{1 + \omega^2\lambda^2} \tag{2-137}$$

and

$$G'' = \frac{G\omega\lambda}{1 + \omega^2\lambda^2} \tag{2-138}$$

Note that both G' and G'' are frequency dependent. The variation of (G'/G) and (G''/G) with λ is shown in Figure 2-17. Note that (G''/G) passes through a maximum when $\omega\lambda = 1$, and becomes small at very low or very high frequencies. On the other hand, G', the real part of the shear modulus, increases as frequency increases, indicating that the faster the sample is deformed, the "stiffer" it becomes.

The frequency dependence of the properties of real viscoelastic materials is obviously more complex than that of the simple Maxwell liquid. The temperature and frequency dependence of μ' and G' for polyisobutylene of molecular weight 1,200,000 are shown in Figure 2-18, a logarithmic plot of μ' and G' as a function of frequency at various temperatures. Note that G', and therefore the stiffness of the material, increases with increasing frequency and decreasing temperature so that at high frequencies and low temperatures, the material is very rigid. Note, also, that μ' increases with decreasing temperature and decreasing frequency so that at low temperatures and, in particular, at low frequencies, the viscous nature of the material predominates.

Ferry[19] has shown that such data can be reduced to a single curve through the use of reduced variables. He represents the behavior of real materials by a generalized Maxwell model having n elements in parallel, as shown by Figure 2-16*c*. He assumes that each G_i is proportional to both absolute temperature and density and that each μ_i has the same temperature dependence. He further assumes that when the temperature changes from some reference temperature T° to any temperature T, each λ_i of the model changes by the same factor a. Under these assumptions

$$G_i = G_i^\circ\left(\frac{\rho T}{\rho^\circ T^\circ}\right) \tag{2-139}$$

$$\lambda_i = a\lambda_i^\circ \tag{2-140}$$

It is now necessary to determine the frequency dependence of the generalized Maxwell model. This is done by determining the frequency

[19] J. D. Ferry et al., *Ind. Eng. Chem.*, **44**, 703 (1952).

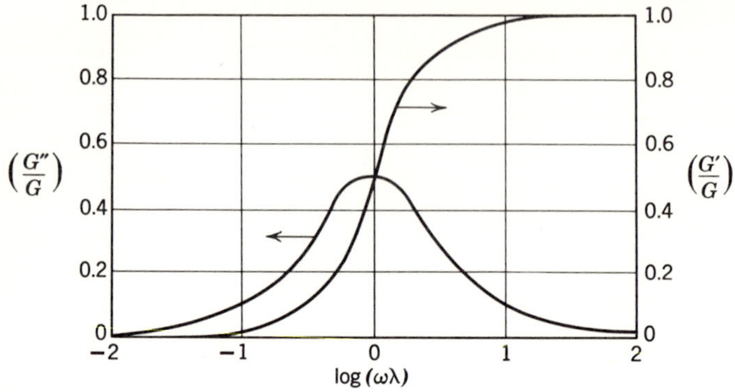

Figure 2-17. Frequency dependence of the dynamic mechanical properties of the Maxwell liquid.

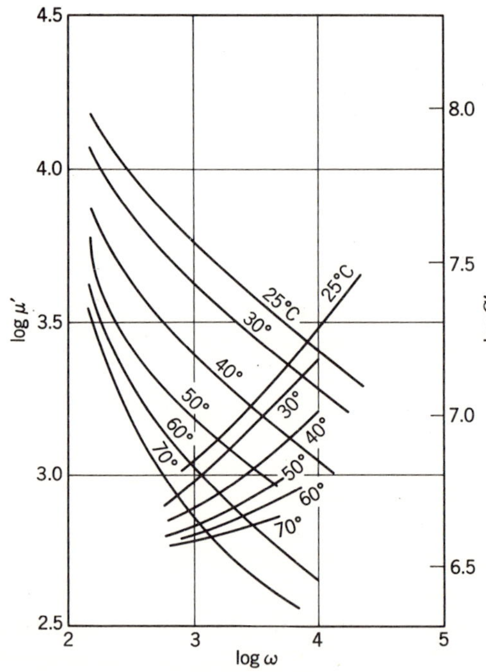

Figure 2-18. Dynamic mechanical properties of polyisobutylene as a function of temperature and frequency (Ref. 19).

dependence of the two-element model shown in Figure 2-16*b* and then generalizing the results to *n* elements.

For the two-element model, let τ_1 represent the stress in the first element and τ_2 the stress in the second element. The total stress τ is the sum of the two, thus

$$\tau = \tau_1 + \tau_2 \tag{2-141}$$

The rheological equation for the first element is written

$$\tau_1 = \mu_1\dot{\gamma} - \lambda_1\left(\frac{d\tau_1}{dt}\right) \tag{2-142}$$

and for the second element,

$$\tau_2 = \mu_2\dot{\gamma} - \lambda_2\left(\frac{d\tau_2}{dt}\right) \tag{2-143}$$

In this system of equations, τ_1, τ_2, τ, and $\dot{\gamma}$ are dependent variables. By combining the three equations to eliminate τ_1 and τ_2, one obtains the rheological equation

$$\ddot{\tau} + (\lambda_1 + \lambda_2)\dot{\tau} + (\lambda_1\lambda_2)\tau = (G_1 + G_2)\ddot{\gamma} + (G_1\lambda_2 + G_2\lambda_1)\dot{\gamma} \tag{2-144}$$

Note that this equation is of precisely the same form as (2-110). The behavior of the model in Figure 2-16*b* (two Maxwell elements in parallel) is exactly the same as the model of Figure 2-13 (Maxwell and Voigt elements in series).

When there are *m* Maxwell elements connected in parallel, the rheological equation becomes a linear differential equation of the *m*th order. It can be written in the form

$$P\tau = Q\dot{\gamma} \tag{2-145}$$

where *P* and *Q* are linear differential operators defined by the equations

$$P = \frac{d^m}{dt^m} + p_{m-1}\frac{d^{m-1}}{dt^{m-1}} + \cdots + p_0 \tag{2-146}$$

$$Q = q_n\frac{d^n}{dt^n} + q_{n-1}\frac{d^{n-1}}{dt^{n-1}} + \cdots + q_0 \qquad (m = n + 1) \tag{2-147}$$

the parameters q_i and p_i determine the properties of the material.

Equation (2-145) is called the operator equation. The Maxwell model and the operator equation are fundamental methods of specifying the viscoelastic properties of materials. Alfrey[20] describes seven methods that can be used to specify the viscoelastic properties of materials and shows the relationships between them.

The response of the model of Figure 2-16*b* to the sinusoidal strain

$$\gamma = \gamma_0 \sin(\omega t) \tag{2-148}$$

[20] T. Alfrey, *J. Appl. Phys.*, **16**, 700 (1945).

is now examined. To obtain $\tau(t)$, introduce equation (2-148) into (2-144) and solve for τ. Again, a solution consisting of a transient and a periodic part is obtained. The periodic part of the solution has the same form as before, and the value of G' and G'' are given by the equations

$$G' = \frac{G_1\lambda_1{}^2\omega^2}{1 + (\omega\lambda_1)^2} + \frac{G_2\lambda_2{}^2\omega^2}{1 + (\omega\lambda_2)^2} \tag{2-149}$$

$$G'' = \frac{G_1\lambda_1\omega}{1 + (\omega\lambda_1)^2} + \frac{G_2\lambda_2\omega}{1 + (\omega\lambda_2)^2} \tag{2-150}$$

Compare equations (2-137) and (2-138) with (2-149) and (2-150). The similarity suggests that the formulas for G' and G'' can be generalized for the case of a model containing n elements in parallel, as shown in Figure 2-16*c*. This is the generalized Maxwell model having values of G' and G'' given by the equations

$$G' = \sum_{i=1,2,\ldots}^{n} \frac{G_i(\lambda_i\omega)^2}{1 + (\omega\lambda_i)^2} \tag{2-151}$$

$$G'' = \sum_{i=1,2,\ldots}^{n} \frac{G_i\lambda_i\omega}{1 + (\omega\lambda_i)^2} \tag{2-152}$$

Furthermore, because of the relationship $G'' = \mu'\omega$, the real part of the complex viscosity is given by the equation

$$\mu' = \sum_{i=1,2,\ldots}^{\infty} \frac{G_i\lambda_i}{1 + (\omega\lambda_i)^2} \tag{2-153}$$

It should also be noted that for steady flow

$$\tau = \sum_{i=1,2,\ldots}^{n} \mu_i\dot{\gamma}_i \tag{2-154}$$

so that the steady flow viscosity is the sum of the individual viscosities. Therefore

$$\mu = \sum_{i=1,2,\ldots}^{n} \mu_i = \sum_{i=1,2,\ldots}^{n} G_i\lambda_i \tag{2-155}$$

To derive the reduced variables, equation (2-139) is now introduced into (2-151). One obtains

$$\frac{G'T^\circ\rho^\circ}{T\rho} = \sum_{i=1,2,\ldots}^{n} G_i{}^\circ\left[\frac{(a\omega\lambda_i{}^\circ)^2}{1 + (a\omega\lambda_i{}^\circ)^2}\right] \tag{2-156}$$

The reduced frequency ω_r and the reduced modulus G_r' are now defined by the equations

$$\omega_r = a\omega \tag{2-157}$$

$$G_r' = G'\left(\frac{T^\circ\rho^\circ}{T\rho}\right) \tag{2-158}$$

so that (2-156) becomes

$$G'_r = \sum_{i=1,2,\cdots}^{n} G_i^\circ \left[\frac{(\omega_r \lambda_i^\circ)^2}{1 + (\omega_r \lambda_i^\circ)^2} \right] \tag{2-159}$$

showing that the reduced modulus G'_r is a function only of the reduced frequency ω_r because the λ_i° and the G_i° in the above equation are constants. If G'_r is plotted against ω_r, all of the points will fall on a single curve regardless of the temperature at which the measurements are made.

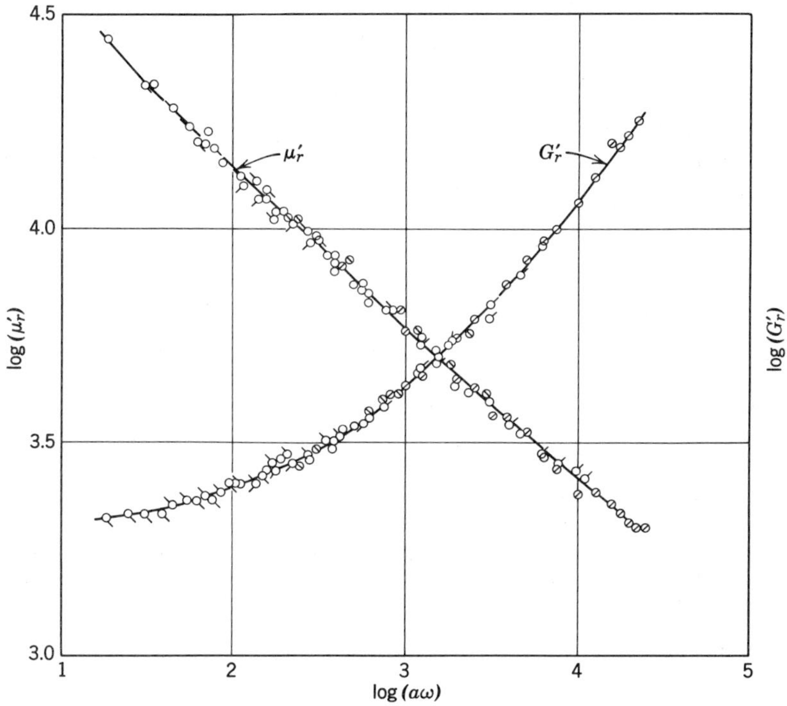

Figure 2-19. Data of Figure 2-18 replotted as reduced variables (Ref. 19).

Other reduced properties can be defined in a similar manner. For example

$$G''_r = G'' \left(\frac{T^\circ \rho^\circ}{T\rho} \right) \tag{2-160}$$

and

$$\mu'_r = a\mu' \left(\frac{T^\circ \rho^\circ}{T\rho} \right) \tag{2-161}$$

Figure 2-19 shows the data of Figure 2-18 recalculated in terms of reduced variables, thus resolving the data into two temperature dependent curves.

The temperature coefficient a is itself temperature dependent. Its values can be found by plotting log μ_r' versus log ω and shifting the curves along a line with a slope of -1 until they coincide. The distance moved in either the horizontal or vertical direction is equal to log a.

Values of a can also be determined from the steady flow viscosities. From equation (2-155),

$$\mu = a\left(\frac{T\rho}{T^\circ\rho^\circ}\right) \sum_{i=1,2,\cdots}^{n} G_i^\circ\lambda_i^\circ = a\left(\frac{T\rho}{T^\circ\rho^\circ}\right)\mu^\circ \tag{2-162}$$

from which one can calculate a values if μ is known as a function of temperature.

Reference should be made to the book by Ferry[21] for a more detailed treatment of the viscoelastic behavior of polymers. He discusses in detail experimental techniques and results obtained on a wide variety of polymeric systems and interprets the results in terms of molecular theory.

Problems

2-1 Show that the two forms of the equation of continuity

$$\frac{D\rho}{Dt} = -\rho(\nabla \cdot \boldsymbol{v})$$

$$\frac{\partial\rho}{\partial t} = -(\nabla \cdot \rho\boldsymbol{v})$$

are equivalent Also, show that both equations reduce to

$$(\boldsymbol{\Delta} \cdot \boldsymbol{v}) = 0$$

for incompressible fluids.

2-2 Show that the two forms of the equation of motion (momentum equation)

$$\rho\frac{D\boldsymbol{v}}{Dt} = -\nabla P + [\nabla \cdot \boldsymbol{\tau}] + \rho\boldsymbol{g}$$

$$\frac{\partial(\rho\boldsymbol{v})}{\partial t} = -[\nabla \cdot \rho\boldsymbol{v}\boldsymbol{v}] - \nabla P + [\nabla \cdot \boldsymbol{\tau}] + \rho\boldsymbol{g}$$

are equivalent.

2-3 What modifications should be made to equation (2-3) if an exothermic chemical reaction were occurring in the fluid?

2-4 Derive the equation of the temperature profile for the problem discussed in Example 2-1 when the upper plate is at temperature T_1 and the lower plate at temperature T_2.

2-5 Show that the position variables in cylindrical and rectangular coordinates are related by the equations

$$x = r\cos\theta$$

$$y = r\sin\theta$$

$$z = z$$

[21] J. D. Ferry, *Viscoelastic Properties of Polymers*, John Wiley, 1961.

and the velocity components by the equations

$$v_x = v_r \cos\theta - v_\theta \sin\theta$$
$$v_y = v_r \sin\theta + v_\theta \cos\theta$$
$$v_z = v_z$$

Then prove the following relationship between the stress components:

$$\tau_{yz} = \tau_{rz} \sin\theta + \tau_{z\theta} \cos\theta$$

2-6 Derive the rheological equation for simple shear for a material that behaves as a Hookean solid below a yield stress S_0 and as a Newtonian fluid above the yield stress. Devise a mechanical model that simulates the behavior of this material.

2-7 Determine the strain history at a point in the material described in Problem 2-6 above when the stress history at the point is that which is shown in Figure 2-6*a*. Assume ($S < S_0 < 2S$).

2-8 Calculate the time required for 50% of the elastic strain of a Voigt solid to be recovered upon removal of the stress. How much time would be required for 90% of the strain to be recovered?

2-9 Find values of the parameters μ_0 and B of the Powell-Eyring equation applicable to the flow data of Figure 2-9.

2-10 Show that the third scalar invariant of the tensor $\boldsymbol{\Delta}$ vanishes for the flow described in Example 2-3.

2-11 Show that the generalized form of the Reiner-Philippoff equation reduces to equation (3) of Table 2-4 for simple shear.

2-12 Determine the values of E and b for the three materials whose viscosity data are plotted in Figures 2-11 and 2-12.

2-13 For the flow described in Example 2-1, calculate the maximum temperature difference in the fluid if $V = 10$ cm/sec, $H = 0.1$ cm, $k = 10^{-3}$ cal/cm-sec-deg C, and $\mu = 10^4$ poise.

2-14 Consider steady isothermal laminar flow of a Newtonian fluid through a tube having a diameter of 0.1 cm and a length of 10 cm. If the viscosity of the fluid is 100 poise and the pressure drop is 10 atm, what is the velocity of the fluid at the tube center?

2-15 Upon integration, equation (2-74) becomes equation (2-75). Show the steps in this integration.

2-16 Derive the rheological equation for a material whose behavior in simple shear is simulated by a mechanical model consisting of a Voigt element and a spring connected in series. What would be the electrical analog of this material?

2-17 For the material described in Problem 2-16, derive an equation relating strain at a point to time for the case where a constant stress S is imposed at the point under consideration at time t_0 and released at time t_1.

2-18 In a constant stress experiment, the viscosity of a non-Newtonian fluid decreased from 100 to 50 poise as the temperature increased from 20° C to 40° C. It is known that the fluid obeys the power law and has a flow index of 0.80. Compute E_τ for the material and estimate its viscosity at 40° C for the case where the change occurs at constant shear rate. Estimate b_τ and $b_{\dot\gamma}$ for the material.

2-19 Derive the relationship

$$b_{\dot{\gamma}} = nb_{\tau}$$

Note that it is analogous to equation (2-103).

2-20 If $G(t)$ represents the ratio of stress to strain in a stress relaxation experiment, show that for the Maxwell liquid

$$G(t) = Ge^{-t/\lambda}$$

and for the Voigt solid

$$G(t) = G$$

If $J(t)$ represents the ratio of strain to stress in a shear creep experiment, show that for the Maxwell liquid

$$J(t) = J + \frac{t}{\mu}$$

and for the Voigt solid

$$J(t) = J[1 - e^{-t/\lambda}]$$

where $J = (1/G)$.

2-21 Show that $|G^*|$ represents the ratio of peak stress to peak strain.

2-22 Show that if a viscoelastic material is subjected to a strain that is a sinusoidal function of time, then the maximum energy stored by the material at any time during a cycle is $G'\gamma_0^2$.

chapter 3

Flow in Tubes and Channels

This chapter deals with fluid flow in channels having circular, rectangular, and annular cross sections—a subject of considerable importance in polymer processing. Because of the extremely high viscosities of most polymeric fluids (10^6 poise being not uncommon), turbulence is seldom encountered in polymer processing. Hence this chapter considers laminar flow only. Internal heat generation, end effects, elastic effects, and capillary tube viscometry are discussed briefly. The chapter concludes with an analysis of wire coating and sheeting dies.

3-1 Isothermal Flow in Tubes—Special Results

The steady isothermal laminar flow of incompressible fluids through straight horizontal tubes of circular cross section is analyzed in this section. Two classes of fluids are considered: power law fluids (which include Newtonian fluids as a special case) and Bingham plastics. A more general but less useful treatment is presented in the next section. In both sections the key assumptions are that the fluid does not "slip" at the wall of the tube and that the fluid viscosity is uniquely determined by its temperature and state of shear.

In the following analysis, z represents the variable distance measured along the tube axis and r is the variable radial distance. The z-axis is so oriented that the pressure increases in the direction of increasing z; hence the gradient $(\partial P/\partial z)$ is positive and the flow is in the negative direction.

The z-component of the momentum equation in cylindrical coordinates, obtained from Table 2-2, is written

$$\frac{\partial P}{\partial z} = \frac{1}{r}\frac{\partial}{\partial r}[r\tau_{rz}] \tag{3-1}$$

which becomes upon integration (and upon noting that because of symmetry $\tau(0) = 0$)

$$\tau_{rz} = \frac{r}{2}\left(\frac{\partial P}{\partial z}\right) \tag{3-2}$$

which is a general relationship independent of the nature of the fluid. It shows that the shear stress is a linear function of radial distance, being zero at the tube center and reaching its largest value at the tube wall.

Equation (3-2) could have been derived directly from a force balance applied to the fluid in the tube rather than through specialization of the general momen-

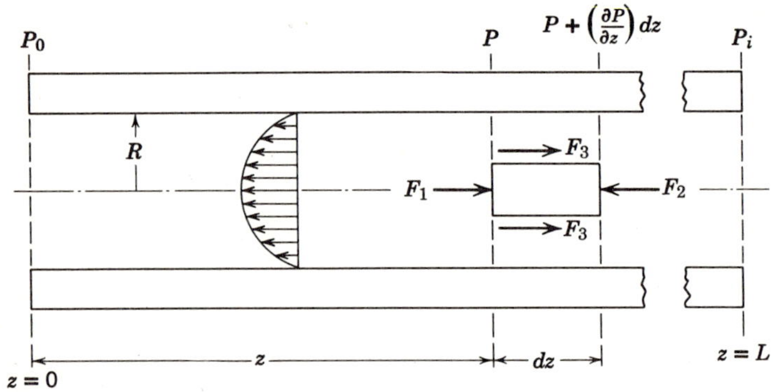

Figure 3-1. Forces acting on a fluid element for steady laminar flow in a tube.

tum equation. Consider the forces acting on the cylindrical element of fluid of radius r and length dz shown in Figure 3-1. Since the motion is steady, there are no unbalanced forces and

$$\sum F_z = 0 = F_1 + F_2 + F_3 \tag{3-3}$$

where the forces F_1, F_2, and F_3 act on the left end, the right end, and the cylindrical surface of the element, respectively. They are given by the equations

$$F_1 = \pi r^2 P$$

$$F_2 = -\pi r^2\left[P + \left(\frac{\partial P}{\partial z}\right) dz\right]$$

$$F_3 = 2\pi r \tau_{rz}\, dz$$

which, when introduced into equation (3-3), yield equation (3-2).

The stress component τ_{rz} (from Table 2-3 with μ replaced by η for non-Newtonian flow) is given by the equation

$$\tau_{rz} = \eta\left(\frac{\partial v_z}{\partial r}\right)$$

which, upon dropping the subscripts and introducing $\dot{\gamma}$, becomes

$$\tau = \eta\dot{\gamma} \tag{3-4}$$

Equation (3-4) is the same as (2-39), the rheological equation for non-Newtonian fluids in simple shear.

The viscosity η of power law fluids (see Table 2-4) is related to $\dot{\gamma}$ by the equation

$$\eta = \eta^\circ \left|\frac{\dot{\gamma}}{\dot{\gamma}^\circ}\right|^{n-1} \tag{3-5}$$

If equations (3-5) and (3-4) are introduced into (3-2) and the resulting equation integrated (the details of this are shown in Example 2-2), one obtains

$$v_z = v_0\left[1 - \left(\frac{r}{R}\right)^{(n+1)/n}\right] \tag{3-6}$$

where v_0, which represents the velocity at the center of the tube, is given by the equation

$$v_0 = -\left(\frac{nR\dot{\gamma}^\circ}{n+1}\right)\left[\frac{R}{2\eta^\circ\dot{\gamma}^\circ}\left(\frac{\partial P}{\partial z}\right)\right]^{1/n} \tag{3-7}$$

The negative sign indicates that the flow is in the negative direction.

The volumetric flow rate Q through a tube of circular cross section is, in general, given by the integral

$$Q = 2\pi\int_0^R rv_z\,dr \tag{3-8}$$

Introducing (3-6) into (3-8) and performing the indicated integration gives

$$Q = \left(\frac{n+1}{3n+1}\right)\pi R^2 v_0 \tag{3-9}$$

The average fluid velocity $\bar{v}_z$, defined as $Q/\pi R^2$, is related to the maximum velocity v_0 by the equation

$$\bar{v}_z = \left(\frac{n+1}{3n+1}\right)v_0 \tag{3-10}$$

and the local velocity v_z is related to the average velocity $\bar{v}_z$ by the equation

$$v_z = \bar{v}_z\left(\frac{3n+1}{n+1}\right)\left[1 - \left(\frac{r}{R}\right)^{(n+1)/n}\right] \tag{3-11}$$

which is obtained from equations (3-6) and (3-10). Velocity profiles for various values of n are shown in Figure 3-2, where the ratio $v_z/\bar{v}_z$ is plotted against (r/R). The curve labeled $n = 1$ is a parabola, a characteristic of

Newtonian flow. For $n < 1$ (pseudoplastic fluids), the viscosity of the fluid is less at the tube wall than at the center and the parabola is flattened. The converse is true for $n > 1$.

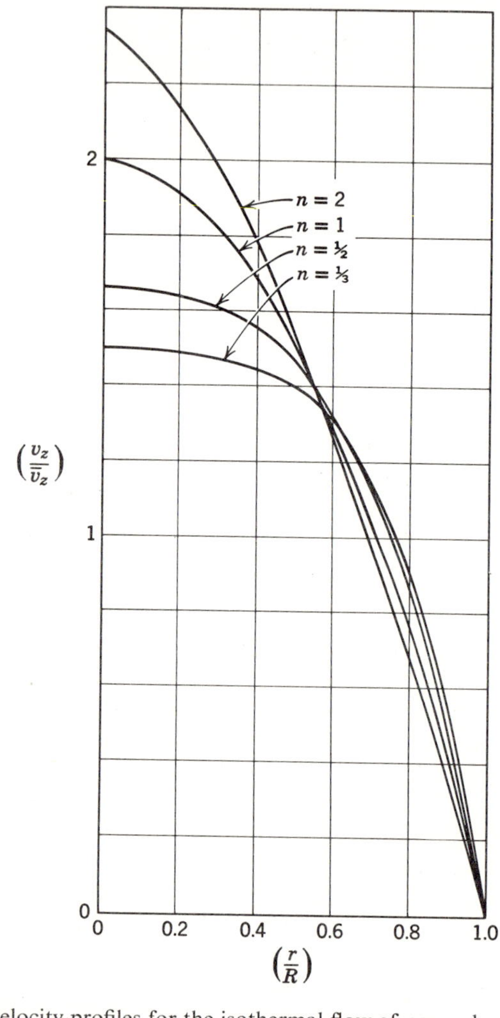

Figure 3-2. Velocity profiles for the isothermal flow of power law fluids through tubes.

The shear stress τ_w at the wall of the tube, obtained directly from equation (3-2), is

$$\tau_w = \frac{R}{2}\left(\frac{\partial P}{\partial z}\right) \tag{3-12}$$

and the shear rate at the wall of the tube, found by differentiating (3-6) and then eliminating v_0 through (3-9), is

$$\dot{\gamma}_w = -\left(\frac{3n+1}{n}\right)\left(\frac{Q}{\pi R^3}\right) \tag{3-13}$$

When an incompressible fluid flows through a tube having a uniform cross section, the volumetric flow rate Q is the same through all cross sections perpendicular to the axis. Examination of equations (3-7) and (3-9) shows that the gradient $(\partial P/\partial z)$ must be independent of z and constant. Hence, for a tube of length L, one can substitute $(\Delta P/L)$ for $(\partial P/\partial z)$ in the above equations, where ΔP represents the total pressure drop over the tube.

For convenience, the formulas derived above for the isothermal flow of power law fluids through tubes are summarized in Table 3-1. Formulas for the special case of Newtonian flow ($n = 1$) are also included in the table.

Table 3-1

Summary of Formulas for the Isothermal Flow of Power Law Fluids Through Tubes

Quantity	Power Law Formulas	Newtonian Formulas
Shear stress at tube wall	$\tau_w = \left(\frac{R\Delta P}{2L}\right)$	$\tau_w = \left(\frac{R\Delta P}{2L}\right)$
Shear stress	$\tau = \tau_w\left(\frac{r}{R}\right)$	$\tau = \tau_w\left(\frac{r}{R}\right)$
Velocity at tube center	$v_0 = -\left(\frac{n\dot{\gamma}^\circ R}{n+1}\right)\left[\frac{R\Delta P}{2\eta^\circ\dot{\gamma}^\circ L}\right]^{1/n}$	$v_0 = -\left(\frac{R^2\Delta P}{4\mu L}\right)$
Shear rate	$\dot{\gamma} = -\left(\frac{v_0}{R}\right)\left(\frac{n+1}{n}\right)\left(\frac{r}{R}\right)^{1/n}$	$\dot{\gamma} = -\left(\frac{2v_0}{R}\right)\left(\frac{r}{R}\right)$
Velocity	$v_z = v_0\left[1 - \left(\frac{r}{R}\right)^{(n+1)/n}\right]$	$v_z = v_0\left[1 - \left(\frac{r}{R}\right)^2\right]$
Average velocity	$\bar{v}_z = \left(\frac{n+1}{3n+1}\right)v_0$	$\bar{v}_z = \left(\frac{v_0}{2}\right)$
Volumetric flow rate	$Q = -\left(\frac{n\pi R^3\dot{\gamma}^\circ}{3n+1}\right)\left[\frac{R\Delta P}{2\eta^\circ\dot{\gamma}^\circ L}\right]^{1/n}$	$Q = -\frac{\pi R^4\Delta P}{8\mu L}$
Shear rate at tube wall	$\dot{\gamma}_w = -\left(\frac{3n+1}{n}\right)\frac{Q}{\pi R^3}$	$\dot{\gamma}_w = -\frac{4Q}{\pi R^3}$

η° represents the non-Newtonian viscosity of the fluid in the standard state defined by the shear rate $\dot{\gamma}^\circ$.

Example 3-1.

In this example, the following data for the steady flow of a fluid, having a density of 0.917 grams/cm³, through a tube having a diameter of 0.30 in. and a length of 6 in.. are analyzed:

ΔP Pressure Drop (psi)	W Flow Rate (lb/hr)
150	15
200	22
255	30
320	40

Reference to Table 3-1 shows that, for power law fluids, the flow rate is proportional to the pressure drop raised to the $(1/n)$th power. The applicability of the power law to the above data is confirmed by the logarithmic plot of the flow data

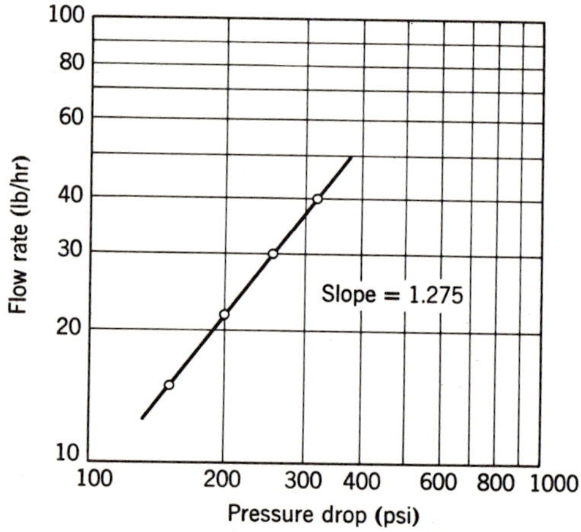

Figure 3-3. Logarithmic plot of flow data.

shown in Figure 3-3. The slope of the line is 1.275 and the reciprocal of the slope, the flow index n of the fluid, is 0.785.

The shear rate and the shear stress at the tube wall, calculated at a flow rate of 30 lb/hr, using the following data

$$n = 0.785$$

$$R = 0.381 \text{ cm } (0.15 \text{ in.})$$

$$Q = W/\rho = 4.13 \text{ cm}^3/\text{sec}$$

$$\Delta P = 17.5 \times 10^6 \text{ dynes/cm}^2 \ (255 \text{ psi})$$

$$L = 15.2 \text{ cm } (6 \text{ in.})$$

are as follows:

$$\tau_w = \frac{R\Delta P}{2L} = 21.9 \times 10^4 \text{ dynes/cm}^2$$

$$\dot{\gamma}_w = \left(\frac{3n+1}{n}\right)\left(\frac{Q}{\pi R^3}\right) = 102 \text{ sec}^{-1}$$

and the viscosity of the fluid at the tube wall is

$$\eta = \left(\frac{\tau_w}{\dot{\gamma}_w}\right) = \frac{21.9 \times 10^4}{102} = 2130 \text{ poise}$$

The value of $\eta°$ can be calculated directly from the rheological equation for the power law fluid

$$\eta = \eta° \left|\frac{\dot{\gamma}}{\dot{\gamma}°}\right|^{n-1}$$

Assigning a value of 1 sec^{-1} to the standard shear rate $\dot{\gamma}°$

$$\eta° = \eta\dot{\gamma}^{1-n} = (2130)(102)^{0.215} = 5730 \text{ poise}$$

and the rheological equation for the fluid becomes

$$\eta = 5730\dot{\gamma}^{-0.215}$$

The steady isothermal flow of a Bingham plastic through a straight horizontal tube is now analyzed. The result of the analysis is the Buckingham-Reiner equation, which relates the volumetric flow rate to the pressure drop.

The following derivation is similar to that of Reiner.[1] In it the rheological equation of the Bingham plastic, (2-38), is combined with the general relation for shear stress, (3-2), giving

$$\frac{\partial v_z}{\partial r} = \frac{r}{2\zeta}\left(\frac{\partial P}{\partial z}\right) - \frac{S_0}{\zeta} \tag{3-14}$$

Note that (3-14) applies only to the region in which the material undergoes shear flow, the region where the shear stress is greater than the yield stress. This is the region ($r_* < r < R$) in Figure 3-4.

In the center region of the tube where $r < r_*$, the shear stress is less than the yield stress and the material behaves like a solid, moving through the tube like a plug. At the radial distance r_*, the shear stress is equal to the yield stress and (3-2) shows that

$$r_* = \frac{2S_0}{p} \tag{3-15}$$

where $p = (\partial P/\partial z)$.

[1] M. Reiner, *Deformation, Strain and Flow*, Interscience Publishers, 1961, p. 117.

The fluid velocity within the region of shear flow, found by integrating (3-14), is given by the equation

$$v_z = -\frac{R^2 p}{4\zeta}\left[1 - \left(\frac{r}{R}\right)^2\right] + \frac{RS_0}{\zeta}\left[1 - \left(\frac{r}{R}\right)\right] \tag{3-16}$$

which satisfies the boundary condition $v_z(R) = 0$.

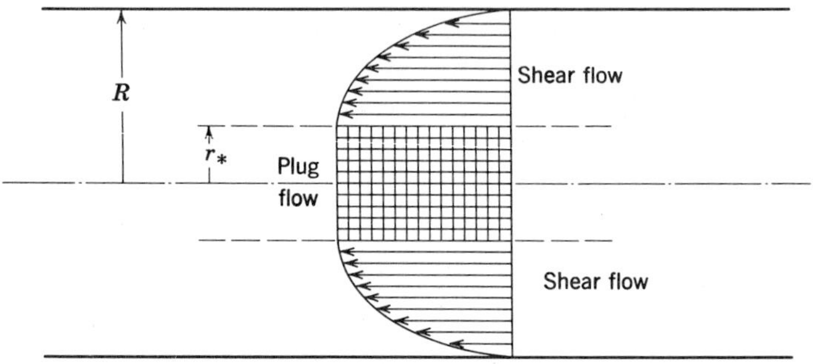

Figure 3-4. Velocity profile for the steady flow of a Bingham plastic through a tube.

The velocity of the plug v_* is obtained from equation (3-16) by replacing r with r_*. Then, by defining a dimensionless variable, $X = (r_*/R) = (2S_0/Rp)$, the plug velocity can be written

$$v_* = -\frac{R^2 p}{4\zeta}[1 - X^2] + \frac{RS_0}{\zeta}[1 - X] \tag{3-17}$$

The volumetric flow rate Q through the tube is given by

$$Q = \pi r_*^2 v_* + 2\pi \int_{r_*}^{R} r v_z \, dr \tag{3-18}$$

where the first term represents the volumetric flow rate of the material in the plug and the second term, the volumetric flow rate of the material that is in shear flow. By introducing the above expressions for velocity into (3-18), one obtains

$$Q = -\frac{\pi R^4 p}{8\zeta}\left[1 - \frac{4X}{3} + \frac{X^4}{3}\right] \tag{3-19}$$

Note that if $S_0 = 0$, the rheological equation of the Bingham plastic reduces to Newton's law and (3-19) becomes the Poiseuille equation.

3-2 Isothermal Flow in Tubes—General Treatment

A more general case of tube flow is now considered, that in which the form of the functional relationship $f(\dot{\gamma})$ in the equation

$$\tau = \eta(\dot{\gamma})\dot{\gamma} = f(\dot{\gamma}) \tag{3-20}$$

is either unknown or cannot be represented accurately by a convenient empirical relationship. Two problems are considered. First, a method is given whereby the *flow curve* can be determined directly from experimental data consisting of flow rate versus pressure drop for flow through a tube, and, second, a rigorous method is given whereby such flow data can be scaled up for the design of pipe lines.

Since a flow curve consists of a plot of τ versus $\dot{\gamma}$, the problem is to derive τ and $\dot{\gamma}$ from $Q - \Delta P$ data. The following general relation for shear stress at the wall of the tube has already been derived:

$$\tau_w = \frac{R\Delta P}{2L} \tag{3-21}$$

If the corresponding shear rate $\dot{\gamma}_w$ at the wall can be calculated, the flow curve is established. The problem, therefore, is to derive a general relationship independent of the nature of the fluid, for the shear rate at the wall.

The solution of this problem, which was originally obtained by Rabinowitsch,[2] can be written

$$-\dot{\gamma}_w = \frac{1}{\pi R^3}\left[3Q + \Delta P \frac{dQ}{d\Delta P}\right] \tag{3-22}$$

where Q is the volumetric flow rate and ΔP the pressure drop. The negative sign indicates that the shear rate $\dot{\gamma}$ is positive if the flow is in the negative direction, and vice versa.

The following development of the Rabinowitsch equation is based on the concise derivation outlined in *Transport Phenomena*.[3] It starts with the general relation for the volumetric flow rate, equation (3-8). Integration by parts yields the following result:

$$Q = 2\pi \int_0^R r v_z \, dr = 2\pi \left[\frac{r^2 v_z}{2} \Bigg|_0^R - \int_0^R \frac{r^2}{2} \, dv_z \right] \tag{3-23}$$

At the wall of the tube, where $r = R$, the value of the first term on the right side of (3-23) is zero because v_z is zero. This term is also zero at the center of the tube where $r = 0$. Therefore (3-23) reduces to

$$Q = -\pi \int_0^R r^2 \, dv_z = -\pi \int_0^R r^2 \left(\frac{\partial v_z}{\partial r}\right) dr \tag{3-24}$$

The shear stress τ can be expressed in the form

$$\tau = \left(\frac{r}{R}\right) \tau_w$$

[2] B. Rabinowitsch, *Z. Physik. Chem.*, **A145,** 1 (1929).

[3] R. B. Bird, W. E. Stewart, and E. N. Lightfoot, *Transport Phenomena*, John Wiley, 1960, p. 67.

by combining equations (3-2) and (3-12). Introducing it as a new variable of integration, equation (3-24) becomes

$$Q = -\frac{\pi R^3}{\tau_w^3}\int_0^{\tau_w} \dot{\gamma}\tau^2\, d\tau \tag{3-25}$$

which can be written in the equivalent form

$$\frac{\tau_w^3 Q}{\pi R^3} = -\int_0^{\tau_w} \dot{\gamma}\tau^2\, d\tau \tag{3-26}$$

Now differentiate both sides of (3-26) with respect to τ_w. The differentiation of the integral follows the Leibnitz rule.[4] One obtains

$$\frac{1}{\pi R^3}\left[\tau_w^3 \frac{dQ}{d\tau_w} + 3\tau_w^2 Q\right] = -\dot{\gamma}_w \tau_w^2 \tag{3-27}$$

but since

$$\tau_w = \frac{R\Delta P}{2L}$$

equation (3-27) can be written

$$-\dot{\gamma}_w = \frac{1}{\pi R^3}\left[3Q + \Delta P \frac{dQ}{d\Delta P}\right] \tag{3-28}$$

which is the Rabinowitsch equation.

For the special case where the fluid obeys the power law, the pressure drop is proportional to the nth power of Q, as shown by the formulas of Table 3-1. Therefore

$$\Delta P = KQ^n$$

where K is the proportionality constant. Differentiating the above expression gives

$$\frac{d\Delta P}{dQ} = KnQ^{n-1}$$

so that

$$\Delta P \frac{dQ}{d\Delta P} = \frac{Q}{n}$$

Hence the Rabinowitsch equation, for the special case of power law fluids, reduces to

$$-\dot{\gamma}_w = \frac{1}{\pi R^3}\left[3Q + \frac{Q}{n}\right] = \frac{Q}{\pi R^3}\left[\frac{3n+1}{n}\right]$$

which is identical to equation (3-13) derived specifically for power law fluids.

The Rabinowitsch equation is the basis of a very useful procedure for

[4] See, for example, I. S. Sokolnikoff and R. M. Redheffer, *Mathematics of Physics and Modern Engineering*, McGraw-Hill, 1958, p. 262.

the scale-up of laboratory flow data in the design of pipe lines for non-Newtonian fluids. Start by writing (3-27) in the form

$$-\dot{\gamma}_w = \frac{3}{4}\left(\frac{4Q}{\pi R^3}\right) + \frac{\tau_w}{4}\frac{d}{d\tau_w}\left(\frac{4Q}{\pi R^3}\right) \tag{3-29}$$

Now introduce a new variable Γ defined by the equation

$$\Gamma = \frac{4Q}{\pi R^3} \tag{3-30}$$

so that (3-29) becomes

$$-\dot{\gamma}_w = \frac{3}{4}\Gamma + \frac{\tau_w}{4}\frac{d\Gamma}{d\tau_w} \tag{3-31}$$

We note that the variables Γ and τ_w are the Reiner[5] consistency variables.

Equation (3-20) applied to the fluid at the tube wall is written

$$\tau_w = f(\dot{\gamma}_w)$$

and upon introducing (3-31) into it, one obtains

$$\tau_w = f\left[-\frac{3}{4}\Gamma - \frac{\tau_w}{4}\frac{d\Gamma}{d\tau_w}\right] \tag{3-32}$$

which shows that there is a unique functional relationship between the variables Γ and τ_w. In other words, if Γ is plotted against τ_w, a single curve that is unique for the fluid at a given temperature will result. This is the basis of the scale-up procedure illustrated in Example 3-2. Note that if the scale-up is made at constant Γ, one laboratory experiment is required. Scale-up problems involving non-Newtonian fluids are discussed in some detail by Alves, Boucher, and Pigford.[6]

Example 3-2.

It is required to determine the diameter of a 500-ft pipe line that will transport a non-Newtonian fluid at the rate of 1.0 gal/min with a maximum pressure drop of 200 psi at room temperature.

The following laboratory flow data on a sample of the fluid were obtained at room temperature by using a tube having a diameter of 0.10 in. and a length of 10.0 in.

ΔP *(psi)*	W *(grams/min)*
2.5	1.15
5.0	2.45
10	5.30
20	11.7
40	27.5

[5] M. Reiner, *op. cit.*, p. 19.

[6] G. E. Alves, D. F. Boucher, and R. L. Pigford, *Chem. Engr. Prog.*, **48,** 385 (1952).

The density of the fluid is 0.90 grams/cm^3.

If these data are plotted on logarithmic graph paper, the points do not fall on a *straight* line. This indicates that the power law is not applicable. The general scale-up procedure described above is used.

The first step is to express the laboratory data as Γ and τ_w, which are defined by the equations

$$\Gamma = \frac{4Q}{\pi R^3} \qquad \tau_w = \frac{R\Delta P}{2L}$$

In this example the data are converted to cgs units although any consistent system of units can be used. The following example shows the calculations of the first point.

$$\Delta P = 2.5 \text{ psi} = 17.2 \times 10^4 \text{ dynes/cm}^2$$

$$Q = \frac{1.15}{(60)(0.90)} = 0.0212 \text{ cm}^3\text{/sec}$$

$$R = 0.05 \text{ in. } (0.127 \text{ cm})$$

$$L = 10.0 \text{ in. } (25.4 \text{ cm})$$

$$\Gamma = \frac{(4)(0.0212)}{(\pi)(0.127)^3} = 13.2 \text{ sec}^{-1}$$

$$\tau_w = \frac{(0.127)(17.2 \times 10^4)}{(2)(25.4)} = 430 \text{ dynes/cm}^2$$

The values for all the points are shown in the table below.

τ_w *(dynes/cm^2)*	Γ *(sec^{-1})*
430	13.2
860	28.2
1720	60.9
3440	134
6880	316

Figure 3-5 shows a plot of these data on logarithmic graph paper. Since τ_w is a unique function of Γ, the curve applies to the flow of the fluid under consideration through *any* tube at room temperature. The operating point for the pipe line must fall on the curve or on an extension of it.

The following values apply to the pipe line:

$$L = 500 \text{ ft} = 1.525 \times 10^4 \text{ cm}$$

$$Q = 1.0 \text{ gal/min} = 63 \text{ cm}^3\text{/sec}$$

$$\Delta P = 200 \text{ psi} = 13.7 \times 10^6 \text{ dynes/cm}^2$$

The unknown quantity is the radius of the pipe line R; thus substituting the above values into the formulas for Γ and τ_w gives

$$\tau_w = 450R \qquad \Gamma = \frac{80.5}{R^3}$$

The problem is to find the particular R that, when substituted into the above expressions, will result in a point falling on the curve of Figure 3-5. The value could be found by a trial-and-error procedure, but it is preferable to use a more direct graphical method. Eliminating R between the above expressions gives

$$\tau_w \propto \Gamma^{-1/3}$$

Therefore the intersection of a straight line having a slope of -0.33 with the curve of Figure 3-5 gives the location of the required point. The position of the

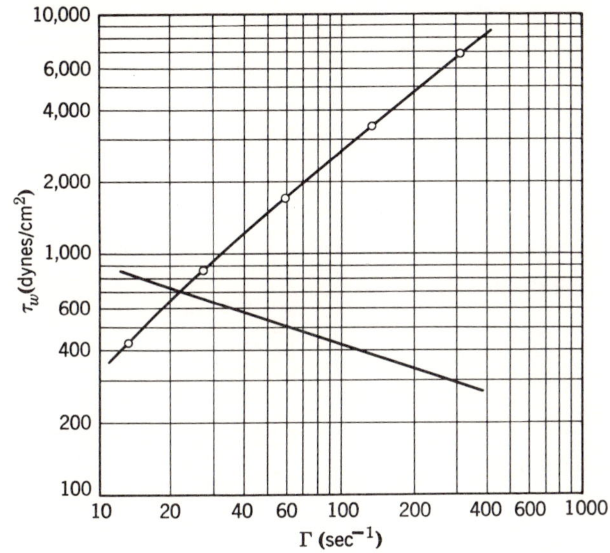

Figure 3-5. Graphical solution of flow problem.

straight line can be fixed by setting $R = 1$ to obtain the point

$$(\tau_w, \Gamma) = (450, 80.5)$$

The value of τ_w at the intersection of the two lines is 700 dynes/cm^2 and the required radius R is given by the calculation

$$R = \frac{\tau_w}{450} = \frac{700}{450} = 1.55 \text{ cm}$$

The required diameter is 3.10 cm or 1.22 in.

It should be emphasized that the preceding discussion is limited to steady isothermal laminar flow. With Newtonian fluids, flow in tubes is laminar until the Reynolds number

$$N_{\text{Re}} = \frac{D\bar{v}\rho}{\mu}$$

exceeds about 2100. This means that the conventional logarithmic plot of friction factor f versus Reynolds number N_{Re} is straight in the region $N_{\text{Re}} < 2100$. This can be demonstrated by solving the Poiseuille equation (see Table 3-1) for ΔP to obtain

$$\Delta P = \frac{8\mu LQ}{\pi R^4} = \frac{8\mu L\bar{v}}{R^2}$$

and then introducing this result into the defining equation for the friction factor

$$\frac{1}{f} = \frac{(\frac{1}{2}\rho\bar{v}^2)}{(R\Delta P/2L)} = \frac{\rho\bar{v}^2}{2\tau_w}$$

to obtain

$$\frac{1}{f} = \frac{1}{16}\left[\frac{D\bar{v}\rho}{\mu}\right]$$

If we retain the same definition of friction factor for non-Newtonian fluids, and if we define a generalized Reynolds number for non-Newtonian fluids by the equation

$$N_{\text{Re}} = \left[\frac{8\,\rho\bar{v}^2}{\tau_w}\right] \tag{3-33}$$

we see that a logarithmic plot of f versus N_{Re} will be straight if the flow is laminar. According to Metzner and Reed,[7] the transition to turbulence occurs at a Reynolds number of about 2100 providing that highly elastic fluids are excluded.

Example 3-3.

Referring to the previous example, we calculate the Reynolds number for non-Newtonian flow when the flow rate is 27.5 grams/min. We have the following data from that example:

$$\tau_w = 6880 \text{ dynes/cm}^2$$
$$W = 27.5 \text{ grams/min}$$
$$\rho = 0.90 \text{ grams/cm}^3$$
$$\bar{v} = \frac{W}{60\pi R^2\rho} = 10.0 \text{ cm/sec}$$

Substituting these values into equation (3-33) gives

$$N_{\text{Re}} = 0.104$$

The polymer melts and solutions encountered in polymer processing almost always have high viscosities. Hence Reynolds numbers are small and the flow is almost always laminar.

---

[7] A. B. Metzner and J. C. Reed, *A.I.Ch.E. Journal*, **1**, 434 (1955).

3-3 Non-Isothermal Flow in Tubes

In this section the effects of heat conduction and heat generation during the laminar flow of a viscous fluid through a straight tube of circular cross section are discussed. A thermodynamic analysis of this problem is presented in Section 8-1, where the first law of thermodynamics is used to derive an over-all energy balance for steady flow. The over-all energy balance enables the total heat and work effects to be calculated, from which the average temperature rise of the fluid as it passes through the tube can be estimated.

In this section a more detailed analysis of the problem is made. The equations of continuity, momentum, energy, and the rheological equation applying to the fluid are used in the mathematical formulation of the problem. Solutions that have been obtained for various special cases are then discussed.

Cylindrical coordinates are used because the flow occurs in a tube. Because steady flow only will be considered, all derivatives with respect to time are zero. Furthermore, we shall assume that v_z is the only nonzero component of the velocity vector (v_θ is zero because of axial symmetry and v_r is assumed to be negligible in comparison to v_z), and we shall neglect body forces. Upon introducing the appropriate expressions for the stress tensor components (from Table 2-3), the equations of continuity, momentum, and energy (from Table 2-2) become (neglecting τ_{zz})

$$\rho\left(\frac{\partial v_z}{\partial z}\right) + v_z\left(\frac{\partial \rho}{\partial z}\right) = 0 \tag{3-34}$$

$$\rho v_z\left(\frac{\partial v_z}{\partial z}\right) = -\left(\frac{\partial P}{\partial z}\right) + \frac{1}{r}\frac{\partial}{\partial r}\left[r\eta\left(\frac{\partial v_z}{\partial r}\right)\right] \tag{3-35}$$

$$\rho c_v v_z\left(\frac{\partial T}{\partial z}\right) = \frac{k}{r}\frac{\partial}{\partial r}\left[r\,\frac{\partial T}{\partial r}\right] + k\left(\frac{\partial^2 T}{\partial z^2}\right) - T\left(\frac{\partial P}{\partial T}\right)_\rho\left(\frac{\partial v_z}{\partial z}\right) + \eta\left(\frac{\partial v_z}{\partial r}\right)^2 \tag{3-36}$$

The total differential of the fluid density ρ, choosing T and P to represent the state of the fluid, is written

$$d\rho = \left(\frac{\partial \rho}{\partial T}\right)_P dT + \left(\frac{\partial \rho}{\partial P}\right)_T dP \tag{3-37}$$

which can be put in the equivalent form

$$d\rho = -\kappa\rho\, dT + \beta\rho\, dP \tag{3-38}$$

by introducing the quantities κ and β, where κ is the coefficient of thermal expansion and β the compressibility of the fluid. These quantities are defined by the equations

$$\kappa = \frac{-1}{\rho}\left(\frac{\partial \rho}{\partial T}\right)_P \qquad \beta = \frac{1}{\rho}\left(\frac{\partial \rho}{\partial P}\right)_T \tag{3-39}$$

from which it can be shown that

$$\left(\frac{\partial P}{\partial T}\right)_\rho = \left(\frac{\kappa}{\beta}\right) \tag{3-40}$$

To consider the variation of the fluid density with the z position coordinate, equation (3-38) is written

$$\left(\frac{\partial \rho}{\partial z}\right) = -\kappa\rho\left(\frac{\partial T}{\partial z}\right) + \beta\rho\left(\frac{\partial P}{\partial z}\right) \tag{3-41}$$

and upon introducing (3-41) into the equation of continuity, one obtains

$$\frac{\partial v_z}{\partial z} = v_z\left[\kappa\left(\frac{\partial T}{\partial z}\right) - \beta\left(\frac{\partial P}{\partial z}\right)\right] \tag{3-42}$$

If one now introduces (3-40) and (3-42) into the equations of momentum and energy, (3-35) and (3-36), one obtains

$$\rho v_z^2\left[\kappa\left(\frac{\partial T}{\partial z}\right) - \beta\left(\frac{\partial P}{\partial z}\right)\right] = -\left(\frac{\partial P}{\partial z}\right) + \frac{1}{r}\frac{\partial}{\partial r}\left[r\eta\left(\frac{\partial v_z}{\partial r}\right)\right] \tag{3-43}$$

$$\rho c_v v_z\left(\frac{\partial T}{\partial z}\right) = \frac{k}{r}\frac{\partial}{\partial r}\left[r\left(\frac{\partial T}{\partial r}\right)\right] + k\left(\frac{\partial^2 T}{\partial z^2}\right) - T\kappa v_z\left[\frac{\kappa}{\beta}\left(\frac{\partial T}{\partial z}\right) - \left(\frac{\partial P}{\partial z}\right)\right] + \eta\left(\frac{\partial v_z}{\partial r}\right)^2 \tag{3-44}$$

Equations (3-43) and (3-44) constitute a pair of simultaneous, nonlinear, partial differential equations. Solutions of certain special cases of these equations have been obtained. First, consider the temperature profile in a power law fluid at a sufficient distance from the tube inlet so that v_z and T no longer change with the z position coordinate. If the simplifying assumptions of constant fluid density ($\kappa = \beta = 0$) and viscosity independent of temperature are made, equations (3-43) and (3-44) reduce to the following:

$$\left(\frac{\partial P}{\partial z}\right) = \frac{\eta^\circ}{r}\frac{\partial}{\partial r}\left[r\left(\frac{\partial v_z}{\partial r}\right)^n\right] \tag{3-45}$$

$$-\frac{k}{r}\frac{\partial}{\partial r}\left[r\left(\frac{\partial T}{\partial r}\right)\right] = \eta^\circ\left(\frac{\partial v_z}{\partial r}\right)^{n+1} \tag{3-46}$$

The boundary conditions for the problem are as follows:

$$\begin{aligned} v_z(R) &= 0 \qquad & T(R) &= Tw \\ v_z'(0) &= 0 \qquad & T'(0) &= 0 \end{aligned} \tag{3-47}$$

where the prime indicates differentiation with respect to r. Note that equation (3-45) contains only the one independent variable v_z; thus a solution of (3-45) and (3-46) is easily obtained by the method outlined in Example 2-4.

It is easy to remove one of the restrictive simplifying assumptions made above, namely, that the fluid density is not temperature dependent. Then the energy equation becomes

$$\frac{k}{r}\frac{\partial}{\partial r}\left[r\left(\frac{\partial T}{\partial r}\right)\right] + T\kappa v_z\left(\frac{\partial P}{\partial z}\right) + \eta^\circ\left(\frac{\partial v_z}{\partial r}\right)^{n+1} = 0 \tag{3-48}$$

with the applicable boundary conditions being (3-47). Toor[8] has given the following solution to this problem:

$$\frac{(T - T_w)}{(T_0 - T_w)^\circ} = \left[1 + \left(\frac{2n}{n+1}\right)\overline{T\kappa}\right]\left[1 - \left(\frac{r}{R}\right)^{(3n+1)/n}\right] - \left[\frac{(3n+1)^2}{2n(n+1)}\overline{T\kappa}\right]\left[1 - \left(\frac{r}{R}\right)^2\right] \tag{3-49}$$

where $(\overline{T\kappa})$ represents the mean value of the $(T\kappa)$ product across the tube and $(T_0 - T_w)^\circ$ represents the difference in fluid temperature between the center of the tube and the wall for the case where $\kappa = 0$.

Values of the dimensionless temperature ratio defined by equation (3-49) are plotted in Figure 3-6 for $(\overline{T\kappa}) = 0.3$, the upper curve applying

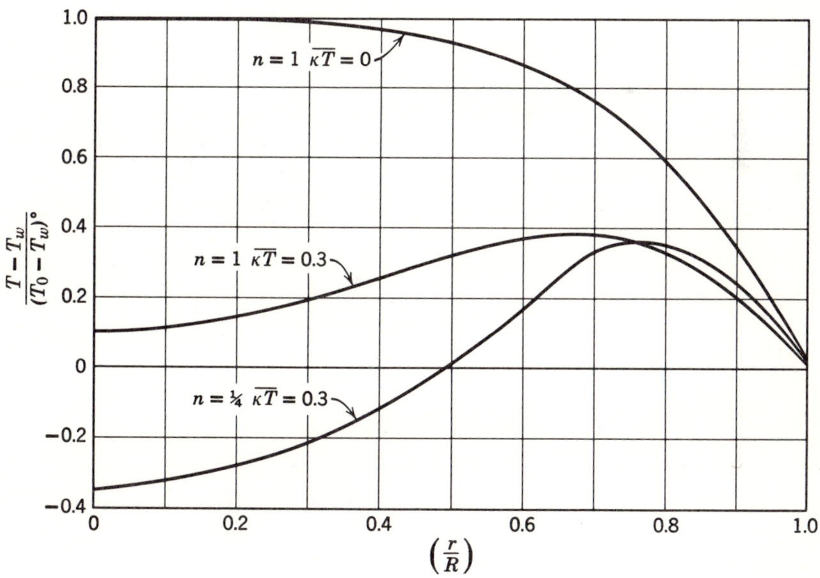

Figure 3-6. Dimensionless temperature distribution across tube radius for power law fluids having different values of $\overline{\kappa T}$.

[8] H. L. Toor, *Ind. Eng. Chem.*, **48,** 922 (1956); *A.I.Ch.E. Journal*, **4,** 319 (1958).

to a Newtonian fluid and the lower to a pseudoplastic fluid having a flow index of 0.25.

For the Newtonian fluid, the temperature profile is depressed in the center of the tube. This is due to the cooling effect of the expansion of the fluid. Since the maximum flow is in the center of the tube, the cooling is a maximum there. With a pseudoplastic fluid having the same value of $\overline{T\kappa}$, the cooling effect remains the same but the rate of frictional heat generation in the center of the tube is reduced because of the flattening of the velocity profile. The net result is to reduce further the temperature of the fluid at the center of the tube. In the example shown, the center temperature is actually lower than the wall temperature.

The assumption of constant fluid properties is, of course, open to question. However, it would appear that if the effect of temperature on viscosity were considered, the fluid viscosity near the tube wall would have to be reduced and the velocity profile flattened still further. The effect would be the same as increasing the pseudoplasticity of the fluid—the center temperature would be lowered. Therefore it would appear that the cooling effects calculated under the assumption of constant viscosity are on the conservative side.

The more general problem in which the fluid temperature is a function of the position variable z, as well as r, is now considered for the case of an incompressible ($\kappa = \beta = 0$) fluid that obeys the power law. The simplifying assumptions that the viscosity η is constant and that heat conduction in the axial direction, represented by the second term on the right side of equation (3-44), can be neglected in comparison to heat convection in the axial direction, represented by the left side of (3-44), are made. This reduces the energy equation to

$$\rho c_v v_z \left(\frac{\partial T}{\partial z}\right) = \frac{k}{r}\frac{\partial}{\partial r}\left[r\left(\frac{\partial T}{\partial r}\right)\right] + \eta^\circ \left(\frac{\partial v_z}{\partial r}\right)^{n+1} \tag{3-50}$$

and the momentum equation, (3-43), to

$$\left(\frac{\partial P}{\partial z}\right) = \frac{\eta^\circ}{r}\left(\frac{\partial v_z}{\partial r}\right)^{n-1} \frac{\partial}{\partial r}\left[r\left(\frac{\partial v_z}{\partial r}\right)\right] \tag{3-51}$$

The solution of equation (3-51) is (2-80). When it is substituted into (3-50), one obtains a linear partial differential equation having T as its dependent variable and r and z as independent variables. Solutions of this equation for the "isothermal tube" and "adiabatic tube" problems have been obtained by Bird.[9] The boundary conditions for these problems are

[9] R. B. Bird, *S.P.E. Journal*, **11**, No. 7, 35 (1955).

written as follows:

Isothermal Tube	Adiabatic Tube
$T(r, 0) = T_i$	$T(r, 0) = T_i$
$T(R, z) = T_w$	$T'(R, z) = 0$
$T'(0, z) = 0$	$T'(0, z) = 0$

where T_i is the initial uniform temperature of the fluid at the entrance to the tube and T_w is the wall temperature of the tube for the isothermal tube problem. The symbol T' represents the derivative of T with respect to r.

Toor[10] analyzed the same problem but included the effect of thermal expansion. In this case the energy equation, (3-44), becomes

$$\rho v_z\left(\frac{\partial T}{\partial z}\right)\left[c_p + \frac{T\kappa^2}{\beta\rho}\right] = \rho v_z c_v\left(\frac{\partial T}{\partial z}\right) = \frac{k}{r}\frac{\partial}{\partial r}\left[r\left(\frac{\partial T}{\partial r}\right)\right] + T\kappa v_z\left(\frac{\partial P}{\partial z}\right) + \eta\left(\frac{\partial v_z}{\partial r}\right)^2 \tag{3-52}$$

The results of Toor were presented in terms of the dimensionless variables

$$\theta = \frac{T - T_w}{\Delta T_\infty}$$

$$X = \left(\frac{r}{R}\right)$$

$$Y = \frac{\alpha z}{v_0 R^2}$$

where α represents the thermal diffusivity of the fluid and ΔT_∞ represents the difference between the wall temperature and the temperature of the fluid at the center of the tube for very large values of z, where T is no longer a function of z. Figure 3-7 shows θ plotted against X with Y as the parameter for the case where the value of the flow index n is 0.50. The curves of Figure 3-7*a* apply when $(\overline{\kappa T}) = 0$, corresponding to an incompressible fluid, while those of Figure 3-7*b* and 3-7*c* are for values of $(\overline{\kappa T})$ of 0.2 and 0.3, respectively.

Toor points out that many polymers have $(\overline{\kappa T})$ values in the range of 0.2 to 0.4; thus significant cooling effects due to thermal expansion should be expected. In the center region of the tube, heat generation due to viscous friction is small and is effectively removed by the thermal expansion process. In the neighborhood of the wall, heat generation is intense, but the generated heat is readily conducted through the short path to the wall and dissipated. In the intermediate region of the tube, conduction losses are reduced; the combined effects of conduction and thermal expansion are

[10] H. L. Toor, *Trans. Soc. of Rheology*, **1**, 177 (1957).

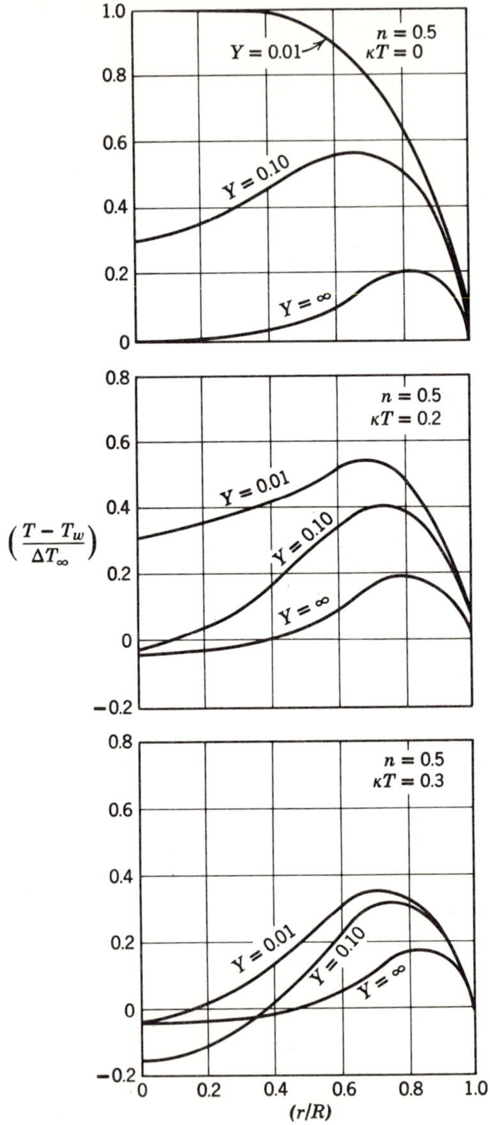

Figure 3-7. Dimensionless temperature across tube radius. The dimensionless parameter Y is proportional to distance measured along tube axis (Ref. 10).

unable to overcome the heat generation, and the temperature rises to its maximum value.

3-4 End Effects

The discussion here is limited primarily to steady isothermal laminar flow through straight tubes having circular cross sections. It will prove convenient in the discussion to refer to the various regions shown in Figure 3-8. Region 3 is designated the region of steady flow. Here the

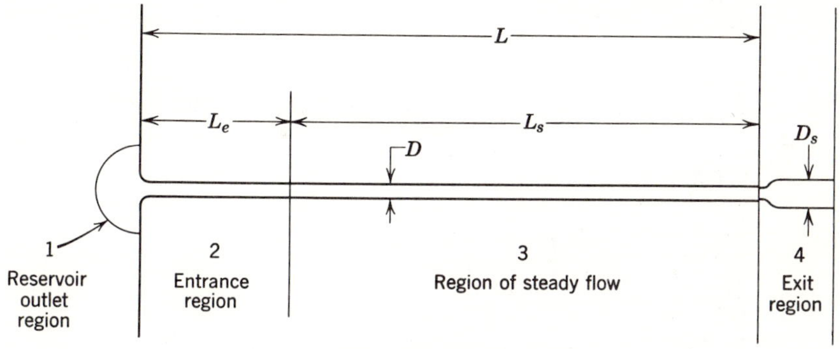

Figure 3-8. Regions associated with flow through tubes.

velocity profile has its characteristic parabolic shape (or distorted parabolic shape for non-Newtonian fluids) and the shear stress τ and the shear rate $\dot{\gamma}$ in the fluid depend only on the radial position coordinate. The analysis of steady isothermal flow in the tubes presented in Section 3-1 applies to the flow in this region.

The development of the steady flow velocity profile occurs in region 2, the entrance region. Here τ and $\dot{\gamma}$ depend on a position coordinate measured along the tube axis as well as on the radial position coordinate. The length of tube required for the complete development of the steady flow velocity profile is designated as L_e. For Newtonian fluids, an approximate solution[11] of the continuity and momentum equations gives the following result for the entrance length:

$$\frac{Le}{D} = 0.03\, N_{\mathrm{Re}} \tag{3-53}$$

where N_{Re} represents the Reynolds number for the flow in the tube.

Consider a fluid flowing at the rate of 1 gram/sec through a tube having an (L/D) ratio of 40 and a diameter of 0.10 cm. If the fluid viscosity is about 1 centipoise, then the value of N_{Re} is about 1300 and $(L_e/D) = 39$. In this case

[11] S. Goldstein, *Modern Developments in Fluid Dynamics*, Oxford University Press, London, 1952, p. 301.

(a)

(b)

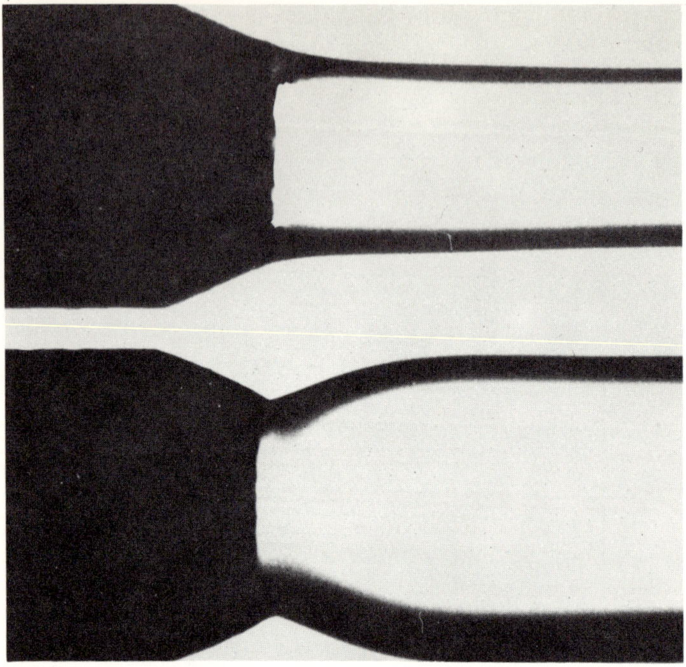

Figure 3-9. Photographs showing expansion and contraction of the free stream in the exit region: (*a*) water and (*b*) 2% carboxy methyl cellulose (Ref. 13).

the entrance region of the tube consists of about 98% of the over-all tube length. On the other hand, if the viscosity should be as high as 1 poise, then the (L_e/D) ratio is reduced to 0.39 and the entrance region amounts to less than 1% of the over-all tube length.

Obviously, if one applies the equations developed for region 3 to the over-all tube, the entrance region will introduce certain errors. An empirical method[12] of correcting for entrance effects has been developed and is discussed in part (*a*) below. With it the effect of the (L/D) ratio of the tube on viscosity values calculated from the over-all pressure drop can be eliminated.

Region 4, which is completely exterior to the tube, is designated the exit region. Here the restraining influence of the tube wall ends, and a second rearrangement of the velocity profile occurs, this time from a parabolic to a flat shape. An approximate analysis, shown in part (*b*) below, assuming conservation of momentum and incompressibility shows that for inelastic Newtonian fluids this velocity rearrangement will result in a 13% reduction in the diameter of the fluid stream. Figure 3-9*a*, which

[12] E. B. Bagley, *J. Appl. Phys.*, **28**, 624 (1957).

is a photograph of a stream of water emerging from a tube, shows that contractions of the order predicted by the theoretical analysis are observed.

It is well known that, under certain conditions, gravity, surface tension, and frictional drag forces acting between the stream surface and the surrounding fluid are factors that play a key role in the exit region. They must be considered in analyzing instabilities, drop formation, wave motion and other phenomena occurring at relatively large distances from the end of the tube. However, McIntosh[13] concluded that in the immediate vicinity of the tube end, these factors are usually of little significance in the extrusion of polymer melts and solutions.

Entrance and exit effects for fluids exhibiting elastic as well as viscous behavior become quite complex. In the reservoir outlet region, region 1, the flow may become drastically unsteady, with large elastic stresses forming and decaying and causing major disturbances in the flow. If the tube is relatively short, these disturbances may persist to the exit region and cause the extrudate surface to become highly irregular. With high stresses and short tubes, the extrudate may actually emerge from the tube in a broken, twisted, or ruptured condition, hence the name "melt fracture" applied by Tordella[14] to this phenomenon.

Figure 3-10 is a photograph that shows the characteristic increase in irregularity of the extrudate as the shear stress increases. The exact cause or nature of these melt fracture effects is still a matter of controversy. Reference should be made to the paper of Metzner,[15] which is a comprehensive review of the pertinent work in this developing field.

If the tube is relatively long or the stresses low, the unsteady conditions introduced in region 1 will damp out and the stream emerging from the tube will again have a smooth and regular surface. However, a characteristic "swelling" is often observed in the exit region, indicating the existence of another phenomenon. Figure 3-9*b* shows a stream of 2% carboxy methyl cellulose in water emerging from a tube. The swelling amounts to about 44%. Recent work,[13] discussed in part (*c*) below, indicates that a possible explanation of this effect is the recovery of elastic strains imparted to the fluid during its travel through the tube.

(*a*) *Entrance Region Correction*

In the entrance region, fluid particles entering the tube from the reservoir are accelerated to their final steady flow velocities. The energy consumed in this process causes the pressure to drop more rapidly in this

[13] D. L. McIntosh, "Elastic Effects in the Extrusion of Polymer Solutions," doctoral thesis, Washington University, St. Louis, 1960.

[14] J. P. Tordella, *J. Appl. Phys.*, **27**, 454 (1956).

[15] A. B. Metzner, *Ind. Eng. Chem.*, **50**, 1577 (1958).

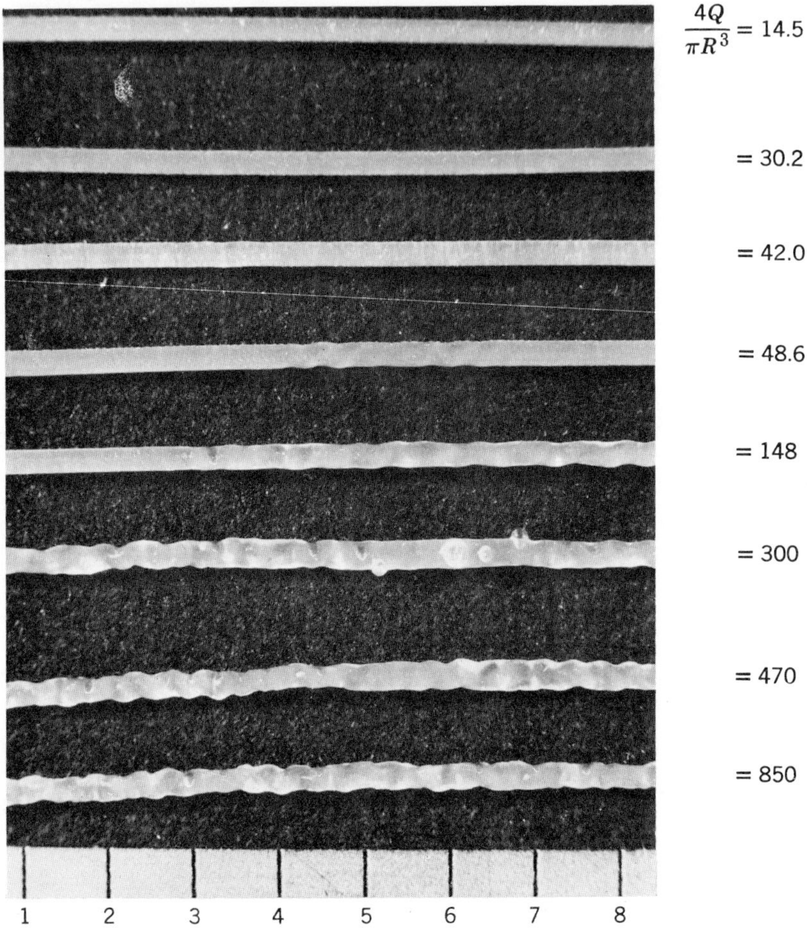

Figure 3-10. Polyethylene (melt index 2.0) filaments extruded at 150°C through a capillary having a length of 0.40 inches and a diameter of 0.02 inches. Scale divisions are 0.01 inches apart. (Courtesy Dr. John P. Tordella, Polychemicals Department, E. I. du Pont de Nemours and Co.)

region than in the steady flow region. As shown by Figure 3-11, the pressure gradient decreases throughout the length L_e of the entrance region and eventually reaches a constant value that is maintained over the length L_s of the steady flow region.

In the steady flow region the shear rate $\dot{\gamma}_w$ at the tube wall can be calculated with equation (3-31) if the shear stress τ_w at the tube wall is known.

Hence the entrance correction problem is essentially one of determining the true value of τ_w from the over-all measurements of flow rate and pressure drop. The procedure developed by Bagley,[12] which is outlined below, enables this to be done.

Noting from equation (3-32) that τ_w is a unique function of the parameter Γ, we have

$$\tau_w = \frac{D}{4} p_s = g(\Gamma) \tag{3-54}$$

where

$$\Gamma = \frac{4Q}{\pi R^3} \tag{3-55}$$

and p_s is the pressure gradient in the region of steady flow.

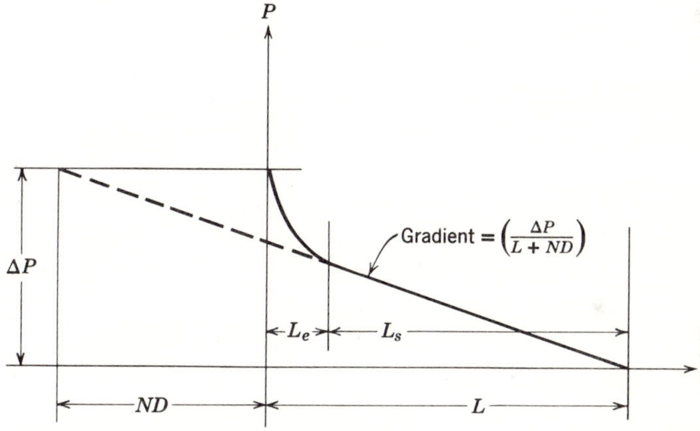

Figure 3-11. Correction of tube length L by addition of length ND in order to calculate the pressure gradient in the steady flow region.

It is clear that

$$p_s = \left(\frac{\Delta P}{L + ND}\right) \tag{3-56}$$

where ND, as shown by Figure 3-11, represents a fictitious tube length that, when added to the actual length, enables one to use the over-all pressure drop in calculating the gradient in the steady flow section.

The expression for the shear stress at the wall in the steady flow section of the tube can now be written

$$\tau_w = \frac{D}{4}\left(\frac{\Delta P}{L + ND}\right) \tag{3-57}$$

and referring to equation (3-54), we see that

$$\frac{D}{4}\left(\frac{\Delta P}{L + ND}\right) = g(\Gamma) \tag{3-58}$$

Upon rearranging, equation (3-58) becomes

$$\frac{L}{D} = -N + \frac{\Delta P}{4g(\Gamma)} \tag{3-59}$$

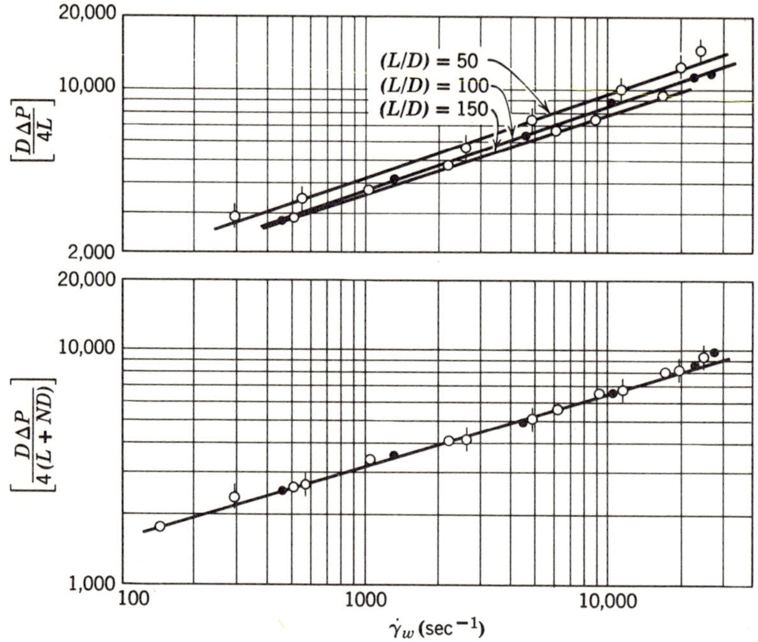

Figure 3-12. Logarithmic flow curves for 2% carboxy methyl cellulose solutions. Entrance correction eliminates the effect of (L/D) ratio (Ref. 13).

which suggests that the value of N might be determined from flow measurements made with a series of capillary tubes having different (L/D) ratios. For each tube, the pressure drop ΔP giving some specific value of Γ would be determined. Then, by plotting (L/D) versus ΔP, a straight line having $-N$ as an intercept would be obtained.

The flow data of McIntosh[13] for 2% carboxy methyl cellulose solutions in water in tubes having (L/D) ratios of 50, 100, and 150 illustrate the use of the entrance correction. Figure 3-12*a* is a logarithmic flow curve in which the shear stress is calculated with a gradient obtained by dividing the over-all pressure drop by the length of the tube. Note that three distinct lines result, showing that the calculated stress is a function of the (L/D) ratio. This indicates the presence of an entrance effect. Applying the entrance correction technique described above, the true stress can be determined and the three curves are reduced to a single one, as shown in Figure 3-12*b*.

The error involved in calculating the stress by using the uncorrected gradient can be quite large. For example, the data of Figure 3-12 show that at a shear rate of 1000 $\sec^{-1}$, the true shear stress at the wall is 3200 dynes/cm^2. With a tube having an (L/D) ratio of 50, the shear stress calculated with the uncorrected gradient is 4100 dynes/cm^2, a deviation of nearly 30% from the true value.

Bagley[12] has tested this method of making entrance corrections with molten polyethylene in capillaries having (L/D) ratios from zero to 15. He found that N was a function of shear rate, but that the method was successful in eliminating the effect of (L/D) on the flow curves.

(*b*) *Velocity Rearrangement in the Exit Region*

If one assumes that there is conservation of the momentum associated with the bulk motion of the fluid between the end of the tube and the point at which the velocity rearrangement is complete, then

$$\bar{m}_t = \bar{m}_s \tag{3-60}$$

where $\bar{m}_t$ is the average momentum per unit mass of fluid at the end of the tube and $\bar{m}_s$ is the average momentum per unit mass of the fluid in the free stream. These quantities can be calculated from the following equations, which follow directly from the definition of momentum:

$$\bar{m}_t = \frac{1}{\bar{v}R^2}\int_0^R 2rv^2\,dr \tag{3-61}$$

$$\bar{m}_s = \bar{v}_s \tag{3-62}$$

For a power law fluid, the local velocity in the tube is related to the mean velocity by the equation

$$v = \bar{v}\left(\frac{3n+1}{n+1}\right)\left[1 - \left(\frac{r}{R}\right)^{(n+1)/n}\right] \tag{3-63}$$

Introducing (3-63) into (3-61) and performing the indicated integration gives the following result:

$$\bar{m}_t = \left(\frac{3n+1}{2n+1}\right)\bar{v} \tag{3-64}$$

Comparing the above equation with (3-62) shows that

$$\frac{\bar{v}}{\bar{v}_s} = \left(\frac{2n+1}{3n+1}\right) \tag{3-65}$$

If the fluid is incompressible, the volumetric flow rate in the tube must

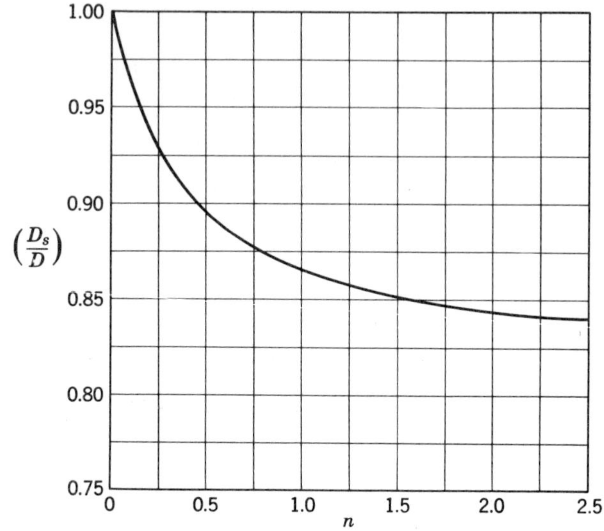

Figure 3-13. Contraction of inelastic power law fluids in exit region due to velocity rearrangement as a function of the flow index n. The diameter of the free stream is D_s and the diameter of the tube is D.

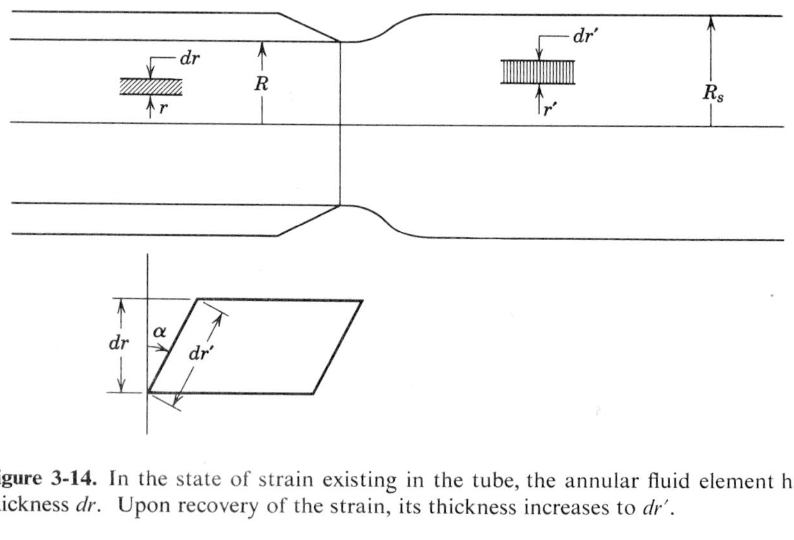

Figure 3-14. In the state of strain existing in the tube, the annular fluid element has thickness dr. Upon recovery of the strain, its thickness increases to dr'.

be exactly the same as in the free stream, and the mean velocities are related to the diameters by the equation

$$\frac{\bar{v}}{\bar{v}_s} = \left(\frac{D_s}{D}\right)^2 \tag{3-66}$$

A comparison of the above two equations shows that

$$\frac{D_s}{D} = \sqrt{\frac{2n+1}{3n+1}} \tag{3-67}$$

For a Newtonian fluid, $n = 1$ and $(D_s/D) = 0.87$, showing that there will be a contraction of about 13%. For pseudoplastic fluids, the velocity profile is more pluglike and a less severe velocity rearrangement is required; hence $(D_s/D) > 0.87$ and for dilatant fluids $(D_s/D) < 0.87$. The variation of (D_s/D) with the flow index n for power law fluids is shown in Figure 3-13.

(c) Swelling in the Exit Region

This section outlines the analysis proposed by McIntosh[13] to explain the swelling of viscoelastic fluids in the exit region. The key concept of the theory is that a recoverable elastic shear strain is imparted to the fluid during its passage through the tube. The recovery of the shear strain in the exit region is the cause of the swelling.

Consider an annular fluid element of thickness dr and located at radial distance r in the tube, as shown in Figure 3-14. The fluid of this element is in a state of strain characterized by the angle α of the parallelogram. Strain recovery is assumed to cause the radial expansion in the exit region, with the angle α becoming zero, the thickness of the element becoming dr', and its position coordinate becoming r'.

From the parallelogram it can be seen that

$$\frac{dr'}{dr} = \sec \alpha = \sqrt{1 + \tan^2 \alpha} \tag{3-68}$$

and defining the shear strain $\gamma = \tan(\alpha)$, the above equation becomes

$$\frac{dr'}{dr} = \sqrt{1 + \gamma^2} \tag{3-69}$$

If strain recovery were the only process operative in the exit region, the radius of the stream would increase by the factor

$$\frac{R_s}{R} = \frac{1}{R}\int_0^{R_s} dr' = \frac{1}{R}\int_0^{R} \sqrt{1 + \gamma^2}\, dr \tag{3-70}$$

which can be written in the following form by introducing the dimensionless variable x, defined as (r/R):

$$\frac{R_s}{R} = \int_0^1 \sqrt{1 + \gamma^2}\, dx \tag{3-71}$$

Superimposing the contraction effect due to velocity profile rearrangement, discussed in part (b) above, on the expansion process and designating the resultant expansion ratio as β, one obtains,

$$\beta = \sqrt{\frac{2n+1}{3n+1}} - 1 + \int_0^1 \sqrt{1+\gamma^2}\,dx \tag{3-72}$$

for fluids that obey the power law in steady tube flow.

Integration of equation (3-72) requires specification of the radial distribution of elastic shear strain. This requires the specification of a rheological model relating shear stress and shear strain. The Voigt element, the simplest rheological model exhibiting recoverable retarded elasticity, was used in the derivation. It can be written

$$\gamma(x) = \frac{\tau}{G}[1 - e^{-t/\lambda}] \tag{3-73}$$

where G is the shear modulus, λ the retardation time, and t represents the time of application of the stress τ. We note that (for a given experiment) in the region of steady flow the strain $\gamma(x)$ depends only upon the radial position variable x, because shear stress τ and residence time t are functions only of x.

If an average shear strain γ_a is defined by the equation

$$\sqrt{1+\gamma_a^2} = \int_0^1 \sqrt{1+\gamma^2}\,dx \tag{3-74}$$

the expansion formula, equation (3-72), becomes

$$\beta = \sqrt{\frac{2n+1}{3n+1}} - 1 + \sqrt{1+\gamma_a^2} \tag{3-75}$$

In principle, one could determine γ_a through equation (3-74) and insert it into equation (3-75) to evaluate β in terms of the fluid properties and the flow rate or pressure drop across the tube. However, McIntosh used an approximate value of γ_a calculated with the equation

$$\gamma_a = \frac{\tau_a}{G}[1 - e^{-t_a/\lambda}] \tag{3-76}$$

where the average shear stress τ_a and the average residence time t_a are defined by the equations

$$\begin{aligned} \tau_a &= \beta(x = \tfrac{1}{2}) = \frac{R\Delta P}{2L} \\ t_a &= \frac{\pi R^2 L}{Q} = 2\left(\frac{L}{D}\right)\left(\frac{3n+1}{n}\right)\left(\frac{\tau^\circ}{2\gamma^\circ \tau_a}\right)^{1/n} \end{aligned} \tag{3-77}$$

Noting that τ_a is directly proportional to ΔP and therefore that t_a is proportional to $\Delta P^{-1/n}$, we see that β should be a unique function of ΔP. Furthermore, examination of the above equations shows that

$$\lim_{\Delta P \to 0} \beta = \lim_{\Delta P \to \infty} \beta = \sqrt{\frac{2n+1}{3n+1}}$$

and also that the β versus ΔP curve will pass through a maximum. The maximum in the curve arises because the stress-strain relationship used in the derivation provides for retarded elasticity. Given sufficient time, the fluid strain would increase in direct proportion to the pressure (stress) until it reached its maximum allowable value—and then it would remain constant. However, the pressure also has another effect. As it increases, the residence time of the fluid particles in the tube (the time of application of the stress) decreases; thus, as the pressure increases, the strain becomes more and more removed from its equilibrium value. These competing processes then account for the maximum in the curve.

The results of the experiments of McIntosh[13] are in good agreement with the general features of his theory of swelling. He extruded 2% carboxy methyl cellulose solutions through long tubes and photographed the free stream in the exit region. Figure 3-15 is a typical sequence of photographs. Note that as the pressure increases, the degree of swelling increases, passes through a maximum, and then decreases.

Figure 3-16 is a plot of β versus ΔP_e for one of these experiments, where

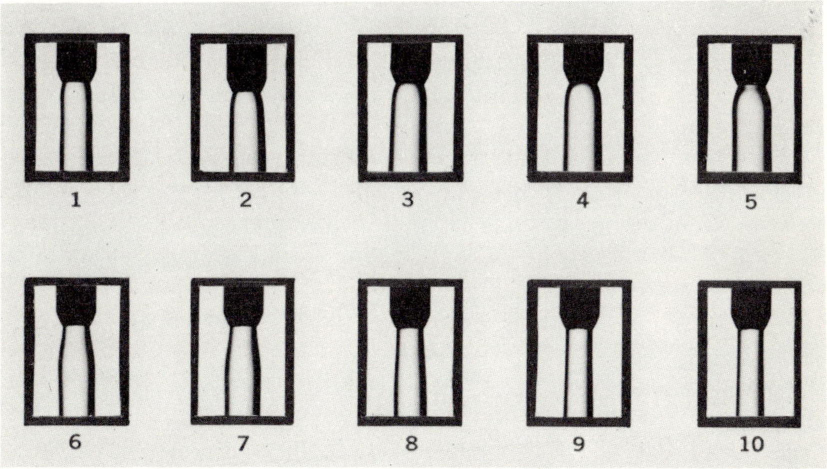

Figure 3-15. Sequence of photographs showing swelling of the free stream in the exit region. Flow rate increasing from 1 through 10 (Ref. 13).

ΔP_e represents the total pressure drop multiplied by $L/(L + ND)$ to correct for the entrance losses. The solid line represents the theoretical curve calculated with the above equations. Values of G and λ were obtained by requiring the theoretical curve to pass through the critical point—the point of maximum expansion. Knowing the coordinates of the critical point and also that the slope of the curve is zero at the critical point, one has two simultaneous equations from which G and λ can be determined. Note that the theory provides for a broader peak than is

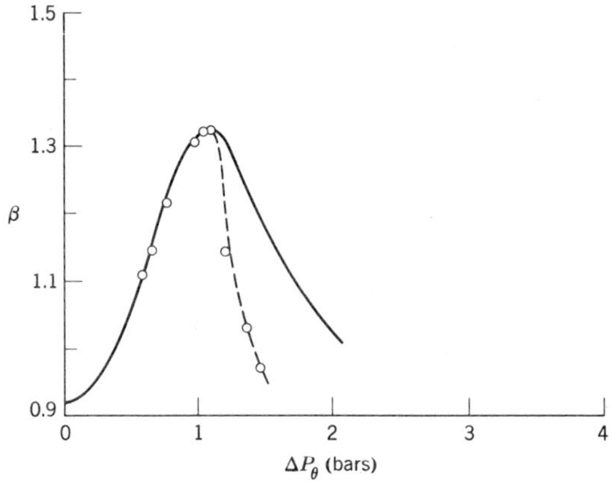

Figure 3-16. Plot of swelling index versus pressure drop for data shown in Figure 3-15. Broken line represents equation (3-76) (Ref. 13).

observed experimentally. The two curves could probably be brought closer together by replacing equation (3-73) with a rheological equation that provides for a more realistic (nonlinear) relationship between stress and strain.

It should be noted that phenomena other than velocity profile rearrangement and elastic strain recovery are operative in the exit region and may have a role in the expansion process. Gravitational and surface tension effects undoubtedly exist, but with highly viscous polymer solutions or melts, they are probably not important. On the other hand, with such liquids normal stresses[16] (probably both axial and radial) are generated in the shear field existing within the capillary. In the exit region these normal stresses undoubtedly have a role in the expansion process. Metzner[17] has derived an expression for the expansion ratio β that contains both axial

[16] See, for example, H. Markovitz, "The Normal Stress Effect in Polymer Solutions", in *Viscoelasticity*, edited by J. T. Bergen, Academic Press, 1960.

[17] A. B. Metzner, W. T. Houghton, R. A. Sailor, and J. L. White, *Trans. Soc. Rheology*, **5**, 133 (1961).

and radial normal stresses. Middleman and Gavis[18] have pointed out that axial normal stresses can be generated in the exit region itself as a result of the velocity rearrangement process and have derived an expression for the expansion ratio β in terms of a single axial tension that includes axial stresses developed both within the capillary and in the exit region.

3-5 Isothermal Flow in Rectangular Channels

Steady isothermal flow of an incompressible Newtonian fluid through a straight horizontal channel having a rectangular cross section is analyzed in

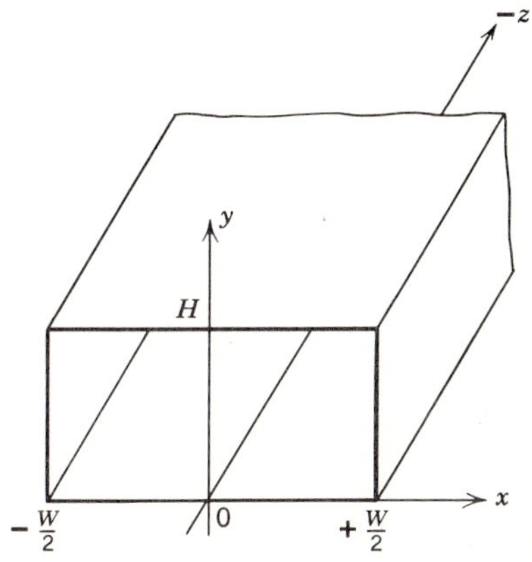

Figure 3-17. Straight horizontal channel of rectangular cross section having width W and height H.

this section. The solution of this problem proves useful in the development of the hydrodynamic analysis of screw pumps presented in Chapter 10.

The cross section of the channel and the position of the coordinate axes, a conventional right-handed set of cartesian coordinates, are shown in Figure 3-17. The width of the channel is W and its height is H. Because the only nonzero component of the velocity vector is v_z, the momentum equation (Table 2-2) reduces to

$$\left(\frac{\partial P}{\partial z}\right) = \left(\frac{\partial \tau_{xz}}{\partial x}\right) + \left(\frac{\partial \tau_{yz}}{\partial y}\right) \tag{3-78}$$

which becomes

$$\left(\frac{\partial P}{\partial z}\right) = \mu\left[\frac{\partial^2 v_z}{\partial x^2} + \frac{\partial^2 v_z}{\partial y^2}\right] \tag{3-79}$$

upon introducing the stress tensor components from Table 2-3.

[18] S. Middleman and J. Gavis, *Physics of Fluids*, **4**, 355 (1961).

The boundary conditions for the problem, established from the assumption that the fluid at the walls of the channel is stationary, are written

$$v_z(x, 0) = 0 \qquad v_z(x, H) = 0 \tag{3-80}$$

$$v_z\left(\pm \frac{W}{2}, y\right) = 0$$

The solution of the boundary value problem consisting of the partial differential equation (3-79) and the boundary conditions (3-80) was first found by Boussinesq.[19] It can be written

$$v_z = \frac{1}{\mu}\left(\frac{\partial P}{\partial z}\right)\left[\frac{y}{2}(y - H) + \frac{4H^2}{\pi^3}\sum_{n=1,3,\cdots}^{\infty}\frac{1}{n^3}\sin\left(\frac{n\pi y}{H}\right)\frac{\cosh\left(\dfrac{n\pi x}{H}\right)}{\cosh\left(\dfrac{n\pi W}{2H}\right)}\right] \tag{3-81}$$

and is obtained by the method of separation of variables[20] outlined below.

The variables of the nonhomogeneous differential equation, (3-79), can be separated by introducing the equation

$$v_z = V(x, y) + \phi(y) \tag{3-82}$$

where $V(x, y)$ is a function of both x and y and $\phi(y)$ depends only upon y, into equation (3-79). One obtains

$$\frac{\partial^2 V}{\partial x^2} + \frac{\partial^2 V}{\partial y^2} + \frac{d^2\phi}{dy^2} = \frac{1}{\mu}\left(\frac{\partial P}{\partial z}\right) \tag{3-83}$$

If the function $\phi(x)$ is defined by the equation

$$\frac{d^2\phi}{dy^2} = \frac{1}{\mu}\left(\frac{\partial P}{\partial z}\right) \tag{3-84}$$

then (3-83) reduces to the linear homogeneous equation

$$\frac{\partial^2 V}{\partial x^2} + \frac{\partial^2 V}{\partial y^2} = 0 \tag{3-85}$$

The solution of equation (3-84) is

$$\phi = \frac{y^2}{2\mu}\left(\frac{\partial P}{\partial z}\right) + C_1 y + C_2 \tag{3-86}$$

where C_1 and C_2 are arbitrary constants. If they are assigned the values

$$C_1 = -\frac{H}{2\mu}\left(\frac{\partial P}{\partial z}\right) \qquad C_2 = 0$$

[19] M. J. Boussinesq, *Math. pures appl.*, **13,** 377 (1868).

[20] See, for example, R. V. Churchill, *Fourier Series and Boundary Value Problems*, McGraw-Hill, 1941.

then the required boundary conditions for the function $V(x, y)$ are

$$V(x, 0) = 0$$
$$V(x, H) = 0$$
$$V\left(\pm \frac{W}{2}, y\right) = \frac{y}{2\mu}\left(\frac{\partial P}{\partial z}\right)(H - y) = f(y)$$

It can be shown that each member of the set of functions

$$\left\{C_n \sin\left(\frac{n\pi y}{H}\right) \cosh\left(\frac{n\pi x}{H}\right)\right\} \qquad n = 1, 2, \cdots$$

is a solution of equation (3-85) and satisfies the first two of the above boundary conditions. Because of the orthogonality of the sine functions, the third boundary condition can be satisfied by forming the Fourier sine series expansion of the function $f(y)$. Hence

$$V = \frac{4H^2}{\mu\pi^3}\left(\frac{\partial P}{\partial z}\right) \sum_{n=1,3,\cdots}^{\infty} \frac{1}{n^3} \sin\left(\frac{n\pi y}{H}\right) \frac{\cosh\left(\frac{n\pi x}{H}\right)}{\cosh\left(\frac{n\pi W}{2H}\right)} \tag{3-87}$$

and the solution of the original boundary value problem is obtained by adding (3-87) and (3-86).

A variation of the preceding problem is obtained when there is no pressure gradient in the channel and the flow occurs under the influence of a moving boundary. This type of flow has been called *drag flow* in contrast to *pressure flow*, which occurs under the influence of a pressure gradient. For the drag flow problem, the momentum equation becomes

$$\frac{\partial^2 v_z}{\partial x^2} + \frac{\partial^2 v_z}{\partial y^2} = 0 \tag{3-88}$$

We consider the case where the top of the channel is in motion and the other three sides are stationary. To facilitate solution of the problem, the origin of the coordinate system is moved to the corner of the channel, as shown in Figure 3-18. The boundary conditions for this problem are written

$$\begin{aligned} v_z(0, y) &= 0 \qquad & v_z(x, 0) &= 0 \\ v_z(W, y) &= 0 \qquad & v_z(x, H) &= V_z \end{aligned} \tag{3-89}$$

where V_z represents the speed at which the top of the channel moves in the $+z$-direction.

Equation (3-88) is a homogeneous linear equation, and it is easily verified that each member of the set of functions

$$\left\{C_n \sin\left(\frac{n\pi x}{W}\right) \sinh\left(\frac{n\pi y}{W}\right)\right\} \qquad n = 1, 2, \cdots$$

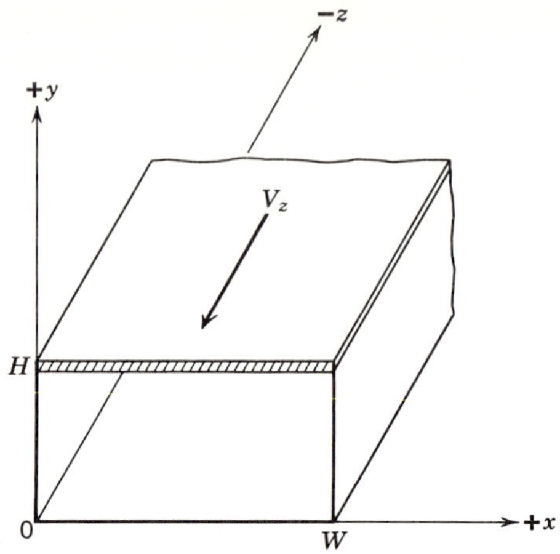

Figure 3-18. Straight horizontal channel of rectangular cross section. Top of channel moves with speed V_z in the $+z$-direction.

is a solution of equation (3-88) and also satisfies the first three of the above boundary conditions. To satisfy the fourth boundary condition one makes use of the orthogonality of the sine functions to form a Fourier sine series. Upon evaluating the coefficients one obtains

$$v_z = \frac{4V_z}{\pi} \sum_{n=1,3,\cdots}^{\infty} \frac{1}{n} \sin\left(\frac{n\pi x}{W}\right) \frac{\sinh\left(\dfrac{n\pi y}{W}\right)}{\sinh\left(\dfrac{n\pi H}{W}\right)} \tag{3-90}$$

which is the solution of the drag flow problem.

3-6 Isothermal Flow Between Parallel Plates

An analysis of steady isothermal drag and pressure flow of power law fluids between parallel plates is presented in this section. The coordinate axes are so oriented, as shown in Figure 3-19, that the only nonzero component of the velocity vector is v_z and the momentum equation (Table 2-2) reduces to

$$\left(\frac{\partial P}{\partial z}\right) = \frac{\partial \tau_{yz}}{\partial y} \tag{3-91}$$

Because its left side is not a function of y, equation (3-91) can be integrated to yield

$$\tau_{yz} = \left(\frac{\partial P}{\partial z}\right) y + C_1 \tag{3-92}$$

where C_1 is the constant of integration. Because pressure flow is symmetrical about the mid-plane, the shear rate and hence the shear stress are zero at the mid-plane, and the above equation becomes

$$\left(\frac{\partial P}{\partial z}\right) = \frac{\tau_{yz}}{y} \tag{3-93}$$

We note that the same result could have been obtained from a force balance applied to an element of the fluid. Referring to Figure 3-19, consider the forces acting on an element of fluid of unit width, height $2y$, and length dz. For steady flow, the forces are balanced and

$$\sum F_z = 0 = F_1 + F_2 + F_3 \tag{3-94}$$

where

$$F_1 = -2y\left[P + \left(\frac{\partial P}{\partial z}\right) dz\right]$$

$$F_2 = +2yP$$

$$F_3 = 2\tau_{yz}\, dz$$

The above expressions for the F's, when introduced into (3-94), reduce it to equation (3-93).

The stress component τ_{yz} is given by the equation

$$\tau_{yz} = \eta\left(\frac{\partial v_z}{\partial y}\right) \tag{3-95}$$

which is obtained from Table 2-3. Introducing equation (3-95) and the rheological equation applicable to power law fluids (obtained from Table 2-4) into equation (3-93) reduces it to (for $\dot{\gamma}^\circ = 1$)

$$\frac{1}{\eta^\circ}\left(\frac{\partial P}{\partial z}\right) = \frac{1}{y}\left(\frac{\partial v_z}{\partial y}\right)\left|\frac{\partial v_z}{\partial y}\right|^{n-1} \tag{3-96}$$

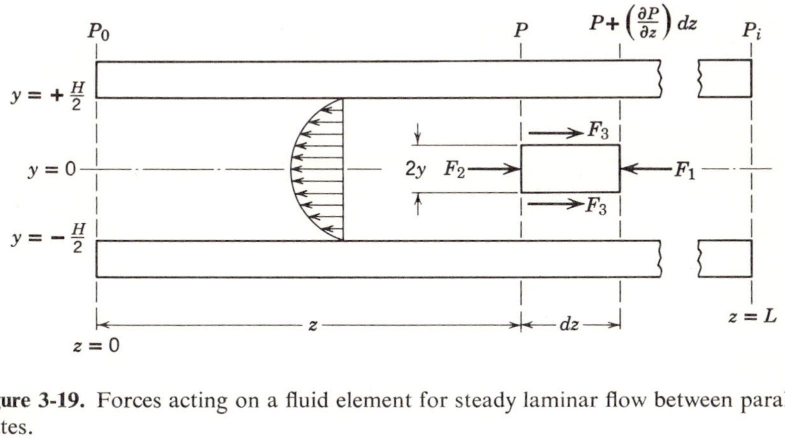

Figure 3-19. Forces acting on a fluid element for steady laminar flow between parallel plates.

which becomes, upon integrating with respect to y

$$v_z = \left(\frac{n}{n+1}\right)\left[\frac{1}{\eta^\circ}\left(\frac{\partial P}{\partial z}\right)\right]^{1/n} |y|^{\frac{n+1}{n}} + C_2 \tag{3-97}$$

The constant of integration is evaluated from the condition $v_z(H/2) = 0$, and (3-97) becomes

$$v_z = v_0\left[1 - \left|\frac{2y}{H}\right|^{(n+1)/n}\right] \tag{3-98}$$

where v_0, the fluid velocity at the mid-plane is given by

$$v_0 = -\frac{H}{2}\left(\frac{n}{n+1}\right)\left[\frac{H}{2\eta^\circ}\left(\frac{\partial P}{\partial z}\right)\right]^{1/n} \tag{3-99}$$

The flow is symmetrical about the mid-plane; hence the absolute value sign must be used in equation (3-98).

The volumetric rate at which the fluid flows between the plates in the z-direction (per width W measured in the x-direction) is given by the integral

$$Q = 2W\int_0^{H/2} v_z\, dy \tag{3-100}$$

Introducing the velocity function into the above integral and performing the indicated integration gives

$$Q = \left(\frac{n+1}{2n+1}\right) v_0 WH \tag{3-101}$$

It is easily shown that the average velocity $\bar{v}_z$ defined by the equation

$$\bar{v}_z = \frac{Q}{WH} \tag{3-102}$$

is related to the maximum velocity v_0 by the equation

$$\bar{v}_z = \left(\frac{n+1}{2n+1}\right) v_0 \tag{3-103}$$

The shear stress at the walls τ_w, obtained directly from (3-93), is

$$\tau_w = \pm\left(\frac{H}{2}\right)\left(\frac{\partial P}{\partial z}\right)$$

where the positive sign applies to the upper surface and the negative sign to the lower surface. The shear rate at the wall, obtained by differentiating the velocity function, is

$$\dot{\gamma}_w = \pm\left(\frac{n+1}{n}\right)\left(\frac{2v_0}{H}\right) \tag{3-104}$$

where the positive sign applies to the lower surface and the negative to the upper. Because of equation (3-103), $\dot{\gamma}_w$ is also given by the relation

$$\dot{\gamma}_w = \pm\left(\frac{2n+1}{n}\right)\left(\frac{2Q}{WH^2}\right) \tag{3-105}$$

For the special case of Newtonian flow $n = 1$, the above equations reduce to the following

$$\bar{v}_z = \frac{2v_0}{3} \tag{3-106}$$

$$v_0 = -\frac{H^2}{8\mu}\left(\frac{\partial P}{\partial z}\right) \tag{3-107}$$

$$v_z = v_0\left[1 - \left|\frac{2y}{H}\right|^2\right] \tag{3-108}$$

$$Q = -\frac{WH^3}{12\mu}\left(\frac{\partial P}{\partial z}\right) \tag{3-109}$$

$$\dot{\gamma}_w = \pm\frac{4v_0}{H} = \pm\frac{6Q}{WH^2} \tag{3-110}$$

$$\tau_w = \pm\frac{H}{2}\left(\frac{\partial P}{\partial z}\right) \tag{3-111}$$

We now consider the drag flow resulting from the motion of the upper plate (shown in Figure 3-19) moving in the z-direction at the constant speed V_z, with the lower plate stationary. Because the pressure gradient is zero, equation (3-91) becomes

$$\left(\frac{\partial \tau_{yz}}{\partial y}\right) = 0 \tag{3-112}$$

Upon introducing the expression for the stress component, equation (3-95), the above equation becomes

$$\frac{\partial}{\partial y}\left[\eta\left(\frac{\partial v_z}{\partial y}\right)\right] = 0 \tag{3-113}$$

or, upon integrating,

$$\eta\left(\frac{\partial v_z}{\partial y}\right) = C_1 \tag{3-114}$$

Upon introducing the rheological equation of the power law fluid, (3-114) becomes (for $\dot{\gamma}^\circ = 1$)

$$\eta^\circ\left(\frac{\partial v_z}{\partial y}\right)^n = C_1 \tag{3-115}$$

The boundary conditions for the flow are written

$$v_z\left(-\frac{H}{2}\right) = 0 \qquad v_z\left(+\frac{H}{2}\right) = V_z$$

and the solution of equation (3-115) satisfying these conditions is

$$v_z = \frac{V_z}{H}\left[y + \frac{H}{2}\right] \tag{3-116}$$

The volumetric flow rate is given by the integral

$$Q = W\int_{-H/2}^{+H/2} v_z\, dy \tag{3-117}$$

which, upon introducing (3-116) into it and performing the indicated integration, becomes

$$Q = \frac{V_z W H}{2} \tag{3-118}$$

The shear rate, found by differentiating (3-116), is

$$\dot{\gamma} = \frac{\partial v_z}{\partial y} = \left(\frac{V_z}{H}\right) \tag{3-119}$$

Note that it is independent of the y position coordinate. The shear stress in the fluid is given by

$$\tau = \tau^\circ\left(\frac{V_z}{H}\right)^n \tag{3-120}$$

Note that the shear rate, the velocity function, and the flow rate are independent of the nature of the fluid. Only the shear stress depends upon the flow index n.

3-7 Isothermal Flow Between Coaxial Cylinders

This section presents an analysis of the motion of a fluid contained in the annulus formed by coaxial cylinders. The radius of the inner cylinder is designated R_i and that of the outer cylinder R_0. The analysis is limited to steady isothermal flow of incompressible fluids.

We shall be concerned with both the v_z and v_θ components of the velocity vector. For these components, the momentum equation becomes

$$\frac{\partial P}{\partial \theta} = \frac{1}{r}\frac{\partial}{\partial r}\left[r^2 \tau_{r\theta}\right] \tag{3-121}$$

$$\frac{\partial P}{\partial z} = \frac{1}{r}\frac{\partial}{\partial r}\left[r \tau_{rz}\right] \tag{3-122}$$

Two types of *drag flow* are considered. In the first, the particle paths are straight lines parallel to the axis of the cylinders. It arises whenever either cylinder moves along its axis. In the other type, the particle paths are concentric circles. This type of flow arises whenever one or the other cylinder rotates about its axis.

For the first type of drag flow, v_θ is zero and the momentum equation (3-122) becomes

$$\frac{\partial}{\partial r}[r\tau_{rz}] = 0 \tag{3-123}$$

The stress component τ_{rz} (obtained from Table 2-3) is

$$\tau_{rz} = \eta\left(\frac{\partial v_z}{\partial r}\right) \tag{3-124}$$

and the rheological equation for power law fluids is

$$\eta = \eta^\circ\left(\frac{\partial v_z}{\partial r}\right)^{n-1} \tag{3-125}$$

Introducing equations (3-125) and (3-124) into (3-123) gives

$$\frac{\partial}{\partial r}\left[r\left(\frac{\partial v_z}{\partial r}\right)^n\right] = 0 \tag{3-126}$$

The boundary conditions for the problem are written

$$v_z(R_i) = V_z \qquad v_z(R_0) = 0$$

and the solution of (3-126) satisfying these boundary conditions can be written

$$\frac{v_z}{V_z} = \frac{\beta^{(n-1)/n} - (r/R_i)^{(n-1)/n}}{\beta^{(n-1)/n} - 1} \tag{3-127}$$

where

$$\beta = \frac{R_0}{R_i}$$

Equation (3-127) is valid for all values of n except $n = 1$. For this special case, the solution of (3-126) satisfying the above boundary conditions is

$$\frac{v_z}{V_z} = \frac{\ln(R_0/r)}{\ln\beta} \tag{3-128}$$

The shear rate obtained through differentiation of (3-127) is

$$\dot{\gamma} = \left(\frac{\partial v_z}{\partial r}\right) = -\left(\frac{n-1}{n}\right)\left(\frac{V_z}{R_i}\right)\left[\frac{1}{\beta^{(n-1)/n} - 1}\right]\left[\frac{r}{R_i}\right]^{-1/n} \tag{3-129}$$

and for the case where $n = 1$, differentiation of equation (3-128) yields

$$\dot{\gamma} = \left(\frac{\partial v_z}{\partial r}\right) = -\frac{V_z}{r\ln\beta} \tag{3-130}$$

The volumetric flow rate Q is given by the integral

$$Q = 2\pi \int_{R_i}^{R_0} r v_z \, dr \tag{3-131}$$

and when the velocity function, equation (3-127), is introduced into it and the integration performed, one obtains

$$Q = \frac{\pi V_z}{2}\left[R_0^2 - R_i^2\right] g(\beta, n) \qquad (n \neq 1) \tag{3-132}$$

where the function $g(\beta, n)$ is defined by the equation

$$g(\beta, n) = \left(\frac{2n-2}{3n-1}\right)\left(\frac{1}{1-\beta^{(1-n)/n}}\right) - \left(\frac{4n}{3n-1}\right)\left(\frac{1}{\beta^2-1}\right) \qquad (n \neq 1)$$

For Newtonian fluids, the flow rate is found by introducing equation (3-128) into the integral. One obtains

$$Q = \frac{V_z \pi}{2}\left[R_0^2 - R_i^2\right] g(\beta, 1) \qquad (n = 1) \tag{3-133}$$

where the function $g(\beta, 1)$ is defined by the equation

$$g(\beta, 1) = \frac{1}{\ln \beta} - \frac{2}{\beta^2 - 1}$$

The limiting values

$$\lim_{\beta \to 1} g(\beta, n) = 1 \qquad \lim_{\beta \to 1} g(\beta, 1) = 1$$

obtained by applying L'hospital's rule, show that for the special case where the radii of the inner and outer cylinders are nearly equal, the volumetric flow rate equation becomes identical to the flow rate equation derived for flow between parallel plates. The function $g(\beta, n)$ can be considered to represent the error involved in using the parallel plate representation of the annulus. Figure 3-20 is a plot of the function versus β, with n as the parameter.

We now consider the motion of the fluid when the outer cylinder is stationary and the inner cylinder is rotating in a counterclockwise direction. The momentum equation reduces to

$$\frac{\partial}{\partial r}\left[r^2 \tau_{r\theta}\right] = 0 \qquad r^2 \tau_{r\theta} = C_1 \tag{3-134}$$

where the stress component $\tau_{r\theta}$ from Table 2-3, is

$$\tau_{r\theta} = \eta r \frac{\partial}{\partial r}\left(\frac{v_\theta}{r}\right) \tag{3-135}$$

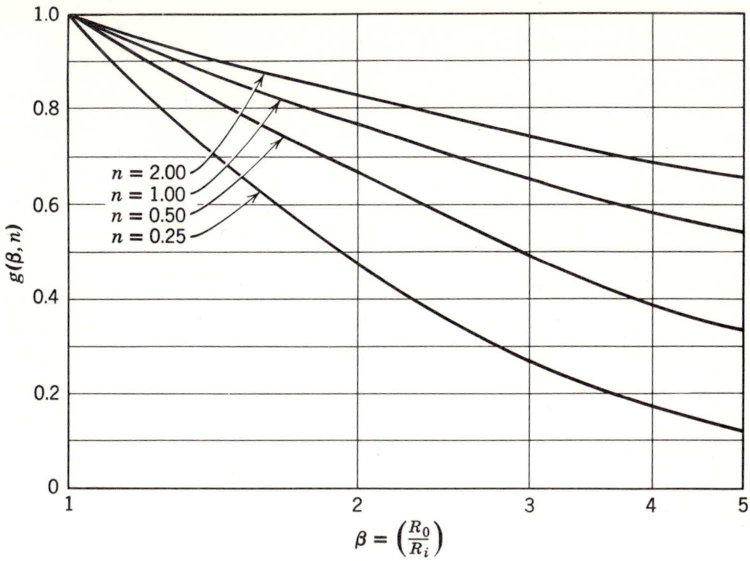

Figure 3-20. Plot of the function $g(\beta, n)$ versus β with n the parameter.

The rheological equation is:

$$\eta = \eta^\circ \left(\frac{\mathbf{\Delta} : \mathbf{\Delta}}{2}\right)^{(n-1)/2} \tag{3-136}$$

where the quantity $(\mathbf{\Delta} : \mathbf{\Delta})$ is

$$(\mathbf{\Delta} : \mathbf{\Delta}) = 2r^2\left[\frac{\partial}{\partial r}\left(\frac{v_\theta}{r}\right)\right]^2 \tag{3-137}$$

as can be seen by referring to Table 2-5.

Combining the above equations gives

$$\eta^\circ r^{n+2}\left[\frac{\partial}{\partial r}\left(\frac{v_\theta}{r}\right)\right]^n = C_1 \tag{3-138}$$

which becomes upon integration

$$v_\theta = C_2 r^{(n-2)/n} + C_3 r \tag{3-139}$$

The boundary conditions for the problem can be written

$$v_\theta(R_i) = V_\theta$$
$$v_\theta(R_0) = 0$$

and upon evaluating the constants of integration, equation (3-139) becomes

$$\frac{v_\theta}{V_\theta} = \left(\frac{r}{R_i}\right)\left[\frac{1 - (R_0/r)^{2/n}}{1 - \beta^{2/n}}\right] \tag{3-140}$$

If the inner cylinder should be stationary and the outer cylinder rotates at speed V_θ, the boundary conditions are

$$v_\theta(R_i) = 0$$

$$v_\theta(R_0) = V_\theta$$

and the velocity distribution is

$$\frac{v_\theta}{V_\theta} = \left(\frac{r}{R_0}\right)\left[\frac{(R_0/r)^{2/n} - (\beta)^{2/n}}{1 - \beta^{2/n}}\right] \tag{3-141}$$

The flow properties of fluids are often investigated with coaxial cylinder viscometers. If one cylinder is driven at a constant speed and the other is free to rotate, then the shear stress and the shear rate can be determined by measuring the speed of the driven cylinder and the moment required to restrain the free cylinder. For example, if the inner cylinder is driven at constant angular speed and the moment M is required to restrain the outer cylinder, then the shear stress in the fluid at the outer cylinder is

$$\tau_{R_0} = \frac{M}{2\pi R_0{}^2 L} \tag{3-142}$$

where L is the length of the cylinder.

Equation (3-135) shows that the shear rate $\dot{\gamma}$ in the fluid at any position r is

$$\dot{\gamma} = r\frac{\partial}{\partial r}\left(\frac{v_\theta}{r}\right) \tag{3-143}$$

so that the shear rate at the outer cylinder wall can be obtained by differentiating equation (3-140) to obtain

$$\dot{\gamma} = \left(\frac{2V_\theta}{nR_i}\right)\left(\frac{r}{R_0}\right)^{-2/n}\left(\frac{1}{1 - \beta^{2/n}}\right) \tag{3-144}$$

and upon specializing by placing $r = R_0$, one gets

$$\dot{\gamma}_{R_0} = \frac{2}{n}\left(\frac{V_\theta}{R_i}\right)\left(\frac{1}{1 - \beta^{2/n}}\right) \tag{3-145}$$

We now consider pressure flow in the axial direction through the annulus. The z-component of the momentum equation, (3-122), is the applicable equation, where the stress component is given by equation (3-124). Integrating (3-122) gives

$$\tau_{rz} = \frac{r}{2}\left(\frac{\partial P}{\partial z}\right) + \frac{C_1}{r} \tag{3-146}$$

The same result can be obtained from a force balance on a fluid element. Referring to Figure 3-21, the pressure increases in the $+z$-direction and the flow is in the $-z$-direction. The shear stress at any radial distance r is τ_{rz} and the pressure at any axial distance z is P. The annular element of fluid having radius r, thick-

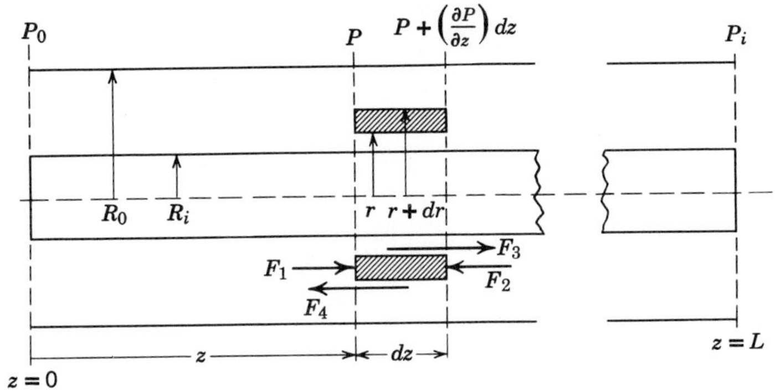

Figure 3-21. Forces acting on a fluid element for steady laminar pressure flow through an annulus.

ness dr, and length dz is in steady motion so that the forces acting upon it are balanced. Therefore

$$\Sigma F_z = 0 = F_1 + F_2 + F_3 + F_4 \tag{3-147}$$

where

$$F_1 = (2\pi r\, dr)P$$
$$F_2 = -(2\pi r\, dr)\left[P + \left(\frac{\partial P}{\partial z}\right) dz\right]$$
$$F_3 = (2\pi r\, dy)\tau_{rz}$$
$$F_4 = (2\pi\, dy)(r + dr)\left[\tau_{rz} + \left(\frac{\partial \tau_{rz}}{\partial r}\right) dr\right]$$

Substituting the above equations into (3-147) and rearranging give the following result

$$r\left(\frac{\partial P}{\partial z}\right) = r\left(\frac{\partial \tau_{rz}}{\partial r}\right) + \tau_{rz} = \frac{\partial}{\partial r}[r\tau_{rz}] \tag{3-148}$$

which is the z-component of the momentum equation, (3-122).

Replacing the stress component in equation (3-122) with (3-124) gives

$$\eta\left(\frac{\partial v_z}{\partial r}\right) = \frac{r}{2}\left(\frac{\partial P}{\partial z}\right) + \frac{C_1}{r} \tag{3-149}$$

and after introducing the rheological equation for the power law fluid, we have

$$\left(\frac{\partial v_z}{\partial r}\right) = \left[\frac{r}{2\eta^\circ}\left(\frac{\partial P}{\partial z}\right) + \frac{C_1}{r\eta^0}\right]^{1/n} \tag{3-150}$$

The special case of Newtonian flow is considered first. In this case

$\eta^\circ = \mu$ and $n = 1$. Upon integrating (3-150), the following result is obtained:

$$v_z = \frac{r^2}{4\mu}\left(\frac{\partial P}{\partial z}\right) + \frac{C_1}{\mu}\ln r + C_2 \tag{3-151}$$

The constants of integration are evaluated from the boundary conditions

$$v_z(R_i) = 0$$
$$v_z(R_0) = 0$$

and the velocity function becomes

$$v_z = -\frac{R_0^2}{4\mu}\left(\frac{\partial P}{\partial z}\right)\left[1 - \left(\frac{r}{R_0}\right)^2 + \left(\frac{\beta^2 - 1}{\beta^2 \ln \beta}\right)\ln (r/R_0)\right] \tag{3-152}$$

The volumetric flow rate through the annulus is obtained by introducing the velocity function, (3-152), into the integral

$$Q = 2\pi\int_{R_i}^{R_0} v_z r\, dr \tag{3-153}$$

After performing the integration and rearranging, one obtains

$$Q = -\frac{\pi R_0^4}{8\mu}\left(\frac{\partial P}{\partial z}\right)\left[\left(\frac{\beta^4 - 1}{\beta^4}\right) - \left(\frac{\beta^2 - 1}{\beta^2}\right)^2\left(\frac{1}{\ln \beta}\right)\right] \tag{3-154}$$

The negative sign indicates that the flow is in the $-z$-direction. Note that when $R_i = 0$, $\beta = \infty$ and equation (3-154) reduces to the Poiseuille equation for flow through a tube.

Fredrickson and Bird[21] obtained the solution of the above problem for power law fluids. Their result can be written in the form

$$Q = \left(\frac{n}{2n + 1}\right)\pi R_0^3\left(\frac{R_0 \Delta P}{2\tau^\circ L}\right)^{1/n}\left(\frac{\beta - 1}{\beta}\right)^{(2n+1)/n} F(n, \beta) \tag{3-155}$$

where the function F depends only upon n and β. Values of the function were obtained by machine computation and are plotted in Figure 3-22.

Note that the value of F becomes nearly independent of n when $(1/\beta)$ takes on values greater than about 0.3. In this region the function can be approximated by the equation

$$F = \left(\frac{\beta + 1}{2\beta}\right)$$

and equation (3-155) becomes

$$Q = \left(\frac{n}{2n + 1}\right)\left(\frac{\pi R_0^3}{2}\right)\left(\frac{R_0 \Delta P}{2\tau^\circ L}\right)^{1/n}\left[\left(\frac{\beta - 1}{\beta}\right)^{(2n+1)/2}\left(\frac{\beta + 1}{\beta}\right)\right] \tag{3-156}$$

[21] A. G. Fredrickson and R. B. Bird, *Ind. Eng. Chem.*, **50**, 347 (1958).

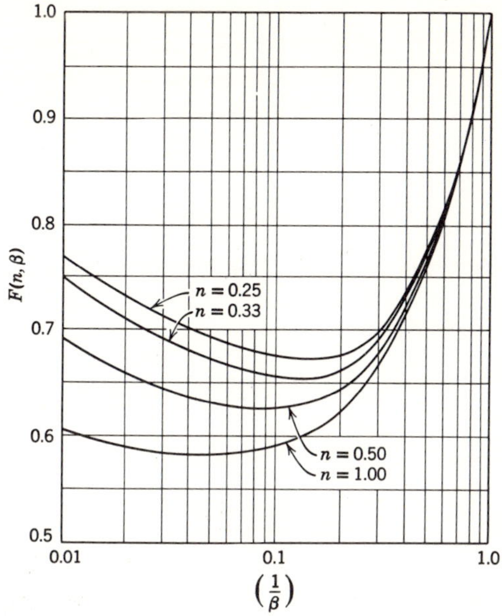

Figure 3-22. The function $F(n, \beta)$ of equation (3-155) versus the reciprocal of β with n as a parameter (data from Ref. 21).

3-8 Wire Coating and Sheeting Dies

Figure 3-23 is a schematic representation of a wire coating die. It consists of a reservoir, containing the coating fluid under a hydrostatic pressure P, connected to a hollow cylinder of diameter D and length L through which the wire is pulled. Because of the hydrostatic pressure of the fluid and because of the dragging action of the moving wire, the fluid flows through the annular space and forms a coating of thickness h on the wire, where h may or may not be equal to H.

The following analysis establishes the relationship between the design variables (the radial clearance H and the length L), the operating variables

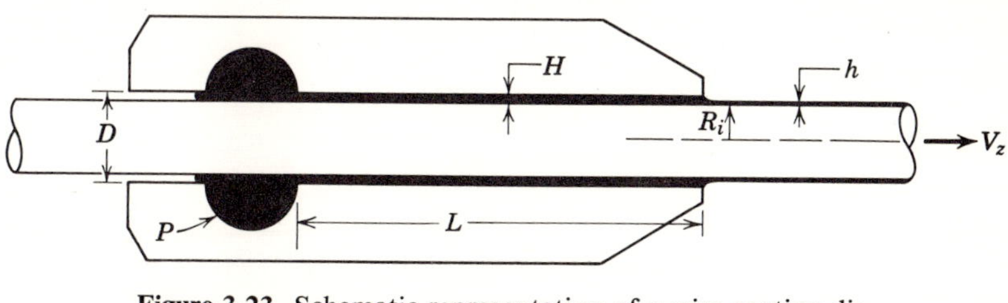

Figure 3-23. Schematic representation of a wire coating die.

(the speed V_z and the pressure P), the fluid viscosity, and coating thickness h. The derivation applies to the special case where the radial clearance H is small in comparison to the bore of the die. Then the equations derived for flow between parallel plates apply to the annular space. In addition, it is assumed that the fluid is Newtonian.

If Q represents the volumetric rate at which fluid is supplied to the die, then

$$Q = Q_d + Q_p \tag{3-157}$$

where Q_d is the volumetric drag flow rate in the annulus and Q_p, the volumetric pressure flow rate. The size of the wire, its speed, and the required coating thickness determine the value of Q, given by the equation

$$Q = \pi V[(R_i + h)^2 - R_i^2] = \pi Vh(2R_i + h) \tag{3-158}$$

The volumetric drag flow rate, given by (3-118), can be written

$$Q_d = \frac{V_z WH}{2} \tag{3-159}$$

where W represents a mean value of the circumference of the annular space. The volumetric pressure flow rate, given by (3-109), can be written

$$Q_p = \frac{WH^3 \Delta P}{12\mu L} \tag{3-160}$$

where ΔP is the difference between the pressure in the reservoir and atmospheric pressure.

If the above four equations are combined and then solved for ΔP, one obtains

$$\Delta P = \frac{6\mu V_z L}{H^2}\left[\frac{4h\pi}{WH}\left(R_i + \frac{h}{2}\right) - 1\right] \tag{3-161}$$

The following approximation for the mean circumference

$$W = 2\pi\left(R_i + \frac{h}{2}\right)$$

reduces equation (3-161) to

$$\Delta P = \frac{6\mu V_z L(2h - H)}{H^3} \tag{3-162}$$

which shows that the required pressure in the wire coating die is directly proportional to the viscosity of the fluid, the linear speed of the wire, and the length of the die.

It is also noted from equation (3-162) that zero pressure is required in the reservoir when the condition $H = 2h$ exists. This is the optimum design condition because changes in wire speed or in the viscosity of the coating fluid will have no effect on the thickness of the coating.

Figure 3-24 is a schematic diagram of an end-fed film die. It consists of a delivery tube of length L and radius R that serves as a reservoir for the fluid entering the die lips. The die lips are parallel plates of length Y

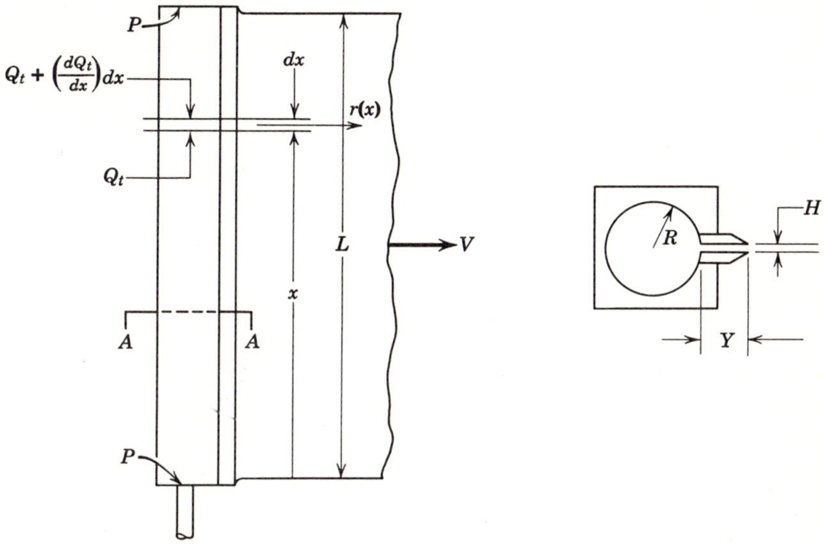

Figure 3-24. Schematic diagram of end-fed film die.

separated by distance H. Flow occurs in two directions. In the tube, the fluid, entering at the point $x = 0$, flows in the x-direction while the flow in the die lips is perpendicular to the axis of the tube.

The center-fed film die and the tubing die are two important variations of this design. In the center-fed film die, the feed port of the delivery tube is located at the center ($x = L/2$) rather than at the end ($x = 0$). In the tubing die, the axis of the delivery tube is a circle.

The following analysis[22] relates the uniformity of the thickness of the film produced by the die to the design and operating variables. Let $r(x)$ represent the extrusion rate through the lips at any distance x from the inlet to the delivery tube. If the quantity $r(x)$ were independent of x, then the film thickness would be uniform—assuming that elastic effects causing strain recovery are negligible or are uniform across the width of the film.

[22] J. F. Carley, *J. Appl. Phys.*, **25,** 1118 (1954); *Modern Plastics*, p. 127, August, 1956.

A uniformity index E can be defined by the equation

$$E = \frac{r_{\text{min}}}{r_{\text{max}}} \tag{3-163}$$

where r(min) represents the minimum extrusion rate and r(max) the maximum. When r is independent of x, the film is uniform and the value of E is unity. If the flow should be blocked at some point, then r(min) is zero and so is the index E.

Formulas for the uniformity index E are derived here for two special cases. In the first case, the flow is isothermal and the decay in pressure along the length of the delivery tube causes $r(x)$ to decrease as x increases. In the second case, the pressure $P(x)$ is constant, but variations in the temperature $T(x)$ cause the rate $r(x)$ to vary.

At any point in the delivery tube, the volumetric flow rate Q_t is given by the equation

$$Q_t = -\frac{\pi R^4}{8\mu}\left(\frac{dP}{dx}\right) \tag{3-164}$$

and the volumetric flow rate $r(x)$ through the lips at any point x is given by the equation

$$r(x) = \frac{H^3 P\,dx}{12\mu Y} \tag{3-165}$$

where P is the hydrostatic pressure relative to the atmosphere at any point along the delivery tube. These equations are valid for isothermal Newtonian flow.

Assuming that the fluid is incompressible, a material balance over an element of the tube leads to the equation

$$Q_t = r(x) + \left[Q_t + \left(\frac{dQ_t}{dx}\right) dx\right] \tag{3-166}$$

If equations (3-164) and (3-165) are introduced into (3-166), one obtains the differential equation

$$\frac{d^2P}{dx^2} - \beta^2 P = 0 \tag{3-167}$$

where the parameter β is defined by the equation

$$\beta^2 = \frac{2H^2}{3\pi R^4 Y}$$

The boundary conditions are

$$P(0) = P_0$$

$$P'(L) = 0 \qquad \text{where} \qquad P' = \left(\frac{dP}{dx}\right)$$

and the solution of equation (3-167) satisfying these boundary conditions is written

$$\frac{P}{P_0} = e^{-\beta x} + e^{-\beta L}\left[\frac{\sinh(\beta x)}{\cosh(\beta L)}\right] \tag{3-168}$$

Because the minimum and maximum extrusion rates occur at the far and near ends of the delivery tube

$$r_{max} = r(0)$$
$$r_{min} = r(L)$$

Reference to equation (3-165) shows that r is directly proportional to P. Therefore the expression for the uniformity index E becomes

$$E = \frac{r_{min}}{r_{max}} = \frac{r(L)}{r(0)} = \frac{P(L)}{P(0)} \tag{3-169}$$

The pressure ratio $P(L)/P(0)$ can be obtained directly from equation (3-168) by replacing x with L. It reduces to the simple form

$$\frac{P(L)}{P(0)} = e^{-\beta L}[1 + \tanh(\beta L)] = \frac{1}{\cosh(\beta L)} \tag{3-170}$$

and the uniformity index becomes

$$E = \frac{1}{\cosh(\beta L)} \tag{3-171}$$

Equation (3-171) can also be written in the approximate form

$$E = \frac{1}{1 + \dfrac{(\beta L)^2}{2} + \cdots} \approx 1 - \frac{(\beta L)^2}{2} \approx 1 - 0.106\left(\frac{H}{Y}\right)\left(\frac{LH}{R^2}\right)^2 \tag{3-172}$$

which gives results agreeing within a few percent of equation (3-171) under the condition $E > 0.8$.

The above formulas show that the uniformity is independent of the viscosity of the fluid and that it is most sensitive to the radius of the delivery tube. For example, if

$$R = 0.35 \text{ in.} \qquad H = 0.010 \text{ in.}$$
$$L = 66 \text{ in.} \qquad Y = 0.40 \text{ in.}$$

the value of the uniformity index E, calculated with equation (3-172), is 0.92. By increasing the radius of the delivery tube to 0.50 in., the uniformity index increases to 0.98.

Carley[22] has also solved this problem for the case where the fluid obeys the power law. He gives the following approximate solution for the uniformity index:

$$E = 1 - 0.525\,(\lambda L)^2 \tag{3-173}$$

where

$$\lambda = \left[\left(\frac{1}{2\pi}\right)\left(\frac{3n+1}{2n+1}\right)\left(\frac{H^{(2n+1)/n}}{Y^{1/n}R^{(3n+1)/n}}\right)\right]^{n/(n+1)}$$

If, in the above example, the flow index of the fluid is 0.50, the uniformity index, as calculated with equation (3-173), is 0.59 when $R = 0.35$ and 0.88 when $R = 0.50$.

A formula for the uniformity index is now derived for the case where the pressure in the delivery tube is constant but where the fluid temperature $T(x)$ may vary along the die lips. Reference to equation (3-165) shows that the extrusion rate $r(x)$ is inversely proportional to the viscosity of the fluid. Therefore equation (3-163) becomes

$$E = \frac{r_{min}}{r_{max}} = \frac{\mu_{min}}{\mu_{max}} \tag{3-174}$$

The viscosity-temperature relationship of equation (2-87), written

$$\mu = ae^{-b_\tau T}$$

reduces (3-174) to

$$E = e^{-b_\tau \Delta T} = 1 - b_\tau \, \Delta T - \cdots \tag{3-175}$$

where

$$\Delta T = T_{max} - T_{min}$$

Note that the temperature coefficient b_τ, which applies to viscosity changes occurring at constant shear stress, is used because the pressure drop $P(x)$ is constant. For power law fluids, the quantities b_τ and $b_{\dot\gamma}$ are related by the equation

$$nb_\tau = b_{\dot\gamma} \tag{3-176}$$

A useful approximate expression for the uniformity index can be obtained from the series of equation (3-175).

$$E = 1 - b_\tau \Delta T = 1 - \frac{b_{\dot\gamma} \Delta T}{n} \tag{3-177}$$

Values of $b_{\dot\gamma}$ for many of the common thermoplastics are listed in Table 2-6.

Equation (3-177) enables the effect of temperature variation on the uniformity to be calculated. Conversely, if a limiting acceptable value for E is established, the required degree of temperature control can be estimated with equation (3-177). For example, if the maximum allowable variation in thickness is $\pm 2.5\%$, the value of E is 0.95. If the temperature coefficient $b_{\dot\gamma}$ is about 0.05 deg. C^{-1} and the flow index n for the fluid is about 0.5, a control system capable of maintaining the fluid temperature at ± 0.25 deg. C is required. For polyethylene, the value of $b_{\dot\gamma}$ is about $\frac{1}{4}$ as large. Therefore, for polyethylene, a control system providing for variations within ± 1 deg. C would give the same uniformity in the extruded film.

Problems

3-1 Show that for isothermal laminar flow of a Newtonian fluid through a tube

$$f = \frac{8}{N_{\mathrm{Re}}}$$

where the friction factor f is a dimensionless quantity defined by the equation

$$f = \frac{\tau_w}{\rho u^2}$$

where u is the average velocity and τ_w is the shear stress at the wall of the tube.

Make a plot of f versus N_{Re}, using logarithmic graph paper. Show that the slope of this line is -1 and that it is a universal plot applying to the isothermal laminar flow of any Newtonian fluid through any tube of circular cross section.

3-2 Prove that the universal plot developed in Problem 3-1 is applicable to power law fluids if the Reynolds number is defined by the equation

$$N_{\mathrm{Re}} = \left[\frac{2n}{n+1}\right]\left[\frac{2(3n+1)}{n}\right]^{1-n}\left[\frac{D^n u^{2-n} \rho (\dot{\gamma}^\circ)^n}{\tau^\circ}\right]$$

Show that this Reynolds number is dimensionless and that it reduces to the conventional one for the special case where $n = 1$.

3-3 A power law fluid having a flow index of 0.8 and a density of 1 gram/cm^3 flows at a rate of 10 cm^3/sec through a tube having a diameter of 0.5 cm and a length of 100 cm. The pressure drop is 10 atm. Calculate the friction factor (as defined in Problem 3-1) for the flow and then calculate the value of the Reynolds number and τ°, taking the standard shear rate as 1 sec^{-1}. Then show that a 50% increase in the flow rate will cause the pressure drop to increase by 38%.

3-4 Show that the Buckingham-Reiner equation can be transformed to

$$u = \frac{RS_0}{4\zeta}\left[\frac{1}{X} - \frac{4}{3} + \frac{X^3}{3}\right]$$

where u is the average velocity. The above equation indicates that u should pass through a minimum at $X = 1$. Is this result correct? Explain.

3-5 The following flow data were obtained with a material known to be a Bingham plastic:

Run	*Pressure Drop (atm)*	*Average Velocity (cm/sec)*
1	10	10
2	15	20

The diameter of the tube is 0.5 cm and its length is 100 cm. Compute S_0 and ζ for the material.

3-6 Show that a plot of $(4Q/\pi R^3)$ versus $[1/X - \frac{4}{3} + X^3/3]$ will be a straight line if the material is a Bingham plastic. What will be the slope and intercept of the line? How could one determine the value of S_0 for a Bingham plastic if $Q - \Delta P$ data for flow through a tube were available?

3-7 A non-Newtonian fluid flows at the rate of 1 cm^3/sec under a pressure drop of 2 atm through a tube having a diameter of 2 cm and a length of 100 cm. What would be the pressure drop if the same fluid, at the same temperature, were pumped through a tube 100 meters long and 20 cm in diameter at a rate of 1000 cm^3/sec? Explain clearly all assumptions made in obtaining your answer.

3-8 Verify equation (3-49).

3-9 A power law fluid reaches steady state flow in a long tube having a radius of 1 cm. The wall of the tube is maintained at 300° K. The pressure gradient is 10^5 dynes/cm^2. The fluid properties are as follows:

$$n = 0.8$$
$$\kappa = 10^{-3} \text{ deg. C}^{-1}$$
$$\eta^\circ = 10^3 \text{ poise (standard shear rate 1 sec}^{-1}\text{)}$$
$$k = 10^{-3} \text{ cal/cm-sec-deg. C}$$

Make a plot of temperature T versus the dimensionless radius (r/R). You may neglect the effect of temperature on viscosity and the effect of pressure on the density of the fluid.

3-10 Consider the drag flow of a Newtonian fluid in a channel whose cross section is a semicircle. The curved surface is stationary and the flat surface moves with speed V_z in the direction of the channel axis. Find the drag flow velocity as a function of position.

3-11 Compare the volumetric drag flow rate in the semicircular channel described in Problem 3-10 with the drag flow in a rectangular channel whose width is equal to the diameter and whose depth is equal to the radius of the semicircle.

3-12 Consider the flow in a rectangular channel when the drag and pressure flow rates are equal but of opposite sign. Make a sketch showing lines of constant velocity (isovels) in a plane perpendicular to the channel axis. (see "Theory of Screw Extruders," p. 596, by W. L. Gore and J. M. McKelvey, ch. 16 of *Rheology, vol. III*, edited by F. R. Eirich, Academic Press, 1960).

3-13 Consider two flat circular disks located in parallel planes and separated by distance H. Both disks have the same radius R. The space between the disks is

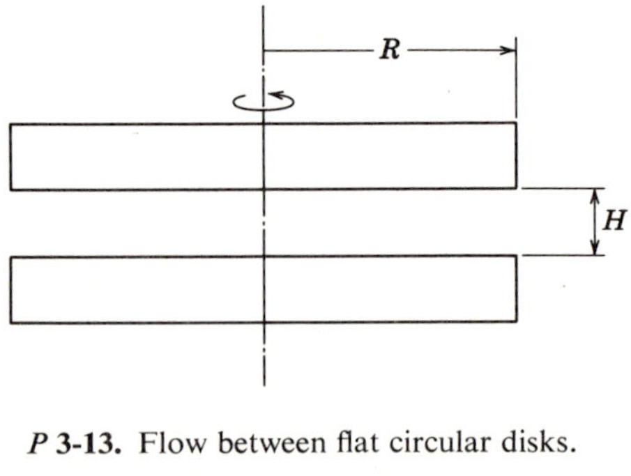

P **3-13.** Flow between flat circular disks.

filled with an incompressible Newtonian fluid. The lower disk is stationary and the upper disk rotates at a constant frequency N. Formulate for this flow problem the partial differential equation that gives the tangential velocity v_θ as a function of the position coordinates r and z, where r is measured out from the axis of the disks and z is measured up from the lower disk. State the boundary conditions for this flow problem.

3-14 Solve the boundary value problem formulated above in Problem 3-13. That is, find a function $v_\theta(r, z)$ that is a solution of the partial differential equation and also satisfies the boundary conditions.

3-15 Derive an equation for the cone and plate viscometer that relates the torque applied to the cone to the dimensions of the instrument, the frequency of rotation of the cone, and the viscosity of the fluid. Assume isothermal Newtonian flow. (see Example 3.5-3 and Problem 3.T in *Transport Phenomena*, by R. B. Bird, W. E. Stewart, and E. N. Lightfoot, John Wiley and Sons, 1960).

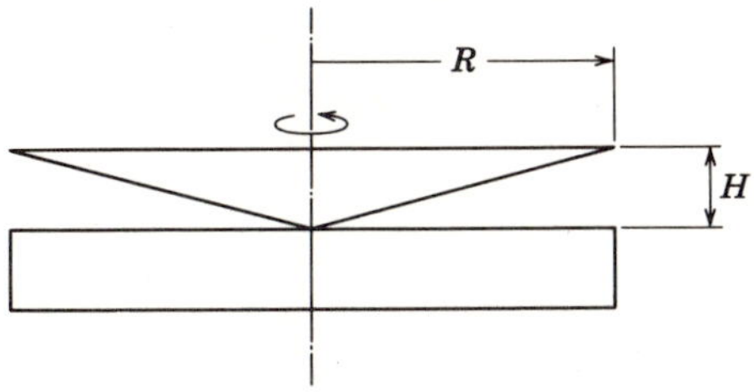

P **3-15.** Flow in the cone and plate viscometer.

3-16 Consider drag flow of a Newtonian fluid in the annular space between coaxial cylinders, where the only nonzero velocity component is $v_\theta(r, t)$. The initial and boundary conditions for a transient or unsteady state flow problem are as follows:

$$v_\theta(r, 0) = 0$$
$$v_\theta(R_i, t) = 0$$
$$v_\theta(R_0, t) = V_\theta$$

Formulate the partial differential equation for this flow problem and then find a solution satisfying it and the above boundary conditions. (See R. B. Bird and C. F. Curtiss, *Chem. Engr. Sci.* **11,** 108 (1959).)

3-17 A molten polymer is to be coated on a cable at a speed of 200 ft/min. The cable diameter is 2 cm and the required coating thickness is 0.1 cm. A die similar to the one shown in Figure 3-23 is available. The bore of the die is exactly 2.1 cm and its length L is 5 cm. Assume that the polymer has a Newtonian viscosity of 10^3 poise. What pressure must be maintained in the reservoir to develop the required coating thickness? If there is a variation of $\pm 10\%$ in the reservoir pressure, what effect will this have on the thickness of the coating?

3-18 Determine the economic pipe diameter for a pipe line to transport a highly viscous non-Newtonian fluid at a rate of 1000 lbs/hr a distance of 1000 meters. The density of the fluid is 1 gram/cm^3. Its non-Newtonian viscosity at a shear rate of 10 sec^{-1} is 500 poise and at 100 sec^{-1} it is 200 poise. You may assume that the efficiency of the pump is 30%, that power for the pump costs 0.04 dollars/kwhr, and that the cost of pipe is 0.15 dollars/lb and that this is independent of the size of the pipe. List all other assumptions made in your calculation.

chapter 4

Expansion and Contraction of Amorphous Polymers

4-1 Introduction

Gases are highly compressible. Doubling the absolute pressure approximately doubles the density and doubling the absolute temperature reduces the density by about half. Process engineering calculations must, therefore, take into account the effects of temperature and pressure on gas volumes. This is usually accomplished by approximating the behavior of gases by simple equations of state, such as the ideal gas law and the van der Waals' equation.

Solids and liquids, on the other hand, are relatively incompressible, and process engineering calculations can often be simplified by assuming that changes in pressure, and sometimes temperature, have no effect on their density. Under most conditions the errors introduced by these assumptions are negligible.

In certain operations of polymer processing, however, small changes in the density of solid and liquid polymers are extremely important and must be carefully considered. In injection molding, for example, if too low an injection pressure is used, the object upon cooling will shrink away from the mold and will have poor dimensional quality. If the injection pressure is too high, the object upon cooling will be in a state of compression and will be removed from the mold only with great difficulty. In the calculation of correct injection conditions, the effects of both temperature and pressure on the volume of the solid and liquid polymer must be considered.

Few equations of state applicable to solid and liquid polymers have been developed. The theoretical aspects of the problem are far more complex than for gases—intermolecular distances are greatly reduced and molecular interactions between polymer chains are extremely complex. Most equations used to calculate volume changes of polymers are interpolation formulas, useful only when experimental data are available.

Other difficulties arise in the study of volume changes of polymers. Below their melting point many polymers have both amorphous regions and regions of high molecular order—crystallites. Since the crystallites have a higher density than the surrounding amorphous regions, the density

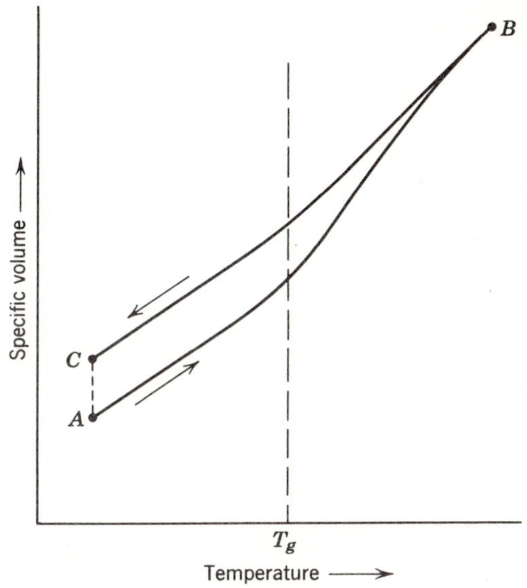

Figure 4-1. Typical volume-temperature curves obtained when an amorphous polymer is heated and cooled through its softening point.

of a given polymer sample depends upon its degree of crystallinity. This, in turn, depends upon its previous history, particularly upon the rate at which it is cooled from the liquid state. Crystallization in polymers is discussed in the next chapter.

Even when crystallization does not occur, complex time effects are observed at temperatures near the softening point, as shown by Figure 4-1. Curve *A-B* is followed when a sample, initially in an equilibrium state, is heated at a constant rate from a temperature below its softening point. Upon reversing the process, curve *B-C* is followed. Over a long period of time, the sample volume decreases, following the line *C-A*, so that the hysteresis loop is eventually closed. For a given material, the extent of hysteresis depends upon the rate of heating and cooling. If the heating and cooling are carried out very slowly, the two curves will coincide.

At the temperature T_g, the glass transition temperature, the material changes from a soft, rubberlike material to a hard, glasslike material. This change does not involve crystallization and is distinctly different from a first-order transition (one characterized by a primary thermodynamic variable, such as specific volume, having a discontinuity).

4-2 Kinetics of Expansion of Amorphous Polymers

Figure 4-2 illustrates the expansion of an amorphous polymer upon an abrupt change of temperature. For $(t < t_0)$, the sample is at a uniform

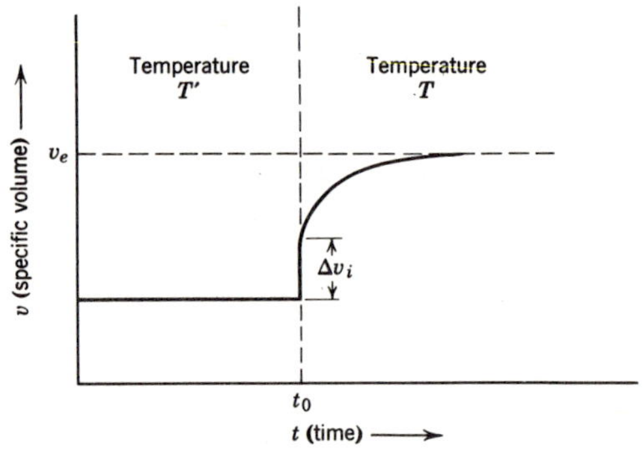

Figure 4-2. Expansion of an amorphous polymer upon an abrupt change of temperature.

temperature T' and at its equilibrium specific volume. At time t_0, the temperature jumps instantaneously to T. The specific volume jumps an amount Δv_i and then gradually increases to its new equilibrium volume v_e.

The first expansion, occurring almost instantaneously, is probably due to an increase in intermolecular spacing arising from increased thermal vibration. This expansion mechanism is similar to that causing the expansion and contraction of ordinary crystalline solids.

The second mechanism occurs at a rate that is directly proportional to the displacement of the specific volume from equilibrium. The rate is also highly temperature dependent. The mechanism of the second expansion process is probably connected with diffusion or viscous flow.

Amorphous polymers have a certain amount of local crystal-like order[1] but no far-reaching structure. Molecules distributed in the space surrounding the ordered regions are imperfectly arranged. The space, therefore, contains holes or gaps. At higher temperatures, the degree of disorder increases, and the percentage of space occupied by holes increases. However, upon a change of temperature, it takes a finite period of time

[1] T. Alfrey, *Mechanical Behavior of High Polymers*, Interscience Publishers, 1948.

for the holes to appear or disappear. This, then, is the explanation of the observed time effect.

Alfrey suggests that the controlling mechanism for the appearance or disappearance of holes is diffusion. On the other hand, Boyer and Spencer[2] believe that the controlling mechanism is one of viscous flow of chain segments. Although the nature of the molecular processes responsible for the observed effects is not completely understood, it is possible to develop an empirical expression that describes the phenomenon.

Noting that the volume of an amorphous polymer (at constant pressure) is a function of time and temperature, its total differential is

$$dv = \left(\frac{\partial v}{\partial T}\right)_t dT + \left(\frac{\partial v}{\partial t}\right)_T dt$$

which can be written in the equivalent form

$$\frac{dv}{dT} = \left(\frac{\partial v}{\partial T}\right)_t + \frac{1}{r}\left(\frac{\partial v}{\partial t}\right)_T \tag{4-1}$$

where

$$r = \left(\frac{dT}{dt}\right)$$

and represents the rate of temperature change.

The experiments of Spencer and Boyer[3] with polystyrene show that the instantaneous (time-independent) volume changes are directly proportional to the temperature change. Hence

$$\Delta v_i = b'\Delta T \qquad \text{or} \qquad \left(\frac{\partial v}{\partial T}\right)_t = b' \tag{4-2}$$

Furthermore, the rate of the isothermal (time-dependent) process is given by the equation

$$\left(\frac{\partial v}{\partial t}\right)_T = k(v_e - v) \tag{4-3}$$

where v_e is the equilibrium volume corresponding to temperature T and k is the rate constant. The temperature dependence of k is given by the equation

$$k = Ke^{-E/RT} \tag{4-4}$$

Furthermore, their experiments showed that if temperature changes occur very slowly, the volume-temperature curve will be straight. Hence it can be represented by the linear equation

$$v_e = a + bT \tag{4-5}$$

where a and b are constants.

[2] R. F. Boyer and R. S. Spencer, *J. Appl. Phys.*, **16,** 594 (1945).

[3] R. S. Spencer and R. F. Boyer, *J. Appl. Phys.*, **17,** 398 (1946).

If equations (4-2), (4-3), and (4-5) are introduced into (4-1), one obtains

$$\frac{dv}{dT} = b' + \frac{k}{r}(a + bT - v) \tag{4-6}$$

The solution of equation (4-6) relates volume to temperature with r, the rate of temperature change, a parameter.

4-3 Calculation of Volume-Temperature Curves

To calculate a volume-temperature curve, one must integrate equation (4-6). Unfortunately, if one takes into account the temperature dependence of the rate constant k, a solution cannot be obtained in closed form. However, numerous numerical techniques can be used to construct the volume-temperature curve.

If one assumes that the rate constant k is constant, (4-6) is readily integrated, but the resulting solution applies only to the narrow range of temperature over which the variation of k is not too large. However, by making step-by-step calculations, using an average value of k for each step, the entire v-T curve can be constructed.

To integrate (4-6) it is convenient to make a change of dependent variable. The new dependent variable ν is defined by the equation

$$\nu = (v_e - v) = (a + bT - v) \tag{4-7}$$

which, upon differentiation, becomes

$$\frac{d\nu}{dT} = b - \frac{dv}{dT} \tag{4-8}$$

The general solution of equation (4-6), for constant k, and in terms of the new variable ν, is written

$$\nu = \alpha + Ce^{-\beta T} \tag{4-9}$$

where C is the constant of integration, and α and β are constants defined by the equations

$$\alpha = \frac{r(b - b')}{k} \qquad \beta = \frac{k}{r}$$

The constant of integration is evaluated from the initial condition that when $T = T_i$, the variable ν has the value ν_i. Equation (4-9) becomes

$$\nu = \alpha - (\alpha - \nu_i)e^{-\beta\Delta T} \tag{4-10}$$

where $\Delta T = (T - T_i)$.

This result can be transformed to the original variable v by introducing equation (4-7). The result is

$$v = (a - \alpha + bT) - (a - \alpha + bT_i - v_i)e^{-\beta\Delta T} \tag{4-11}$$

which is the final working equation for the calculation.

The next step is to determine values of the rate constant k as a function of temperature; this is easily done if half-time data are available. The half-time is defined as the time required for a sample, upon an abrupt change of temperature, to expand half the distance to the new equilibrium volume. The following derivation shows the relationship between the half-time and the rate constant:

For isothermal expansions, the applicable rate equation is (4-3). Its general solution can be written

$$(v_e - v) = Ce^{-kt} \tag{4-12}$$

where C is the constant of integration. If the specific volume is initially v_i, then the constant C can be evaluated. Equation (4-12) becomes

$$\frac{v_e - v}{v_e - v_i} = e^{-kt} \tag{4-13}$$

If θ represents the half-time, then when $t = \theta$,

$$v = (1/2)\,(v_e + v_i)$$

and (4-13) can be solved for k. The result is

$$k = \frac{\ln (2)}{\theta} \tag{4-14}$$

which shows that the rate constant k is inversely proportional to the half-time.

Half-times for the isothermal expansion of polystyrene at five different temperatures are available.[3] Rate constants are calculated from these data, using (4-14), and are shown in Table 4-1. Then, as suggested by (4-4), a plot of the logarithm of k versus the reciprocal absolute temperature is made. This is Figure 4-3. The experimental data fall within a temperature range of about 300° K to about 400° K. Values of k up to about 500° K are obtained by extrapolation.

As a specific example, the calculation of the first point on a v-T curve is shown below. The case where the sample is initially in an equilibrium state at 300° K and is heated at the rate of 0.01 deg/sec is considered. The constants of the polystyrene are assumed to have the following values:

$$a = 0.8450 \text{ cm}^3/\text{gram}$$
$$b = 3.75 \times 10^{-4} \text{ cm}^3/\text{gram-deg. C}$$
$$b' = 2.50 \times 10^{-4} \text{ cm}^3/\text{gram-deg. C}$$

The initial volume of the sample is obtained by the following calculation, using equation (4-5):

$$v_i = (a + bT_i) = 0.8450 + 300(3.75 \times 10^{-4})$$
$$= 0.9575 \text{ cm}^3/\text{gram}$$

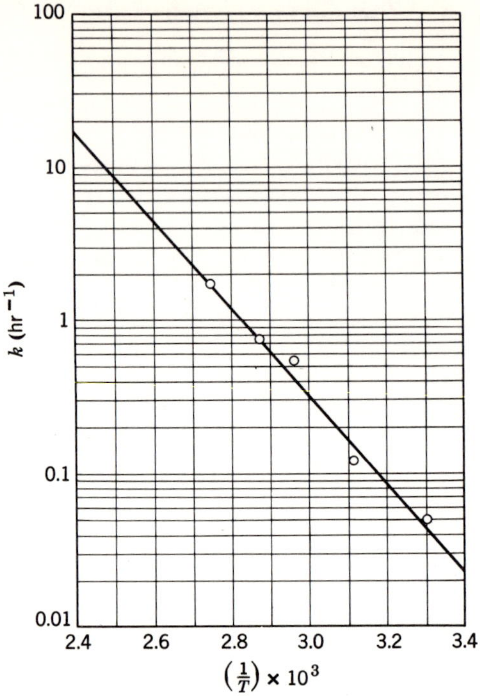

Figure 4-3. Temperature dependence of the rate constant k for the expansion of polystyrene (Data from Ref. 3).

Table 4-1*

Values of the Rate Constant for the Isothermal Expansion of Polystyrene

T(° K)	k (hr^{-1})	T(° C)
310	0.068	37
330	0.26	57
350	0.80	77
370	2.30	97
390	6.0	117
. . . . .	. . . . .	. . . . .
410	13.2	137
430	27.5	157
450	57	177
470	105	197
490	190	217

Note: Values below dotted line obtained by extrapolation.

* Calculated from data of Ref. 3.

Temperature increments of 20 deg were selected for the calculation. The first increment covers the range from 300° K to 320° K. The value of rate constant k at the mid-point of the increment, 310° K, is used in the calculation. The specific volume at 320° K is 0.9653 cm³/gram. It is found by substituting the following values into equation (4-11):

$$k = 0.068 \text{ hr}^{-1} \qquad r = 36 \text{ deg. C/hr}$$
$$\Delta T = 20 \text{ deg. C} \qquad \beta = 0.00189 \text{ deg. C}^{-1}$$

This procedure is repeated until the entire curve is established. The family of $v - T$ curves shown in Figure 4-4 has been calculated by this method.

Figure 4-4 illustrates how the rate of temperature change affects the shape of the volume-temperature curve. With rapid rates of heating or cooling, the curve departs from equilibrium at much higher temperatures than when the rate is low. Hence the glass transition temperature is a function of the rate of cooling or heating.

The glass transition temperature is considered to be that temperature at which the relaxation processes in the polymer occur at a rate that is the same order of magnitude as the cooling (heating) rate. At higher temperatures, relaxation is more rapid. Hence with rapid cooling (heating), the glass transition occurs at higher temperatures.

Factors that promote the relaxation processes, for example, the addition of a plasticizer, tend to reduce the glass transition temperature. Factors that retard relaxation, such as branching and cross-linking, tend to increase the glass transition temperature. Geometric factors and molecular cohesion are also very significant in determining the glass transition temperature, as can be seen in Table 4-2, where the glass transition temperatures of some common thermoplastics are listed.

Table 4-2

Glass Transition Temperatures of Thermoplastics

Polymer	T_g (° C)
Polybutadiene	−85
Polyethylene	−68
Polyvinylidene chloride	−17
Polyamide 66	47
Polystyrene	81

4-4 Annealing for Dimensional Stability

Consider, for example, polystyrene sheet stock being made in a process where molten polystyrene at 500° K is extruded into a water bath at 300° K. The sheet stock, in a subsequent operation, is cut into parts for a particular

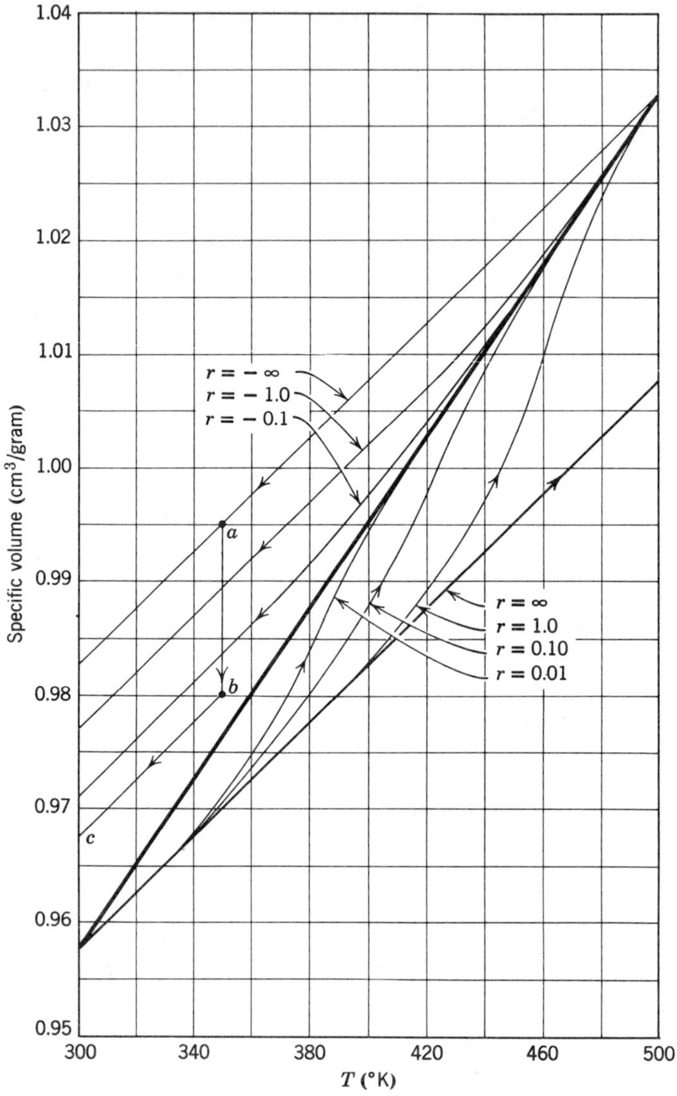

Figure 4-4. Calculated volume-temperature curves for the heating and cooling of polystyrene, with r (the rate of temperature change) expressed in deg C/sec.

application requiring close tolerances. If a maximum shrinkage of 1% in the cut parts can be tolerated and if room temperature is 300° K (81° F), how long should the sheet stock be stored before being cut? If the storage time of the sheet stock is to be reduced by "annealing" in a room at 350° K for a period of time, what annealing time is required?

When the molten polystyrene is solidified by quenching in water, the cooling rate is very high and the material follows the upper curve ($r = -\infty$) of Figure 4-4. At 300° K its specific volume would initially be 0.9825 cm^3/gram. This would eventually decrease to the equilibrium volume of 0.9575 cm^3/gram.

For an allowable shrinkage of 1% in the cut parts, the sheet stock must be allowed to shrink until its specific volume reaches 0.9672 cm^3/gram before it is cut. For an isothermal change, the volume-time relationship is given by equation (4-13), which, when the above volumes are substituted into it, becomes

$$\frac{v_e - v}{v_e - v_i} = \frac{0.9575 - 0.9672}{0.9575 - 0.9825} = 0.388 = e^{-kt}$$

At 300° K, the rate constant k, obtained from Figure 4-3, is 0.035 hr^{-1} and required annealing time is 30 hr.

The line *a-b* in Figure 4-4 shows the path the material would follow upon annealing at 350° K. At point *b*, the volume is 0.9800 cm^3/gram, and, if the temperature were then reduced to 300° K, the material would be at the required point *c*. Because the process *b-c* can occur very rapidly, the total time required is that for process *a-b*. For this process

$$\frac{v_e - v}{v_e - v_i} = \frac{0.9762 - 0.9800}{0.9762 - 0.9950} = e^{-kt}$$

At 350° K the rate constant is 0.80 hr^{-1} and the required annealing time is 2 hr.

4-5 Equation of State

An equation relating volume to temperature and pressure is an equation of state. The simplest equation of state is the ideal gas law, which is written

$$V = \frac{RT}{P} \tag{4-15}$$

where V is the molar volume and R is the universal gas constant. It has the advantage of having no arbitrary constants but gives accurate results only when applied to gases at relatively low densities.

The van der Waals' equation of state gives a more accurate description of the behavior of gases, but it contains two constants that must be determined experimentally. For 1 mole of a gas, it is written

$$(V - B)\left(P + \frac{A}{V^2}\right) = RT \tag{4-16}$$

where A and B are constants. The constant B corrects for the volume occupied by the molecules themselves and the quantity (A/V^2) for the "internal" pressure caused by the attractive forces acting among the molecules. Note that as the molar volume decreases, the internal pressure increases.

The investigations of Spencer and Gilmore[4] showed that the *P-V-T* behavior of many polymers can be represented reasonably well by the following modified van der Waals' equation of state:

$$(v - b)(P + \pi) = \frac{RT}{M} \tag{4-17}$$

where v is the specific volume of the polymer and M the molecular weight of an "interaction unit." The constants π and b must be determined experimentally, as well as the interaction unit M. Note that in equation (4-17), the internal pressure is independent of specific volume and, therefore, of temperature and pressure.

Spencer and Gilmore evaluated the constants π and M from a series of measurements of P and v at a fixed temperature. If equation (4-17) is differentiated and then rearranged to give

$$\left[-RT\left(\frac{\partial P}{\partial v}\right)_T\right]^{1/2} = \pi\sqrt{M} + \sqrt{M}\,P \tag{4-18}$$

it can be seen that π and M can be determined from values of $(\partial P/\partial v)_T$ and P, either by plotting or by the method of least squares. The value of b can then be calculated from a known value of v at atmospheric pressure and a known temperature. Spencer and Gilmore found that the internal pressure π remains relatively constant over a rather wide range of pressures. Their results are summarized in Table 4-3.

Table 4-3*
Constants for Equation of State

	M	π (atm)	b (cm³/gram)
Polystyrene	104	1840	0.822
Polymethyl methacrylate	100	2130	0.734
Ethyl cellulose	60.5	2370	0.720
Cellulose acetate butyrate	54.4	2810	0.688
Polyethylene	28.1	3240	0.875

* From Ref. 4.

At atmospheric pressure, the internal pressure π is much greater than P and equation (4-17) can be written

$$v = b - \left(\frac{RT}{M\pi}\right) \qquad (P \ll \pi) \tag{4-19}$$

[4] R. S. Spencer and G. D. Gilmore, *J. Appl. Phys.*, **21**, 523 (1950).

Upon differentiation, equation (4-19) becomes

$$(M\pi) = \frac{R}{(\partial v/\partial T)_P} \qquad (P \ll \pi) \tag{4-20}$$

which shows that the $(M\pi)$ product can be calculated from isobaric thermal data obtained at low pressures.

The thermal expansion data given previously can be used to calculate the $(M\pi)$ product for polystyrene. Two values of $(\partial v/\partial T)_P$ were given. The value of 3.75×10^{-4} cm³/gram-deg. C applies when volume changes occur so slowly that the sample is always in a state of equilibrium. The value of 2.50×10^{-4} cm³/gram-deg. C applies when the volume changes occur almost instantaneously. Values of $(M\pi)$ calculated with these data, using equation (4-20), are compared below with the value of $(M\pi)$ obtained from the values given in Table 4-2, which were derived from isothermal P-v data:

Method of Calculaton	$(M\pi) \times 10^{-4}$
Equilibrium isobaric v-T data	21.9
Nonequilibrium isobaric v-T data	32.2
Equilibrium isothermal P-v data	19.1

Note that there is good agreement between the values calculated from the equilibrium measurements. Assuming that the value of M does not change drastically, it can be seen that the internal pressure appears to rise sharply as one goes from equilibrium to nonequilibrium conditions. The reason for this is that under nonequilibrium conditions, the volume does not have time to change completely, so $(\partial v/\partial T)_P$ is less than its equilibrium value.

Volume changes occurring at constant pressure or at constant temperature are often a matter of concern. These changes are described by the compressibility β and the coefficient of thermal expansion κ. These quantities are defined by the following equations:

$$\beta = -\frac{1}{v}\left(\frac{\partial v}{\partial P}\right)_T \tag{4-21}$$

$$\kappa = \frac{1}{v}\left(\frac{\partial v}{\partial T}\right)_P \tag{4-22}$$

In general, β and κ are themselves functions of temperature and pressure. For all substances there is a decrease in volume as the pressure increases. Therefore β is always a positive quantity. For most substances an increase in temperature results in an increase in volume, so that κ is usually positive. However, there are some exceptions, for example, water in the range between 0° C and 4° C.

Values of β and κ can be derived from equation (4-17) through differentiation. The following results are obtained:

$$\frac{1}{\beta} = (P + \pi)\left[1 + \left(\frac{Mb}{RT}\right)(P + \pi)\right] \tag{4-23}$$

$$\frac{1}{\kappa} = T - \left(\frac{Mb}{R}\right)(P + \pi) \tag{4-24}$$

Example 4-1

Calculate the decrease in the specific volume of molten polystyrene at 300° F (422° K) when its pressure is increased from atmospheric to 10,000 psi (680 atm).

Since atmospheric pressure is negligible in comparison to π, the equation of state (4-17) can be written

$$\left(\frac{v_1 - v_2}{v_1}\right) = \frac{1 - \pi/(P + \pi)}{1 + (Mb\pi/RT)}$$

where P represents the final pressure of 10,000 psi. Using the constants given in Table 4-3 for polystyrene, the decrease in the specific volume is found to be 4.87%.

If the constants of the equation of state had not been available, the volume change could have been estimated from compressibility data. For example, if β is assumed to be independent of temperature and pressure, equation (4-21) can be integrated to give

$$v = Ce^{-\beta P}$$

where C is the constant of integration. Therefore the volume change is given by the equation

$$\left(\frac{v_1 - v_2}{v_1}\right) = 1 - e^{-\beta \Delta P}$$

The value of β at 300° F and 1 atm (calculated with the equation of state) is 0.98×10^{-4} atm^{-1}, which, when substituted into the above equation, gives a decrease in specific volume of about 6.7%. A better result could be obtained by using the value of β at the mean pressure between the initial and final states.

Problems

4-1 As a rough approximation, the temperature dependence of the rate constant k for the expansion of polystyrene is given by the equation

$$k = \frac{A}{B - T}$$

where A and B are constants. For $B = 410$ and $A = 55$, compare the calculated values of k with those given in Table 4-4 for the temperature range 350° K to 400° K. Then introduce the above expression in equation (4-6) and integrate to obtain an explicit expression for v as a function of T. Compute v for polystyrene heated to 400° K from 300° K at rates of 0.01, 0.10, and 1.0 deg. C/sec. Compare the results with Figure 4-4.

4-2 Show that

$$d \ln (v) = \kappa \, dT - \beta \, dP$$

where v is the equilibrium specific volume, and therefore that

$$\left(\frac{\partial \kappa}{\partial P}\right)_T = -\left(\frac{\partial \beta}{\partial T}\right)_P$$

4-3 Estimate the pressures required to compress, at a temperature of 200° C, the materials listed in Table 4-2 so that their specific volumes at 200° C are equal to their equilibrium specific volumes at room temperature and pressure.

4-4 Calculate the work required for the reversible isothermal compression of 1 kg of molten polyethylene at 140° C when the pressure is increased from 1 to 1000 atm. (See Example 5, p. 78 of "Heat Transfer and Thermodynamics" by J. McKelvey, ch. 2 of *Processing of Thermoplastic Materials*, edited by E. C. Bernhardt, Reinhold, 1959.)

chapter 5

Crystallization of Polymers

by Dr. Eric Baer[1]

Whether a given polymer is amorphous or crystalline depends primarily on geometric factors. In the main, crystalline polymers possess structural regularity.[2] The widely used condensation polymers, which include the polyamides and polyesters, are crystalline. Addition type polymers with identical substituents on a single chain, such as polyethylene and polytetrafluoroethylene, also have a high degree of crystallinity.

Amorphous polymers have no uniform arrangement of the molecules in the solid state. Typical amorphous polymers are the

$$\left[\begin{array}{cc} H & X \\ | & | \\ C\text{—} & C \\ | & | \\ H & Y \end{array}\right]_n$$

where different substituents are attached to alternate carbon atoms of the chain. Examples include polystyrene, polymethyl methacrylate, and polyvinyl acetate. In these noncrystalline polymers, alternate carbon atoms in the chain are asymmetric and the *d* and 1 configurations are randomly distributed along the chain. Some of these materials have been polymerized by using stero-specific catalysts that impart order to the chain geometry and permit crystallization to occur.[4]

[1] Polychemicals Department, E. I. du Pont de Nemours and Co., Wilmington, Del.

[2] C. W. Bunn, *J. Appl. Phys.*, **25,** 820 (1954).

[3] L. Mandelkern, *Chem. Rev*, **56,** 903 (1956)

[4] G. Natta, *J. Poly. Sci.*, **16,** 143 (1955).

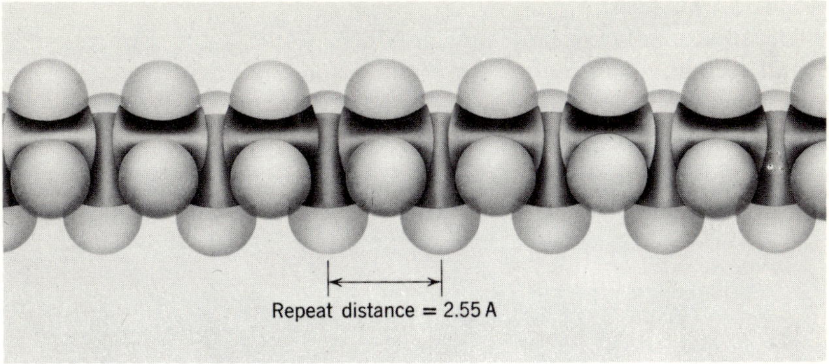

Figure 5–1. Molecular model showing the conformation of the polyethylene chain. (Courtesy Dr. E. S. Clark, Polychemicals Department, E. I. du Pont de Nemours and Co.)

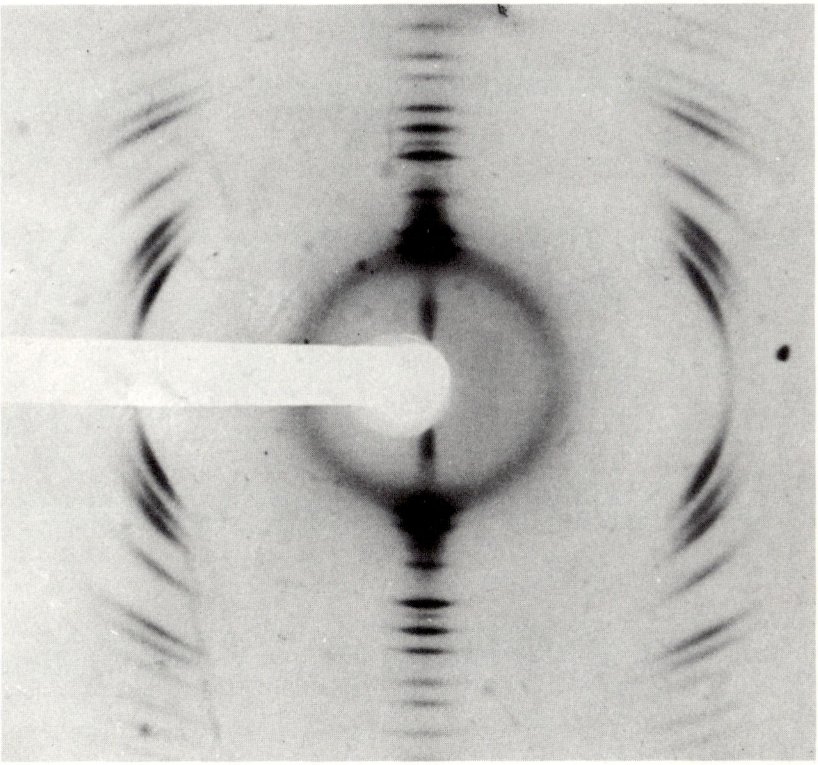

Figure 5–2. X-ray diffraction pattern of a polyethylene fiber. (Courtesy Dr. E. S. Clark, Polychemicals Department, E. I. du Pont de Nemours and Co.)

5-1 The Unit Cell

The structure of crystalline high polymers is revealed by various experimental techniques including x-ray diffraction, infrared absorption, and optical and electron microscopy. The unit cell is the building block from which the crystals are formed. In the unit cell, portions of polymer chains are held in a closely ordered arrangement by relatively weak van der Waals and secondary valence forces.

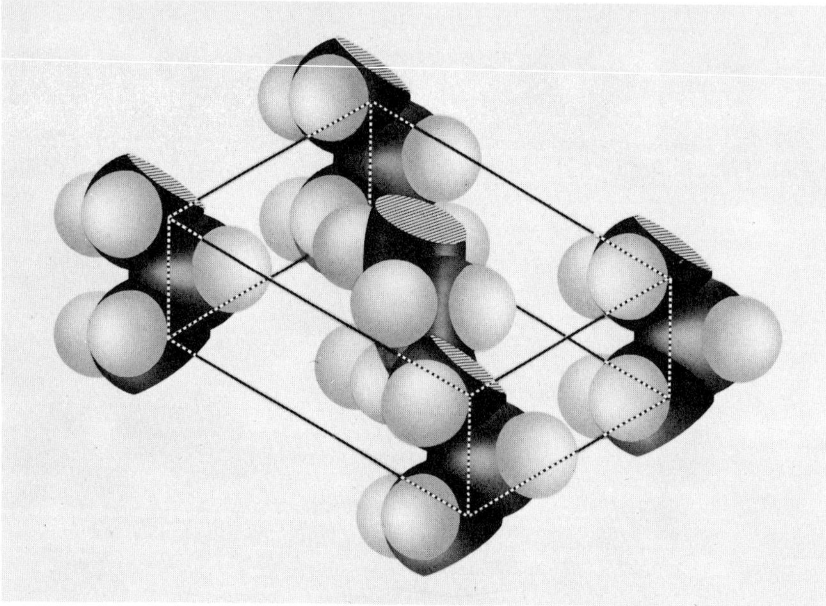

Figure 5–3. The unit cell of polyethylene. (Courtesy Dr. E. S. Clark, Polychemicals Department, E. I. du Pont de Nemours and Co.)

The atoms or molecules in the unit cells of ordinary crystals are discrete units. With polymers, however, only a portion of a polymer chain enters a unit cell, and the structure of the unit cell depends upon chain conformation and packing.

The conformation of a chain molecule relates to the periodicity of atoms along the chain. It is characterized by the nature of the repeat unit and the repeat distance. Figure 5-1 shows that the conformation of a polyethylene chain is a zigzag arrangement with a repeat unit —CH_2—CH_2— and a repeat distance of 2.55 A.

Chain packing determines the nature of the unit cell. Its structure is revealed through x-ray diffraction measurements. A typical x-ray diffraction pattern of a drawn polyethylene fiber is shown in Figure 5-2. The technique of deducing the cell structure from the x-ray diffraction pattern is beyond the scope of this book. It is discussed in some detail by Bunn.[5]

Figure 5-3 represents the unit cell of polyethylene, a parallelepiped having the dimension of 2.55 A measured along the chain axis. Its other dimensions are 7.42 A and 4.94 A, as shown in the figure. Calculations show that the density of the unit cell at room temperature is 1.000 grams/cm^3. Measured densities of polyethylenes range from 0.90 to 0.98 grams/cm^3, the reduction being due to lattice defects and the presence of amorphous regions.

5-2 Morphology of Crystalline Polymers

Currently two models are proposed to describe the morphology of crystalline high polymers. The older concept, the fringed micelle or crystallite model, has been described in the standard texts[6,7] on high polymers. A schematic representation of it is given in Figure 5-4. The parallel lines, which represent the regions of high order, are called crystallites or micelles. These regions are separated by regions of disorder, the amorphous regions. A single polymer molecule can belong to many individual crystallites and have parts of its chain contributing also to amorphous regions.

Lamellar or platelike single crystals of polymers have been grown from dilute solutions, and evidence suggesting the existence of single crystals in the solid state has been obtained. This indicates that the fringed micelle model is probably invalid in some instances, and the lamella or thin platelet model[8–12] has been proposed in its place.

An electron photomicrograph of a single crystal is shown in Figure 5-5. This photograph shows the spiral structure of a polyoxymethylene crystal composed of many platelet layers. The crystal was grown from a dimethyl phthalate solution. It has been shown that although the size and shape of the single crystal depend on the specific growth conditions, some type of lamellar structure is usually found. Electron diffraction experiments

[5] C. W. Bunn, *Chemical Crystallography*, Clarendon Press, Oxford, 1946.

[6] P. J. Flory, *Principles of Polymer Chemistry*, Cornell University Press, Ithaca, New York, 1953.

[7] T. Alfrey, *Mechanical Behavior of High Polymers*, Interscience Publishers, New York, 1948.

[8] A. Keller, in *Growth and Perfection of Crystals*, edited by R. H. Doremus, B. W. Roberts and D. Turnbull, John Wiley, New York, 1958.

[9] P. H. Till, *J. Polymer Sci.*, **24,** 301 (1957).

[10] A. Keller, *Phil. Mag.*, **2,** 1171 (1957).

[11] E. W. Fischer, *Z. Naturforsch.*, **12a,** 753 (1957).

[12] P. H. Geil, "Morphology of an Acetal Resin," paper presented at Am. Physical Soc. meeting, Detroit, March, 1960.

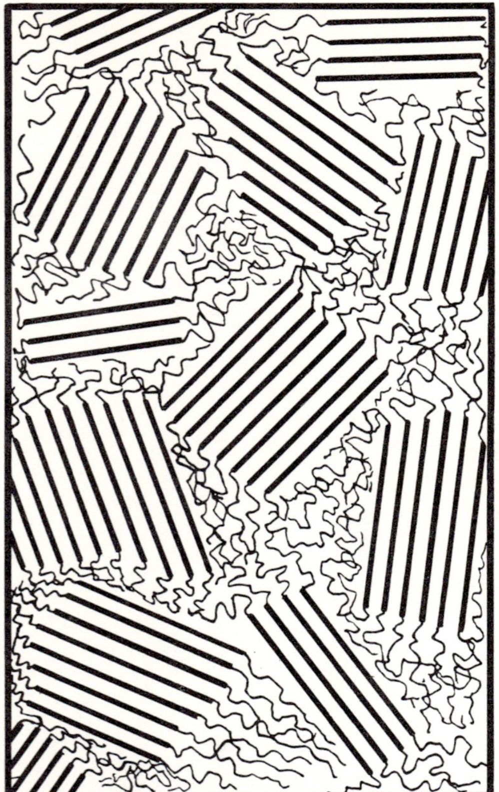

Figure 5–4. Schematic representation of the fringed micelle concept of a crystalline polymer (Ref. 6).

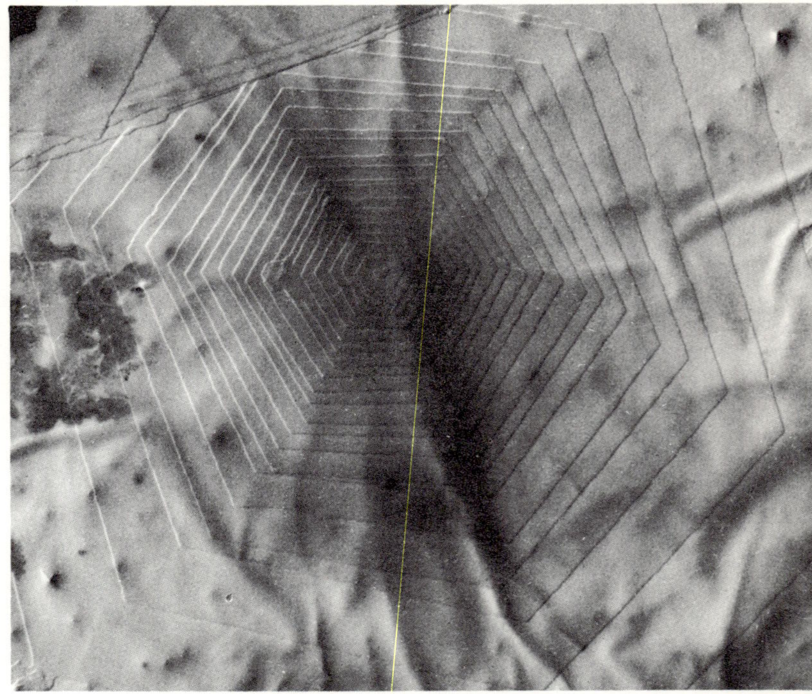

Figure 5–5. Electronmicrograph of a single crystal of polyoxymethylene grown from solution. (Courtesy Dr. P. H. Geil, Polychemicals Department, E. I. du Pont de Nemours and Co.)

have shown that the long chain molecules are folded and oriented normal to the plane of the platelet, which usually has a thickness of about 100 A.

Many polymeric systems also exhibit a spherulitic structure when crystallized from the melt. A characteristic Maltese cross is usually seen when the polymer is viewed between crossed polarizers in the polarizing

Figure 5–6. Spherulites of a low density polyethylene viewed between crossed polarizers in a microscope. (Courtesy Dr. P. H. Geil, Polychemicals Department, E. I. du Pont de Nemours and Co.)

microscope. Figure 5-6 shows some typical spherulites in a low density polyethylene. Spherulites are capable of growing in all three directions and diameters of about 0.1 mm are commonly observed. Electron microscopic observations have indicated that some spherulites have a lamellar structure similar to the single crystals, and there is some evidence[13] showing that spherulites are composed of stacks of these crystals. In some cases it has not been determined if lamellae or fibrils are present within the spherulite. Considerable research remains to be done in this general area, for this

[13] E. W. Fischer, *Z. Naturforsch.*, **12a,** 753 (1957); R. Eppe, E. W. Fischer, and H. A. Stuart, *J. Poly. Sci.*, **34,** 721 (1959).

recent concept of the crystalline structure of polymers has opened a very interesting new area of polymer science.

5-3 Orientation

A certain amount of orientation results when a polymer melt flows through a die. However, much higher degrees of orientation are usually obtained by stretching after crystallization at a temperature below the melting point. The stretching temperature for optimum orientation

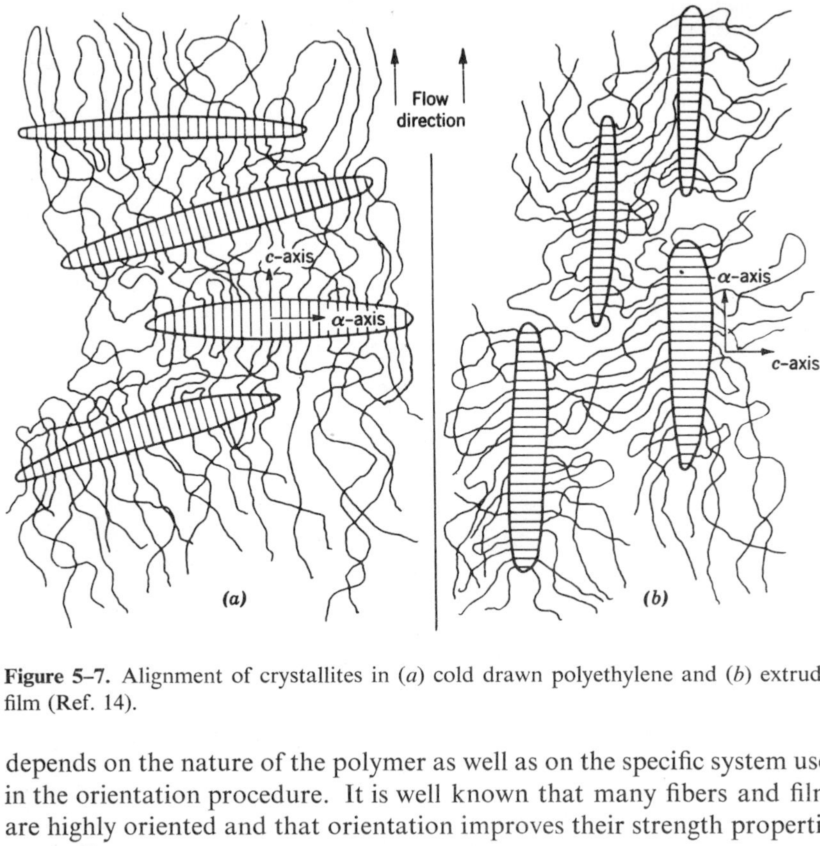

Figure 5–7. Alignment of crystallites in (*a*) cold drawn polyethylene and (*b*) extruded film (Ref. 14).

depends on the nature of the polymer as well as on the specific system used in the orientation procedure. It is well known that many fibers and films are highly oriented and that orientation improves their strength properties markedly.

Studies[14] on polyethylene films using infrared dichroism, x-ray diffraction, and birefringence measurements have revealed some interesting results on the orientation of the molecules and the crystals. The orientation of the crystalline and amorphous regions was discussed on the basis

[14] S. L. Aggarwal, G. P. Tilley, and O. J. Sweeting, *J. Appl. Poly. Sci.*, **1**, 91 (1959).

of the fringed micelle model for extruded polyethylene film as well as for film that was cold drawn 200% after extrusion.

In the case of the extruded films, it was shown that the *a*-axis of the crystallites (which is the long dimension) was oriented, while the *b*- and *c*-axis had no orientation. Furthermore, it was shown that the chains in the crystallites were perpendicular to the machine direction of extrusion. This is illustrated schematically in Figure 5-7*b*. In addition, there was no evidence of any marked orientation in the amorphous regions. In the case of the cold drawn films, both the chains in the crystalline and amorphous regions were oriented parallel to the stretching direction. It was shown that the *c*-axis of the crystal was oriented parallel to the stretching direction, as Figure 5-7*a* illustrates.

5-4 Stiffness and Internal Friction

Glass transitions occur in most commercial amorphous polymers between about 320° K and 420° K. At a glass transition point, an amorphous polymer loses some of its stiffness and brittleness and becomes softer and more rubbery. It becomes better able to undergo large deformations without fracture.

Both torsion pendulum and nuclear magnetic resonance techniques have been used to study transitions in solid polymers of both the amorphous and the crystalline variety. Results of a torsion pendulum experiment with solid polyoxymethylenes of different crystallinities are shown in Figure 5-8. Experiments of this type yield information concerning the stiffness and the internal friction of the material.

The torsion modulus G is a measure of the stiffness of the material. Note that the more crystalline sample is stiffer over the entire range of temperatures covered by the experiment. In the neighborhood of 200° K, the material loses stiffness as it passes through the glass II transition. At 390° K the sample passes through the crystal disordering point. As the temperature approaches the crystalline melting point, the modulus G decreases to a very small value.

The logarithmic decrement δ, a measure of the "internal friction" of the polymer, is determined by measuring the damping of the pendulum. The internal friction depends on the relative amount of strain energy stored as potential energy in the sample compared to that dissipated as heat through molecular motion. Note from Figure 5-8 that the internal friction is lower in the more crystalline sample than in the less crystalline one, indicating that internal friction is strongly associated with the amorphous as well as the crystalline regions of the material.

The sharp peak in the internal friction curve in the neighborhood of the glass II peak indicates that the amorphous regions are undergoing a

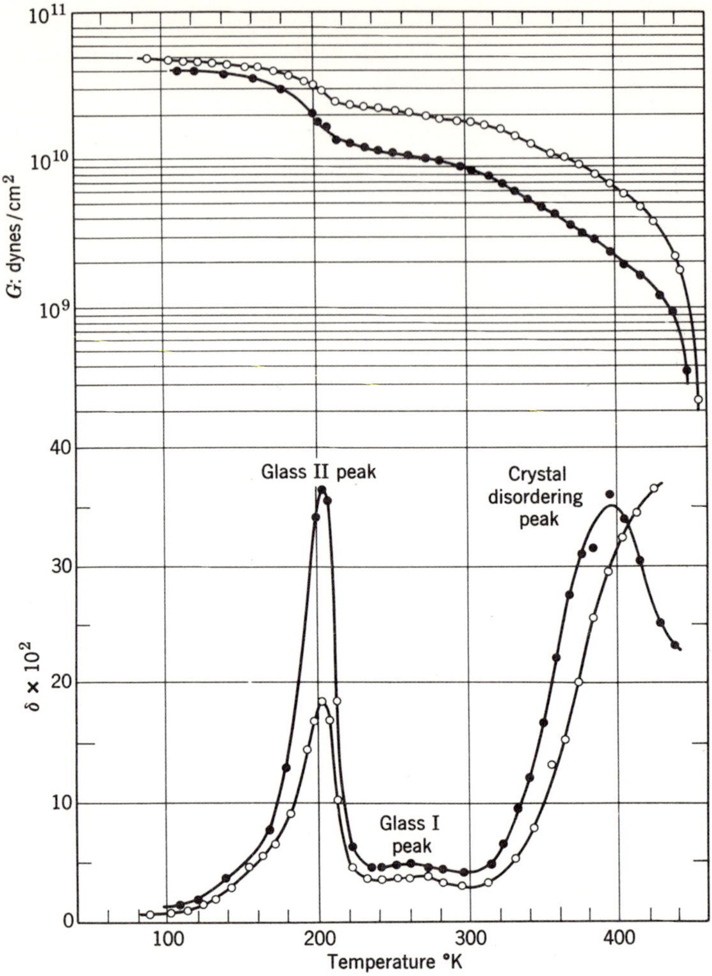

Figure 5–8. Internal friction and torsion modulus of polyoxymethylene. —○— 79.5% crystallinity, —●— 65.0% crystallinity. (Courtesy Dr. N. G. McCrum, Polychemicals Department, E. I. du Pont de Nemours and Co.)

relatively abrupt change in behavior at this temperature, corresponding to the change from a stiff, brittle material to a soft, elastic material.

The glass transition temperatures for most commercial crystalline polymers are usually between 170° K and 270° K, which is considerably lower than those in amorphous polymers. For this reason, crystalline polymers are ductile at normal use temperatures and also stiff because the crystalline aggregates supply rigidity up to temperatures close to the crystalline melting point. This means that polymers with a high crystalline melting point and a low glass transition point will have superior mechanical properties. Polyoxymethylene and various polyamides are crystalline polymers that behave in this manner.

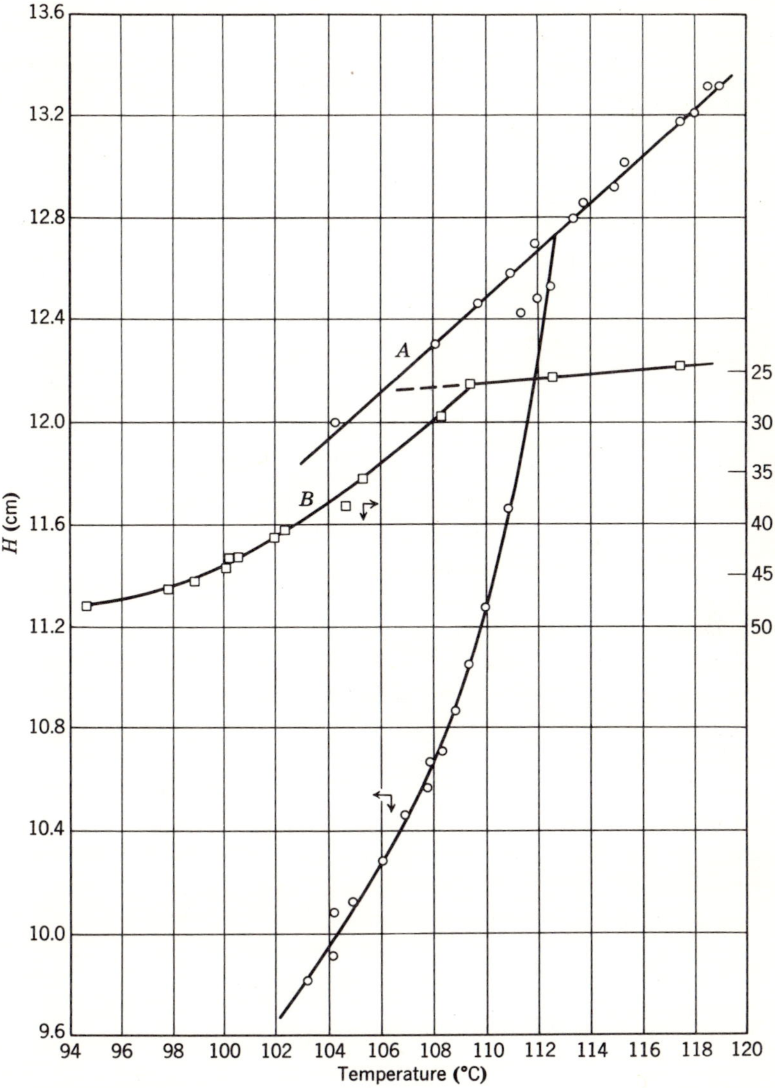

Figure 5–9. Melting curves for polyethylene determined by (*A*) dilatometer measurements and (*B*) infrared measurements (Ref. 15).

Many techniques have been used to measure the melting point of crystalline polymers including optical microscopy, infrared absorption, and dilatometry. In order to obtain reproducible melting points, the sample must be heated very slowly. If this procedure is followed, results have shown that the melting point is independent of the thermal history of the test specimen and the conditions of crystallization.

Figure 5-9 shows the melting curves for a low density polyethylene using two different techniques. A melting point of 112.5° C was observed using the dilatometric method while a value of 109.5° C was observed by infrared absorption measurements, showing that the measured melting point is dependent on the specific technique used.[15]

Although the melting of crystalline polymers is definite, a slight melting range is to be expected because of crystalline imperfections. A classical first order transition requires the melting to be very sharp, and this can be realized only if the crystalline regions are perfectly ordered. Since all crystalline polymers have branches or end groups, crystal imperfections must occur; thus it is not surprising that a crystalline polymer exhibits a melting range.

5-5 Kinetics of Crystallization

In addition to direct observation, the nature of crystalline polymers can be studied through crystallization kinetics. When these two techniques are combined, mutually supporting evidence aiding the over-all understanding of the system is often obtained.

Two techniques are commonly used to study rates of crystallization. A dilatometric method is frequently used to follow changes in volume during the crystallization process. In Figure 5-10, the fractional change in volume ($\Delta V/\Delta V_{\infty}$) at various crystallization temperatures is shown as a function of time for polypropylene.[16] The crystallization rate curves are sigmoidal in shape when the fractional change in volume is plotted against log time and superimposable when shifted along the log time axis. A second method,[17,18] which is frequently used, involves infrared spectrophotometry. Similar curves are obtained when absorption is plotted as a function of time.

A widely used theory[19] in crystallization kinetics relates the fraction of uncrystallized material ϕ to time t with the equation

$$\ln \phi = -Kt^n \tag{5-1}$$

[15] R. Buchdahl, R. L. Miller, and S. Newman, *J. Poly. Sci.*, **36.** 215 (1959).

[16] L. Marker, P. M. Hay, G. P. Tilby, R. M. Early, and O. J. Sweeting, *J. Poly. Sci.*, **38,** 33 (1959).

[17] W. H. Cobbs and R. L. Burton, *J. Poly. Sci.*, **10,** 275 (1953).

[18] F. P. Reding and A. Brown, *J. Appl. Phys.*, **25,** 848 (1954).

[19] M. Avrami, *J. Chem. Phys.*, **7,** 1103 (1939); **8,** 212 (1940).

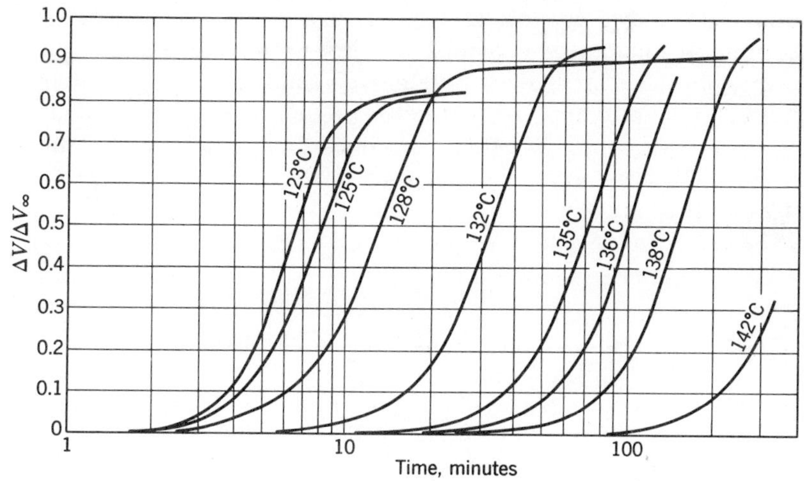

Figure 5–10. Volume changes during crystallization of polypropylene at various temperatures (Ref. 16).

where K is an over-all rate constant and n is a parameter having values of 1, 2, 3, and 4 according to the nature of the nucleation and growth processes, as summarized in Table 5-1.

Table 5-1

Type of Growth	Nucleation	n-Values
Linear	Constant	1
(1 dimensional)	Sporadic	2
Platelike	Constant	2
(2 dimensional)	Sporadic	3
Spherical	Constant	3
(3 dimensional)	Sporadic	4

Constant nucleation means that the density of nucleation sites does not change during the crystallization process, while sporadic nucleation means that the density of nucleation sites continually increases during the process.

If ΔV_∞ represents the total volume change upon complete crystallization and ΔV the volume change upon partial crystallization, then $(\Delta V/\Delta V_\infty)$ represents the fraction crystallized. Hence

$$\phi = 1 - \frac{\Delta V}{\Delta V_\infty}$$

Figure 5-11 shows crystallization rate data for polypropylene plotted according to equation (5-1), with $n = 3$. Note that in the initial stages of the crystallization process, the data are represented quite well by the equation, but after the crystallization process is about half complete, there is an increasing deviation from linearity. This fact is presumably due to a change in the value of n caused by the onset of secondary crystallization processes. These may continue for years.

When equation (5-1) is converted to the form

$$\ln[-\ln \phi] = \ln K + n \ln t \tag{5-2}$$

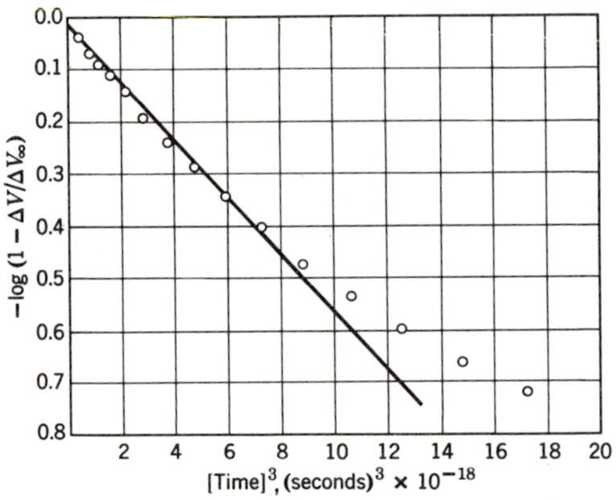

Figure 5-11. Avrami plot of crystallization rate data for polypropylene at 128°C (Ref. 16).

it is clear that a plot of $\ln[-\ln \phi]$ versus $\ln t$ on linear graph paper should be a straight line, provided n does not change. Figure 5-12 shows kinetic data for polyethylene terephthalate (mp 265° C) plotted according to equation (5-2) for crystallization at different quench temperatures.[20] Note that n changes from 2 at the low temperature to 4 at the high temperature, showing that the mechanism of crystallization in this case is dependent upon quench temperature. At the higher quench temperatures, the growth is spherical and the nucleation sporadic. However, at the lower temperature, the growth can either be linear and the nucleation sporadic or platelike and the nucleation constant. By direct observation, using an electron microscope, it should be possible to distinguish between the two mechanisms. This example illustrates the need for both kinetic analysis and direct observation.

The rate constant K is temperature dependent and, for many systems,

[20] L. B. Morgan, *Phil. Trans. Royal Soc.*, **A247,** 21 (1954).

passes through a maximum at some temperature. This is readily shown by plotting K versus T. Reference to equation (5-1) shows that the reciprocal half-time for the process is proportional to $K^{1/n}$ and should show much the same variation with T.

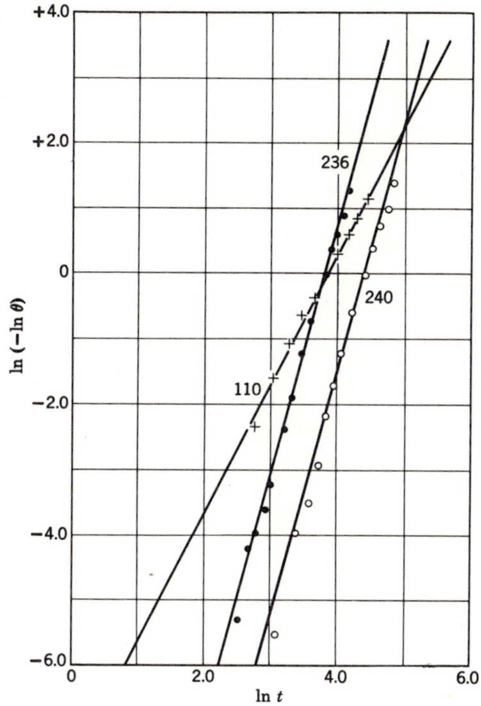

Figure 5–12. Effect of quench temperature on the mechanism of crystallization (Ref. 20).

Figure 5-13 shows data for the reciprocal half-times as a function of temperature for polyethylene succinate.[21] The separate curves A to F are for various molecular weight fractions. The curves show that the temperature at which the rate is a maximum is independent of the molecular weight. However, the reciprocal half-times at the maximum crystallization temperature decrease with increasing molecular weight, which means that the maximum crystallization rate decreases as the molecular weight increases.

Similar results have been obtained in experiments measuring the growth rate of spherulites.[22] Figure 5-14 shows some typical data for the growth of

[21] K. Ueberreiter, G. Kanig, and A. S. Brenner, *J. Poly. Sci.*, **16**, 53 (1955).

[22] L. Mandelkern, *S.P.E. Journal*, **15**, 63 (1959).

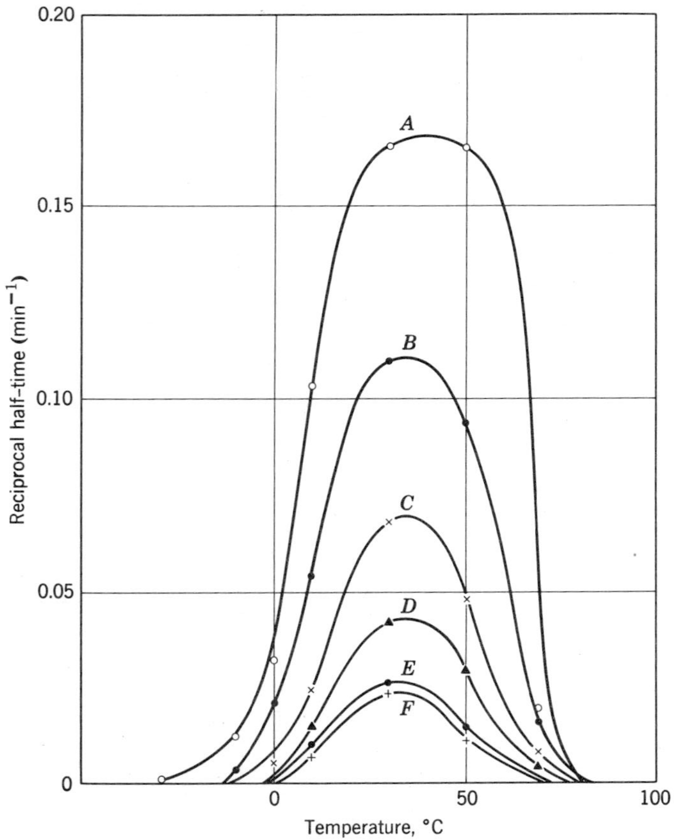

Figure 5–13. Reciprocal half-times for polyethylene succinate. Curves are for various molecular weight fractions with molecular weight increasing from A to F (Ref. 21).

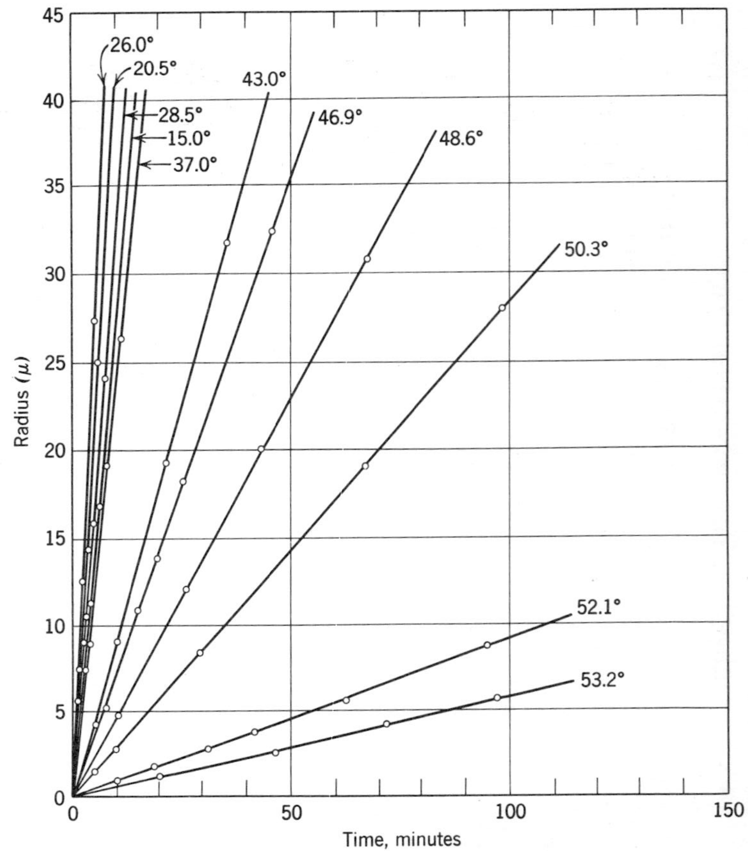

Figure 5–14. Growth rate of spherulites of ethylene adipate at various temperatures (Ref. 22).

spherulites as a function of time at various temperatures below the melting point. The slope of a straight line drawn through a given set of data points is the isothermal growth rate constant. These studies showed that the maximum growth rate occurred around 26° C. Therefore it is concluded that the temperature dependence of spherulitic growth rate is similar to the development of crystallinity as previously described.

When a polymer is cooled below its melting point, crystallization does not occur immediately. The time required for the first signs of crystallization to appear is called the induction time, τ_i. Figure 5-10 shows that the induction time for polypropylene is less than 2 min at a quench temperature of 123° C. The fact that the induction time is very temperature dependent is shown by a direct comparison with data at 142° C. At this temperature, the induction time is about 90 min. It has been shown that temperature dependence of the induction time can be expressed with a relation of the form

$$\tau_i \propto \Delta T^{-\epsilon}$$

where ΔT is the difference between the melting point and the quench temperature. Values of ϵ ranging from 2.5 to 9 have been reported in the literature.

The effect of orientation on the crystallization rate has been shown by observing changes in density with time for stretched rubber.[23] Figure 5-15 clearly shows that the crystallization rate increases markedly as the

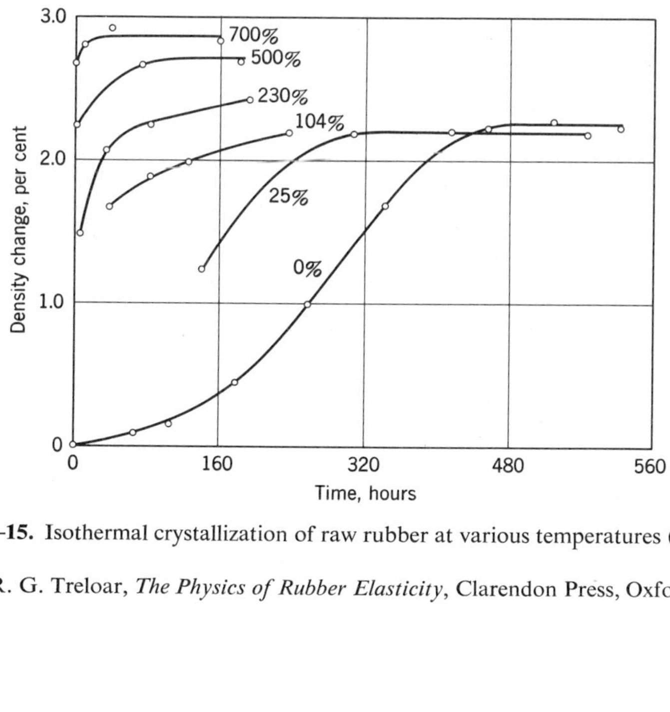

Figure 5–15. Isothermal crystallization of raw rubber at various temperatures (Ref. 23).

[23] L. R. G. Treloar, *The Physics of Rubber Elasticity*, Clarendon Press, Oxford, 1949.

strain is increased. At the higher extensions, the crystallization rate is so fast that the density change method cannot follow the initial crystallization process. These results indicate that stretching promotes chain alignment, apparently an essential step in the crystallization process.

Problems

5-1 Consider the crystallization of polyethylene terephthalate when the molten material is quenched at 110° C. Estimate the time required for the material to become 50% crystallized and for it to become 95% crystallized.

5-2 Calculate the half-time for the crystallization of polyethylene terephthalate at 110° C and for the crystallization of polypropylene at 128° C.

5-3 From the curves of Figure 5-10 determine K, the rate constant for the crystallization of polypropylene, as a function of temperature.

chapter 6

Surface Phenomena

Two dissimilar substances, brought into intimate contact, exert attractive forces on each other across their common interface. Often these forces are of sufficient magnitude so that the substances tenaciously resist separation. This phenomenon is called adhesion. It is essentially a surface phenomenon, being strongly dependent upon the nature of the surfaces and upon the intimacy of contact of the surfaces.

An interface is an essential feature of an adhesive bond. A related phenomenon, cohesion, refers to attractive forces acting among like molecules and is a bulk rather than a surface phenomenon. Cohesive forces are responsible for the close packing of molecules in solids and liquids as contrasted with gases.

Adhesion is an important aspect of such bonding operations as extrusion coating and thermoplastic laminating. In another bonding operation—heat sealing—an adhesive bond is formed initially but because of chain diffusion, the interface gradually disappears and the bond becomes cohesive in nature.

In order to make printing ink adhere to many polymer films, it is necessary to activate the surface of the film. Activation is a surface phenomenon in which a change in the chemical or physical nature of the surface is induced by passing the film through a gas flame or a corona discharge.

This chapter presents an introduction to the theory of adhesion, reviews processes for surface activation of polymers, and presents an analysis of the kinetics of surface activation via the corona discharge, the most important process for activation.

6-1 Thermodynamics of Adhesion

A thermodynamic analysis of adhesion leads to certain conclusions concerning the *work* or energy of adhesion. It makes no assumptions regarding molecular structure or the nature of the intermolecular attractive forces responsible for adhesion. On the other hand, it yields no information concerning the *force* required to separate a pair of adherends, a matter of considerable practical interest. For this, one must postulate a potential energy function based upon certain assumptions concerning the molecular structure at the interface.

For a reversible process in which the initial and final states are at the same temperature and pressure, thermodynamic analysis shows that

$$W' = -\Delta G \text{ (reversible process)} \tag{6-1}$$

where W' is the net work (total work less work of expansion) done by the system on the surroundings and ΔG represents the change of the free energy of the system between the initial and final states.

The above equation is now applied to the process illustrated by Figure 6-1*a*, in which a column of liquid having cross-sectional area A is separated reversibly into two parts. The result is the production of new surface having area $2A$. The work done on the system (the liquid) during the

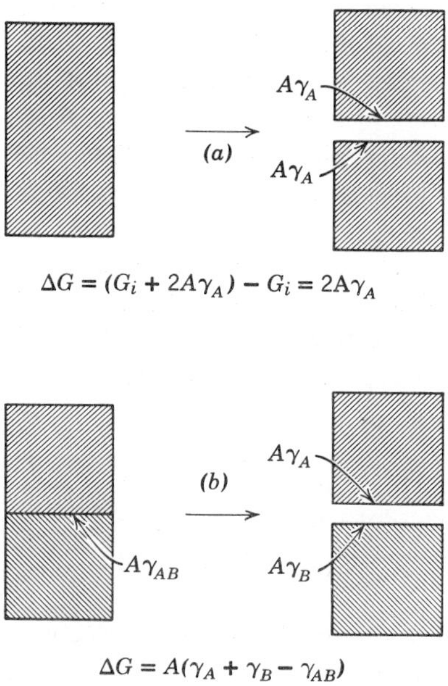

Figure 6–1. Calculation of free energy change for (*a*) process illustrating work of cohesion and for (*b*) process illustrating work of adhesion.

separation process appears as an increase in the free energy of the system, as required by equation (6-1). With the same initial and final temperature and pressure, the only change occurring in the state of the system is the increase of surface, and it is to this that the increase of free energy of the system must be attributed.

Let G_i represent the free energy of the system in the initial state. In the final state, the free energy of the system is $G_i + 2A\gamma_L$, where γ_L is the free energy of the surface expressed on the basis of unit area. The free energy change for the process illustrated by Figure 6-1*a* is, therefore, $2A\gamma_L$, and the total work done *on the system* during the process is $+2A\gamma_L$. When expressed on the basis of unit area, this is called the work of cohesion and is designated here as W_c. Therefore,

$$W_c = 2\gamma_L \tag{6-2}$$

The work of cohesion represents the minimum work required to overcome the intermolecular attractive forces that give the liquid its cohesional strength across a plane surface having an area of 1 cm^2.

The cause of the surface free energy is the preferential attraction of surface molecules toward the interior. This results from the close packing of molecules on the liquid side of the interface and the sparse packing on the vapor side. (Molecules in the interior are, on the average, attracted equally in all directions

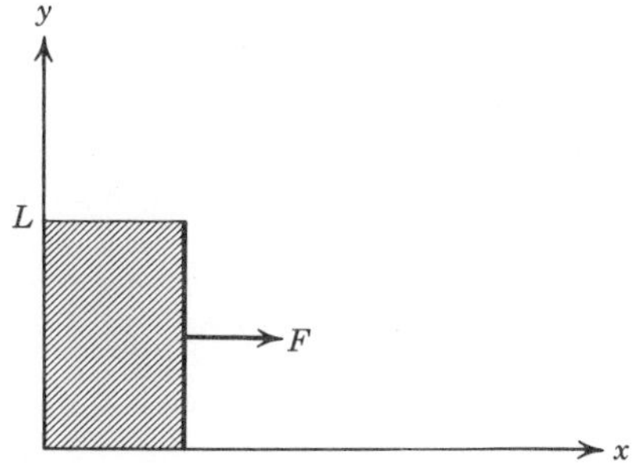

Figure 6–2. Force F moves bar to the right to create new surface.

because the molecular density is the same in all directions.) As a result of the inner-directed force at the surface, liquids tend to assume spontaneously a state of minimum surface corresponding to a state of minimum free energy.

Because of this tendency for spontaneous contraction, a liquid surface can be thought of as being in a state of tension. The tension in a surface can be represented as a force vector acting tangent to the surface at all points and in all directions. Furthermore, as the surface contracts or expands, the force remains constant so that the surface tension is independent of the size of the surface.

Now, consider the process of new surface formation illustrated by Figure 6-2, where the force F is applied to the bar of length L, which forms one boundary of the surface. The work done in moving the bar in the x-direction is given by the calculation

$$W = \int_{x_1}^{x_2} F\,dx = F\,\Delta x$$

The area of the new surface formed by the process is $L\Delta x$ so that the work done per unit area of surface formed, the surface energy, is

$$\frac{W}{L\Delta x} = \frac{F}{L} = \gamma_L$$

The force per unit length (F/L) is called the surface tension. Note that it is equal to the surface energy per unit area and the symbol γ is used for both quantities. The cgs units for surface tension are dynes/cm and the units for surface energy are ergs/cm^2.

Solids also have a free energy associated with their surfaces, designated here by the symbol γ_S. To avoid ambiguity, a double subscript notation is often used. For example, γ_{SL} refers to the surface energy of a solid in contact with a liquid, while γ_{SA} refers to the surface energy of the solid in contact with air. In general, these quantities will be different, and it is more precise to speak of interfacial rather than surface energies in these cases. The symbols γ_S and γ_L imply that the surface is in a vacuum.

The minimum work required to separate a pair of immiscible liquids at their interface, a process illustrated schematically by Figure 6-1b, is given by the Dupré equation

$$W_a = \gamma_A + \gamma_B - \gamma_{AB} \tag{6-3}$$

W_a is the work of adhesion, γ_{AB} the interfacial energy for the pair of liquids, and γ_A and γ_B the surface energies of the pure liquids. The work of adhesion for a solid-liquid system is given by the analogous equation

$$W_a = \gamma_L + \gamma_S - \gamma_{LS} \tag{6-4}$$

There are no simple and direct methods for determining γ_S and γ_{SL} independently. However, they can be related to the contact angle, a quantity that can be directly measured. Consider a liquid drop resting on a flat, horizontal, solid surface and surrounded by air, as shown in Figure 6-3. At the edge of the drop, the three interfaces have a common boundary, represented by a point when a cross section of the drop is considered. Force vectors in the plane of the cross section, and representing the three interfacial tensions at this point, are shown in the figure. Note that the vectors are drawn tangent to the surfaces in which they are operative. For a system in equilibrium, the forces are balanced and

$$\gamma_{SA} = \gamma_{SL} + \gamma_{LA} \cos(\theta) \tag{6-5}$$

The angle θ is called the contact angle of the drop. According to the above equation, it is a function only of the interfacial energies and, therefore,

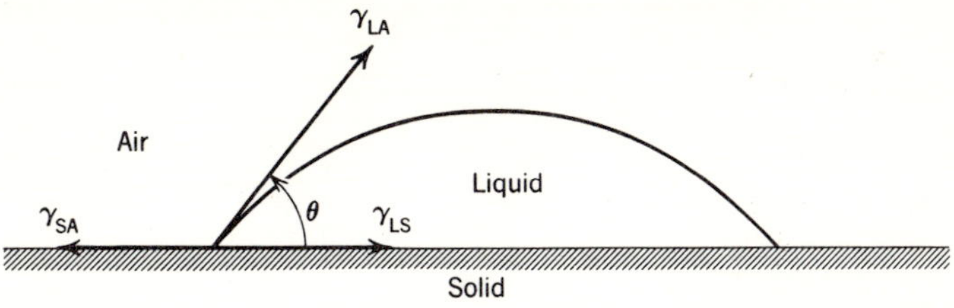

Figure 6–3. Liquid drop resting on horizontal surface. Vectors represent interfacial tensions.

is a property of the system. The contact angle gives an indication of the ability of a liquid to wet a solid, complete wetting occurring only when $\theta = 0$.

A hysteresis effect is usually observed when contact angles are measured. A different angle is obtained when the liquid advances over the surface than when it recedes from the surface. Advancing and receding contact angles may differ by as much as 40°. Other factors discussed by Adam,[1] including surface contamination and roughness, make the measurement of reproducible contact angles a problem of considerable experimental difficulty.

The combination of (6-5) and (6-4) gives the expression

$$W_a = \gamma_{LA}[1 + \cos(\theta_a)] \qquad (\theta > 0) \tag{6-6}$$

for the work of adhesion. Note that it contains only directly measurable quantities but that it is restricted to those cases of solid-liquid adhesion in which less than complete wetting of the solid by the liquid occurs. With complete wetting, drop formation cannot occur and the validity of (6-5) is doubtful.

Kraus[2] has summarized the results of many direct measurements of the work of adhesion. These measurements were made using precision calorimeters with the solid in a finely powdered form having known surface area. The value of W_a for a large class of solid-liquid systems is in the range of 200 to 300 ergs/cm^2, the adhesion in these systems being due, presumably, to dispersion forces. If the liquid has a dipole moment, the work of adhesion is usually increased by 100 to 200 ergs/cm^2 because of the electrostatic attractions resulting from the dipole. Those liquids capable of hydrogen bonding show the largest values of W_a, 700 to 1000 ergs/cm^2 being observed with metal oxides and water.

Allan[3] has measured contact angles of drops of alcohol-water solutions

[1] N. K. Adam, *The Physics and Chemistry of Surfaces*, 3rd ed., Oxford University Press, 1941.

[2] Gerard Kraus, "Energy of Adhesion, Molecular Forces, and Adhesive Rupture," in *Adhesion and Adhesives*, John Wiley, 1954, p. 45.

[3] A. J. G. Allan, *J. Poly. Sci.*, **38**, 297 (1959).

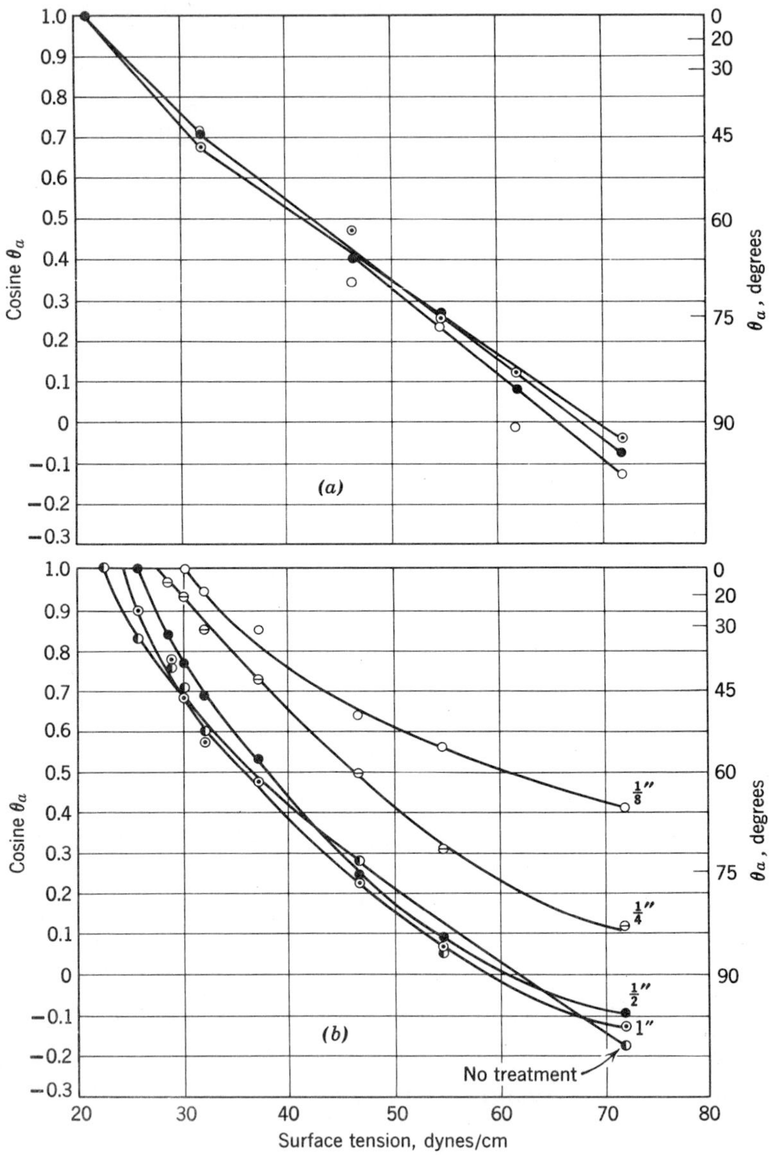

Figure 6–4. Effect of surface tension on contact angle for (*a*) unactivated polyethylenes and (*b*) polyethylene films of various degrees of activation (Ref. 3).

on the surface of polyethylene films. By changing the composition of the solutions, the value of γ_{LA} was varied between 20 and 70 ergs/cm^2. The results obtained with three unactivated polyethylene films are shown in Figure 6-4*a*. The relationship between γ_{LA} and cos (θ_a), where θ_a is the advancing contact angle, is nearly linear. The quantity γ° represents the maximum value of γ_{LA} for which complete wetting ($\theta_a = 0$) can occur. It is a characteristic of the surface. For the unactivated polyethylenes tested by Allan, its value is about 21 ergs/cm^2.

The results of a similar series of experiments using surface activated polyethylene is shown by Figure 6-4*b*. As the degree of activation increases, the value of γ° increases, becoming as large as 30 ergs/cm^2 for the sample with the maximum degree of activation. Ink adhesion measurements, made by sandwiching a thin layer of ink between two films and measuring the force required to peel the films apart, were then correlated with the γ° values. The results are shown in Figure 6-5, where ink adhesion is

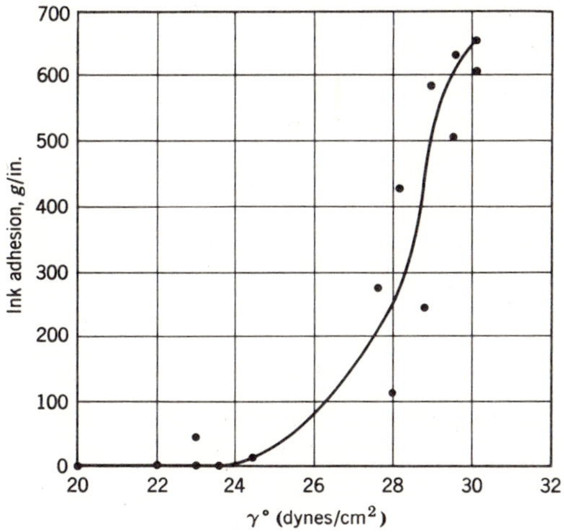

Figure 6–5. Effect of γ° on ink adhesion. For the ink, $\gamma_{LA} = 26.6$ dynes/cm^2 (Ref. 3).

plotted against γ°. Below about 28 ergs/cm^2 adhesion is almost negligible. Above 28 ergs/cm^2 it rises rapidly to about 600 grams/in. The value of γ_{LA} of the ink used in the test was 26.6 ergs/cm^2, and Allan concluded that a necessary condition for adhesion was complete wetting of the surface by the adhesive.

Thermodynamic analysis enables some tentative conclusions to be made concerning adhesive and cohesive separation processes in solid-liquid systems. It is assumed that the separation process, regardless of whether it is cohesive or adhesive, results, initially, in the creation of surfaces uncontaminated with air or adsorbed vapors of any kind. The work of cohesion and adhesion is then given by equations (6-2) and (6-3), respectively. Subtracting the one from the other gives

$$W_a - W_c = (\gamma_S - \gamma_L - \gamma_{LS}) \tag{6-7}$$

The contact angle for the solid-liquid system under consideration, when measured in air, is given by (6-5). Combining it with the above equation gives

$$W_a = W_c - (\gamma_{SA} - \gamma_S) + [\gamma_{LA} \cos(\theta) - \gamma_L] \tag{6-8}$$

The energy of adsorption of air on the solid surface is designated ΔE_s, where

$$\Delta E_s = \gamma_{SA} - \gamma_S$$

Assuming that the adsorption of air on the liquid surface is negligible, $\gamma_{LA} = \gamma_L$, and equation (6-8) becomes

$$W_a = W_c - \Delta E_s - \gamma_{LA}[1 - \cos(\theta)] \tag{6-9}$$

If the surface tension of the liquid is such that it just wets the solid, $\theta = 0$, and (6-9) becomes

$$W_a = W_c - \Delta E_s$$

Since adsorption is an exothermic process, ΔE_s is a negative quantity, and the work of adhesion exceeds the work of cohesion. One concludes that upon separation the system will fail cohesively in the liquid rather than adhesively at the solid-liquid interface. The data shown in Table 6-1 for the adhesion of polyethylene and polystyrene to steel tend to confirm this conclusion.

Table 6-1

Limiting Adhesion Values for Polyethylene and Polystyrene to Steel*

Polymer	Temp (°C)	Limiting Adhesion (kg/cm^2)	Tensile Strength (kg/cm^2)
Polyethylene	22	194	141
	−24	345	282
	−60	444	570
Polystyrene	82	211	268
	64	271	300
	45	338	335
	25	412	377
	−24	560	507

* G. Kraus and J. E. Manson, *J. Poly. Sci.*, **6**, 625 (1952).

Adhesion was measured by bonding thin films of the polymer to the end faces of metal cylinders and then separating the cylinders in a testing machine. The value of limiting adhesion was obtained by testing different film thicknesses and extrapolating to zero thickness. This minimizes edge and other extraneous effects. The agreement between the values of limiting adhesion and the tensile strength of the polymer is remarkably close, indicating that failure occurs cohesively rather than adhesively.

6-2 Intermolecular Forces in Adhesion

The intermolecular forces responsible for adhesion are of the type usually referred to as secondary valence forces and include those arising from hydrogen bonding, permanent dipoles, induced dipoles, and dispersion forces. These forces are a complicated function of the distance of separation and of their mutual orientation. Reference should be made to texts[4] on molecular physics for a discussion of these forces and methods of calculation.

In general, two molecules will repel each other if their distance of separation r is less than some equilibrium value r_0 and will attract each other when $r > r_0$. At the position r_0, the potential energy ϕ of the system has its minimum value $-\phi_0$. Either increasing or decreasing the distance of separation will require work that will be stored in the system as potential energy. For a given separation, the amount of energy stored depends upon the nature of the potential energy function of the system under consideration. Figure 6-6 shows a typical[2] potential energy function for a molecule outside a solid surface located at $r = 0$. The potential energy is taken as zero when the molecules are infinitely far removed. The derivative of the potential function gives the force that must be applied to maintain the molecules at a given separation r. It is shown in Figure 6-6. Note that the force undergoes a change in sign at r_0, indicating a change from repulsive to attractive forces.

The work required to change the molecules from their equilibrium configuration to one in which they are infinitely far separated is given by the integral

$$W = \int_{r_0}^{\infty} F\,dr = \int_{r_0}^{\infty} \left(\frac{d\phi}{dr}\right) dr = \phi(\infty) - \phi(r_0)$$

where

$$\phi(\infty) = 0$$

$$\phi(r_0) = -\phi_0$$

so that the work required for the separation process is ϕ_0.

[4] See, for example, *Molecular Theory of Gases and Liquids*, by J. O. Hirschfelder, J. O. Curtiss, and R. B. Bird, John Wiley, 1954.

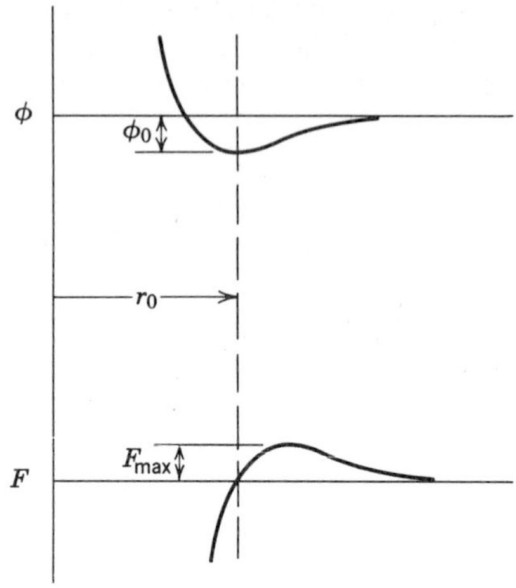

Figure 6–6. Potential and force curves for a molecule outside a solid surface at $r = 0$ (Ref. 2).

Kraus and Manson[5] have calculated the adhesion of polyethylene and polystyrene to steel. They assumed that the adhesion is due almost entirely to dispersion forces and that the potential energy function for the interaction between a monomer unit of the polymer and the surface of the steel is given by the equation

$$\phi = Br^{-9} - A(2r)^{-3} \tag{6-10}$$

where A and B are characteristic constants for the system.

The constants A and B can be expressed in terms of ϕ_0 and r_0 by noting from Figure 6-6 that

$$\phi(r_0) = -\phi_0$$

$$\phi'(r_0) = 0 \qquad \text{where} \qquad \phi' = (d\phi/dr)$$

Equation (6-10) can then be written

$$\phi = -\left(\frac{\phi_0}{2}\right)\left[3\left(\frac{r_0}{r}\right)^3 - \left(\frac{r_0}{r}\right)^9\right] \tag{6-11}$$

The force F, found by differentiating (6-11), is

$$F = \frac{9\phi_0}{2r_0}\left[\left(\frac{r_0}{r}\right)^4 - \left(\frac{r_0}{r}\right)^{10}\right] \tag{6-12}$$

Note that when $r = r_0$, $F = 0$ and that when F passes through the point $(r/r_0) = (10/4)^{1/6}$, it has its maximum value of

$$F_{\max} = 1.46\left(\frac{\phi_0}{r_0}\right)$$

[5] Gerard Kraus and James E. Manson, *J. Poly. Sci.*, **6**, 625 (1951); **8**, 448 (1952).

Taking 100 ergs/cm² as an order of magnitude estimate for the work of adhesion of this system and the equilibrium separation distance as 4 A°, the value of $F_{\max}$ is given by the calculation

$$F_{\max} = \left[\frac{(1.46)(100)}{(4 \times 10^{-8})}\right] = 37 \times 10^8 \text{ dynes/cm}^2$$

$$= 3700 \text{ kg/cm}^2$$

Note that this value is many orders of magnitude greater than the measured limiting adhesion for polyethylene and polystyrene on steel.

Taylor and Rutzler[6] have pointed out that geometrical restrictions imposed by the structure of the polymers and the metal surface may prevent all but a small fraction of the available monomer units from interacting with the surface. Then the value of the work of adhesion used in the above calculation would be much too large. On the basis of studies using molecular models, they concluded that as few as 1% of the available interaction might be realized in the adhesion of polystyrene and polyethylene to steel.

6-3 Surface Activation of Polymers

Although nonpolar polymers, such as polyethylene, do not adhere well to other materials, there are many applications in which they are bonded to paper, metals, cellophane, inks, and other plastics. In order to achieve satisfactory adhesion in these bonding operations, it is necessary to activate the surface of the nonpolar material before it comes in contact with the other adherend.

Various techniques have been used for surface activation. These include sundry chemical treatments involving chlorine gas, ozone, and other oxidizing agents, exposure to hot gases and flames, and exposure to various types of electrical discharges.

According to Allan,[3] complete wetting is a necessary condition for good adhesion. As shown by Figure 6-4, activation increases the wettability (decreases the contact angle) of a surface toward a polar liquid. Because materials such as polyethylene are wet by printing inks only after being activated, surface activation is a necessary prerequisite to the printing of polyethylene.

In addition to making polyethylene compatible with printing inks, activation appears to increase its adhesion to metals and metal oxide surfaces. It has been reported that the adhesion of polyethylene to stainless steel increases with the degree of oxidation of the polyethylene.[7]

[6] D. Taylor and J. E. Rutzler, *Ind. Eng. Chem.*, **50,** 928 (1958).

[7] F. J. Bockhoff, E. T. McDonel, and J. E. Rutzler, *Ind. Eng. Chem.*, **50,** 904 (1958).

Most investigators agree that activation is the result of formation of polar groups in the surface. Rossman[8] activated polyethylene film samples with a flame, a glow discharge, and a brush discharge from a Tesla coil. Spectrophotometric examination indicated that oxidation occurred in all three cases.

Hines[9] conducted similar studies on polyethylene film that had been flame activated. He concluded that hydroxyl and ether groups, probably in the form of formals and hemiformals, were formed in the surface.

Various chemical treatments for the activation of polymer film surfaces have been reported and are reviewed by Bloyer.[10] The simultaneous action of chlorine gas and ultra-violet radiation proved moderately successful. Wolinski[11] reported that films could be activated with a combination of ozone and a hydrogen halide or ozone with nitrous oxide. Horton[12] patented a process for surface activation via oxidation with chromic acid and other strong oxidizing agents. Kreidl[13] reports that activation is obtained by exposing the surface of polyethylene to a high temperature while the bulk of the material is kept at a low temperature. He proposes using hot combustion gases, hot air, exposure to open flame without actual contact, and exposure to electric heating elements.

There are two activation processes of commercial importance. One process, patented by Kritchever,[14] involves the direct exposure of the film surface to an open flame. The other process, patented by Traver,[15] involves passing the film through an electric field in which rapid electron bombardment of the surface occurs.

6-4 Surface Activation by Corona Discharge

Surface activation by corona discharge requires relatively simple equipment, is easy to control, and yields uniform and reproducible results. The process, which has considerable commercial importance, is shown schematically in Figure 6-7.

The equipment consists of a power supply that provides a high voltage alternating potential to a sharp-edged electrode. A rotating metal drum, maintained at ground potential, is located so that there is a narrow gap between its surface and the edge of the electrode. The film is carried by the drum

[8] K. Rossman, *J. Poly. Sci.*, **19,** 141 (1956).

[9] R. A. Hines, "An Investigation of the Chemical Changes Occurring on the Surface of Polyethylene During Treatment to Render the Surface Printable," paper delivered at A.C.S. meeting, September, 1957.

[10] S. Bloyer, *Modern Plastics*, **32,** 105 (1955).

[11] L. E. Wolinski, U.S. Patents 2,715,075, 2,715,076, and 2,715,077 (Aug. 9, 1955).

[12] P. V. Horton, U.S. Patent 2,668,134 (Feb. 2, 1954).

[13] W. H. Kreidl and F. Hartmann, *Plastics Technology*, **1,** 31 (1955).

[14] M. F. Kritchever, U.S. Patent 2,648,097 (Aug. 11, 1953).

[15] G. W. Traver, Brit. Patent 771,234 (Mar. 27, 1957).

through the gap, thus exposing its free surface to the electrical discharge.

An equivalent electrical circuit for the process is also shown in Figure 6-7. It assumes that the plastic film is nonconducting, and thus introduces only capacitance into the circuit. In its unionized state, the air in the gap is essentially nonconducting, but when ions are present, an electric current can be carried across the gap. The air, therefore, is represented as a capacitor and a resistance in parallel, with one or the other predominating, depending upon the condition of the air. Hence, before breakdown of the air (ionization), the equivalent electrical circuit consists of two capacitors in series and, after breakdown, of a capacitor and resistance in series.

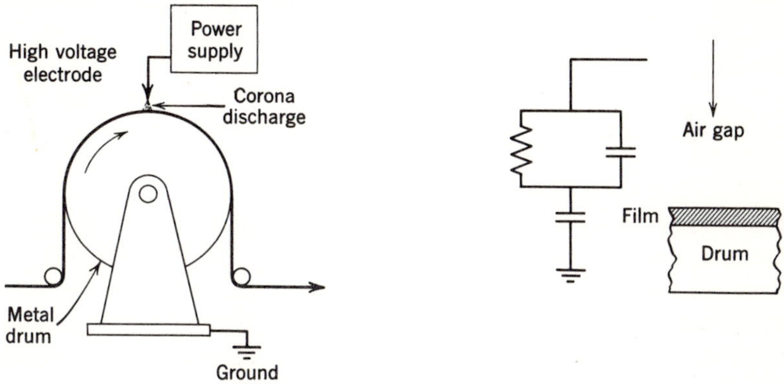

Figure 6–7. Apparatus for surface activation of plastic film by corona discharge with equivalent electrical circuit at right.

Corona discharges, which are characterized by high voltages and low currents, occur at atmospheric pressure when one of the electrodes has sharp edges or points. The discharge begins when the electrode potential increases to the point where partial ionization of the gas takes place. The gap becomes filled with a pale violet glow, a hissing noise is heard, and the odor of ozone can be detected. Bright blue streamers emerge from sharp points on the electrode and terminate in the gap. If the plastic film were not present, or if it should rupture so that the streamers could reach the other electrode, sparking and then arcing would result, and the current would increase drastically.

The current in the corona discharge is carried by electrons and ions. During the first half cycle, charged particles of the same polarity as the electrode collect on the film surface. This reduces the effective field strength and thus quenches the corona before the end of the half cycle. When the polarity of the electrode changes, the now oppositely charged particles on the film surface cause an increase in the gradient in the gap, and the corona starts at a lower applied potential than in the initial half cycle. Therefore the plastic film has the effect of enabling the discharge to be sustained at a lower applied potential.

The primary operating variables in the activation process are the voltage and frequency applied to the electrode, the distance between the electrode and the film surface, and the linear speed of the film. The following kinetic analysis[16] of surface activation via corona discharge attempts to relate these variables to the degree of activation of the film. The key assumptions in the analysis are as follows:

1. Activation is due to the production of polar groups on the surface of the film.
2. Polar groups can be produced only at certain reactive sites on the surface of the film.
3. The key step in the formation of a polar group is the collision of a charged particle with a reactive site. Upon such a collision the probability of forming a polar group is K.

The degree of activation y of the surface is defined as (N/N_0), where N_0 is the total number of reactive sites per unit area of the surface and N is the number that contain polar groups. The degree of activation thus represents the fraction of the reactive sites that contain polar groups. It ranges from zero for a completely unactivated surface to unity for one in which all of the reactive sites contain polar groups.

According to the above assumptions, the rate of activation (dy/dt) of a surface will be proportional to the product of the average rate $\bar{r}$ at which the surface is bombarded with particles and the probability p of a particle striking a reactive site. The proportionality constant in the rate equation represents the probability of the formation of a polar group upon a collision of the required type. The rate equation is therefore written

$$\frac{dy}{dt} = K(\bar{r})(p) \tag{6-13}$$

If α represents the fraction of the surface that contains reactive sites, then

$$\alpha = aN_0 \tag{6-14}$$

where a is the area of a single reactive site. Since y is the fraction of the reactive sites containing polar groups, $(1 - y)$ represents the fraction of the sites available for reaction. Therefore

$$p = \alpha(1 - y) \tag{6-15}$$

[16] J. C. Porter, "Surface Activation of Polyethylene Film by Electric Discharge," doctoral thesis, Washington University, St. Louis, 1960.

If G represents the number of particles swept against the surface of the film during one complete cycle, then the average rate of bombardment is Gf, and the rate equation becomes

$$\frac{dy}{dt} = KGf\alpha(1 - y) \tag{6-16}$$

If the surface is initially (at time zero) completely unactivated, the initial condition is written

$$y(0) = 0$$

and the solution of (6-16) meeting this initial condition is

$$y = 1 - e^{-\alpha KGft} \tag{6-17}$$

When the film passes through the discharge at linear speed U, the exposure time t can be expressed in terms of U and the area of discharge A_d by the equation

$$t = \frac{A_d}{U}$$

so that the rate equation becomes

$$y = 1 - \exp\left(-\frac{\alpha KGfA_d}{U}\right) \tag{6-18}$$

The degree of activation y, representing the fraction of the reactive sites that contain polar groups, must be related to some directly measurable quantity in order to test the kinetic analysis. Porter[16] derived a relationship between the degree of activation and the force required to peel a pressure sensitive tape from the film. He assumed that the adhesion of the tape to the film involved the formation of three types of secondary valence bonds, the first type being those formed between an unreactive site and the adherend, the second between a reactive site and the adherend, and the third between a reactive site containing a polar group and the adherend.

The number of unreactive sites per unit area is designated N_1 and the strength of the bond formed at one of these sites is designated f_1. There are N sites that contain polar groups and $(N_0 - N)$ reactive sites that do not contain polar groups. The strength of a bond involving a polar group is designated f_3, and bonds formed at the free reactive sites have a strength designated f_2. The total adhesive force between the film surface and the adherend is given by the equation

$$F = f_1N_1 + f_2(N_0 - N) + f_3N \tag{6-19}$$

For a completely unactivated film, $N = 0$, and (6-19) reduces to

$$F_0 = f_1N_1 + f_2N_0 \tag{6-20}$$

For a completely activated film, $N = N_0$, and (6-19) becomes

$$F_1 = f_1N_1 + f_3N_0 \tag{6-21}$$

Upon subtracting equation (6-20) from (6-21), one obtains

$$(F_1 - F_0) = N_0(f_3 - f_2) \tag{6-22}$$

When equation (6-19) is written in the form

$$F = f_1N_1 + f_2N_0 + N(f_3 - f_2)$$

it is readily seen that by introducing (6-20) and (6-22), it reduces to

$$\left(\frac{N}{N_0}\right) = y = \left(\frac{F - F_0}{F_1 - F_0}\right) \tag{6-23}$$

By assuming that adhesive forces are proportional to the forces measured in a peel test, equation (6-23) can be written

$$y = \left(\frac{S - S_0}{S_1 - S_0}\right) \tag{6-24}$$

where the S's are directly measurable quantities.

Porter[16] obtained kinetic data for the activation of polyethylene film using an apparatus designed by Grynkiewicz.[17] In the Grynkiewicz apparatus, the film is wrapped around the drum so that, upon rotation of the drum, each element of the film makes repeated passes through the discharge. If A represents the total area of the film and A_d the area of the discharge, the time that any element is exposed to the discharge is obtained by multiplying the total time τ by the area ratio (A_d/A) and the kinetic equation becomes

$$y = 1 - e^{-k\tau} \tag{6-25}$$

where the rate constant k is defined by the equation

$$k = (\alpha KGf)\frac{A_d}{A} \tag{6-26}$$

Equation (6-25) shows that the degree of activation y is a unique function of the quantity $k\tau$. The solid curve labeled *theory* in Figure 6-8 is a plot of y versus kt according to equation (6-25). The degree of activation rises from zero (at time zero) and makes an asymptotic approach to unity, with most of the change occurring over about 2 decades of the abscissa.

Typical kinetic data, consisting of peel strengths S versus time τ, are shown in Table 6-2. The peel strength data were obtained by applying a

[17] J. P. Grynkiewicz, "Apparatus for the Surface Activation of Solid Polyethylene with a Corona Discharge," master's thesis, Washington University, St. Louis, 1960.

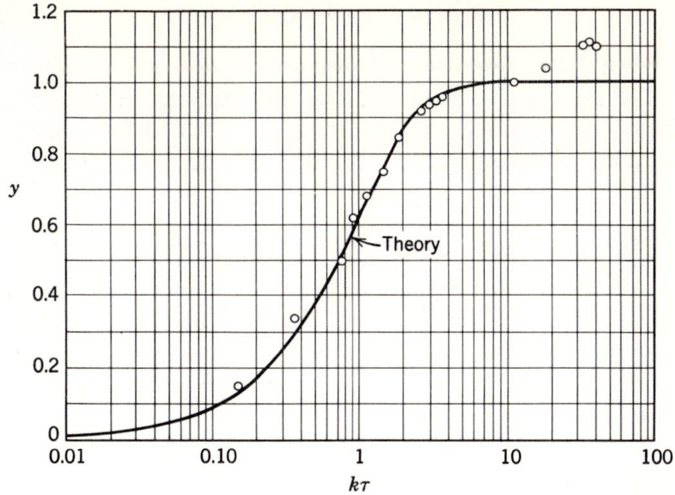

Figure 6–8. Kinetic data for surface activation of polyethylene via corona discharge (Ref. 16).

Table 6-2

Kinetic Data for the Surface Activation of Polyethylene*

Time (sec)	Peel Strength (grams/in.)
4	330
10	530
20	710
25	840
30	910
40	990
50	1100
60	1200
70	1170
80	1190
90	1200
100	1220
300	1270
500	1310
900	1370
1000	1380
1100	1370

$S_0 = 160$ grams/in.	$U = 63$ ft/min
$S_1 = 1260$ grams/in.	$I = 65$ microamperes
$f = 60$ cycles/sec	$V = 8650$ volts
gap $= 50$ mils	$T = 25°$ C

* Data from Ref. 16.

pressure sensitive tape (Minnesota Mining Scotch Brand No. 800) to the film and then peeling the tape from the film with an Amthor Type 272 tensile tester. The value of S_0 is the mean of repeated tests made on the unactivated film. The value of S_1, which was established statistically, gives the best fit of the data to equation (6-25).

If the degree of activation y, calculated from the above data, is plotted against τ, a curve similar in shape to the theoretical curve is obtained. The value of the rate constant k, obtained by shifting the experimental curve along the abscissa until it coincides with the theoretical curve, is 0.037 sec^{-1}. The experimental data points, shown in Figure 6-8, are in excellent agreement with the theoretical curve until values of y approaching unity are obtained.

To translate these results to conventional apparatus in which the film makes a single pass through the discharge, it is noted that

$$t = \left(\frac{A_d}{A}\right)\tau = \left(\frac{L_d}{\pi D}\right)\tau \tag{6-27}$$

where L_d is the effective length of the discharge measured in the direction of motion of the film and D is the diameter of the drum in the experimental apparatus. The linear speed U at which an element of the film receives the required exposure t is (L_d/t), and because of equation (6-27)

$$U = \left(\frac{\pi D}{\tau}\right) \tag{6-28}$$

If the rate equation is solved for τ and substituted into the above equation, one obtains

$$U = -\left[\frac{\pi Dk}{\ln(1 - y)}\right] \tag{6-29}$$

Now, consider the case where the operating conditions are the same as in the above experiment, that is, where $k = 0.037\ \text{sec}^{-1}$. If D is about 6 in. and if the film is to be 90% activated, then equation (6-29) shows that the linear speed of the film through the discharge should be 1.52 ft/min.

Examination of equation (6-26) shows that the rate constant k is directly proportional to the frequency. Therefore, if the frequency should be raised from 60 to 3000 cycles/sec, the rate constant would be increased by a factor of 50 and the allowable linear speed of the film would increase to 76 ft/min.

The degree of activation of the film, with all other variables fixed, decreases as the linear speed at which the film passes through the discharge increases. The relationship between speed and degree of activation is given by equation (6-29). Figure 6-9 is a plot, based on this equation, showing the effect of speed on degree of activation for the above conditions. Note that as the speed increases from zero to 50 ft/min, there is a decrease of only 5% in the degree of activation, but at speeds greater than 50 ft/min, the decrease occurs more rapidly.

The factors that determine the value of the rate constant k are now examined in more detail. If the average charge carried by the ions and

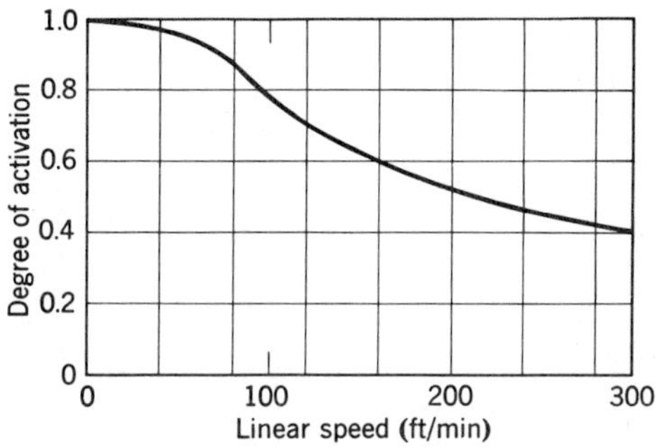

Figure 6–9. Effect of the linear film speed on degree of activation at constant applied potential.

electrons in the corona discharge is $\bar{q}$, then the current I is given by the equation

$$I = Gf\bar{q} \tag{6-30}$$

and the equation for the rate constant, (6-26), becomes

$$k = \frac{\alpha K I A_d}{\bar{q} A} \tag{6-31}$$

which shows that k should be directly proportional to I, provided K, A_d, and $\bar{q}$ remain constant.

Porter[16] determined values of k in a series of ten experiments in which the current ranged from 5 to 850 microamperes. He found that the ratio (k/I) was nearly constant, having the following mean value:

$$\frac{k}{I} = 0.0019 \tag{6-32}$$

An increase in the potential applied to the electrode causes an increase in the current and, therefore, an increase in the rate constant. The data in Table 6-3 illustrate the relationship between voltage, current, and the rate constant.

Table 6-3

Current-Voltage Relationship in a Corona Discharge*

V (volts)	I (microamperes)	k (sec^{-1})	(k/I)
3150	5	0.007	0.0014
6000	25	0.038	0.0015
8650	65	0.109	0.0017

* Data from Ref. 16.

Figure 6-10 illustrates the current voltage relationship. Below about 2000 volts, the breakdown voltage, there is essentially no current flow.

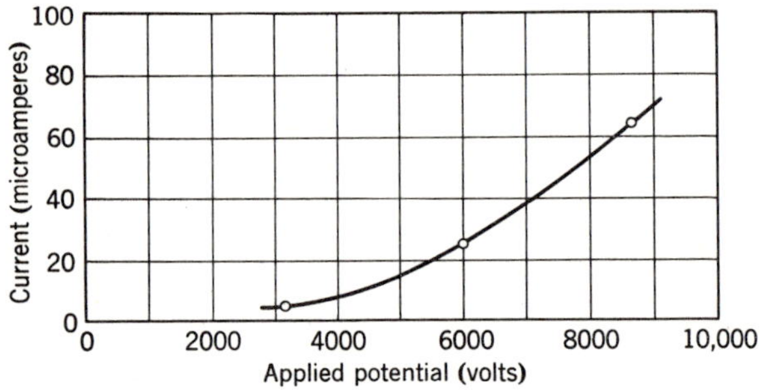

Figure 6–10. Current-voltage curve for corona discharge (Ref. 16).

Above 2000 volts, the current increases as the voltage increases, but the relationship is clearly nonlinear.

The applied potential probably has some effect on the coefficient K and thus exerts a secondary effect on the rate constant. The coefficient K represents the probability that a collision of a particle with a reactive site will result in the formation of a polar group. The following discussion shows that K must be a very small number, but that it should increase as the applied potential increases.

Equation (6-31) is rearranged to give the following explicit expression for K:

$$K = \left(\frac{\bar{q}}{\alpha}\right)\left(\frac{k}{I}\right)\left(\frac{A}{A_d}\right) \tag{6-33}$$

In the above mentioned experiments of Porter, the value of the ratio (k/I) is 1900 $coulombs^{-1}$. Assuming that each charged particle in the discharge carries unit electronic charge, the value of $\bar{q}$ is 1.6×10^{-19} coulombs. The area ratio is unknown, but assuming that the effective length of the discharge L_d is about 0.10 in., the ratio (A/A_d) is about 200, and if $\alpha = 0.1$, the coefficient K is of the order 10^{-12}. Thus, for every collision of a charged particle with a reactive site resulting in the formation of a polar group, there are 10^{12} ineffective collisions.

If one assumes that an effective collision involves the rupture of a bond having an energy of the order of 100 kcal/mole, a colliding particle must have a minimum energy of roughly 4 electron volts to have a chance of being effective. If the mean free path of the charged particles is 10^{-5} cm (that of oxygen at room temperature and pressure), a field strength of 4×10^5 volts/cm is required to impart, on the average, the required kinetic energy to a colliding particle.

In the experiments of Porter shown in Table 6-3, the average field strength ranges from about 25,000 volts/cm to 68,000 volts/cm. For a particle to acquire the minimum kinetic energy required for an effective collision, it must fall through a free path 6 to 16 times longer than the mean free path. In oxygen at room temperature, the fraction of molecules having a free path 10 times greater than the mean free path is about 10^{-6}, which indicates that roughly 10^6 collisions of particles having the minimum energy are required before a polar group is formed.

It is also apparent that the field strength can be changed by varying the gap distance. In another experiment in which the voltage was constant, Porter found that the value of the rate constant k was reduced from 0.11 sec^{-1} when the gap was 50 mils to 0.08 sec^{-1} when the gap was 200 mils. It should be noted, however, that any conclusions regarding the effect of field strength on the coefficient K must be tentative because the quantity obtained by dividing the applied potential by the gap distance may not be indicative of the field strength within a few mean free paths of the film surface.

Problems

6-1 If γ represents the free energy of a surface and Γ the total surface energy, show that

$$\Gamma = \gamma - T\frac{d\gamma}{dT}$$

6-2 The following table shows the surface tension of water as a function of temperature between 20° C and 30° C:

Temperature (°C)	Surface Tension (dynes/cm)
20	72.8
25	72.0
30	71.2

Estimate the total surface energy of water at 25° C. Calculate the work of cohesion for water at 25° C.

6-3 The contact angle of water on graphite at 25° C is 90°. Calculate the work of adhesion of water on graphite at 25° C. Estimate the heat of immersion of graphite in water.

6-4 If W_a is the work of adhesion, show that

$$\frac{dW_a}{dT} = \left(\frac{d\gamma_{LA}}{dT}\right)[1 + \cos\theta] + \gamma_{LA}\sin\theta\left(\frac{d\theta}{dT}\right)$$

and calculate the contact angle of water on graphite at 30° C.

6-5 Polyethylene film is being activated in a continuous process using corona discharge apparatus. The linear speed of the film through the apparatus is 100 ft/min and the degree of activation is estimated to be 0.7. Estimate the degree of activation at 150 ft/min.

chapter 7

Dielectric Behavior of Polymers

This chapter is a brief introduction to the dielectric behavior of high polymers, a subject of interest in polymer processing primarily because dielectric losses can sometimes be used as a source of internal heat generation in polymers. *Dielectric heating*, as it is called, is used in polymer processing chiefly in sheet thermoforming and in heat sealing of polar plastics.

When an electrical insulator—such as a plastic—is placed in a rapidly alternating electric field, an appreciable amount of heat is often generated within it. The rate of heat generation depends upon the strength and frequency of the field and upon the molecular structure and state of the material. In polymers containing polar groups, such as polyvinyl chloride, heat generation becomes intense at high frequencies. This is the basis of dielectric heating.

7-1 Dielectric Constants and Polarization

Consider a pair of electrodes separated by an insulator (the dielectric) with one electrode having charge $+q$ and the other an equal charge of opposite sign. If v is the potential difference between the electrodes, then

$$q = Cv \tag{7-1}$$

where the coefficient C is called the capacitance. Equation (7-1) can be considered to be the defining equation for capacitance. The magnitude of C depends upon the geometry of the electrodes, their spacing, and upon the nature of the dielectric.

For the special case where the electrodes are in a vacuum and consist of closely spaced parallel plates, the capacitance C_0 can be calculated with the formula

$$C_0 = K\frac{A}{b} \tag{7-2}$$

where A is the area of one of the plates, b is the distance of separation, and K is a fundamental constant having the value of 8.85×10^{-12} coulomb^2/newton-meter^2 in the rationalized mks system of units.

Now, consider parallel plates separated by a material dielectric rather than a vacuum. In this case the strength of the field in the space between the electrodes is reduced. Therefore, in order to maintain the field at the strength it would have with the electrodes in a vacuum, it would be necessary to place a larger charge on the electrodes. This effect is due to the *polarization* of the dielectric.

One type of polarization occurs because the electric field distorts the symmetric charge distribution in molecules, thereby inducing dipoles. The induced dipoles then orient themselves in the applied field and set up their own internal field. This opposes the applied field and thus reduces its strength. With molecules that have permanent dipoles, the process is much the same except that the initial distortion step is not required. The first type of polarization is called distortion polarization and the second, orientation polarization.

Polarization is always accompanied by an increase in capacitance. If C_0 is the capacitance when the electrodes are separated by a vacuum and C the capacitance when the same electrodes are separated by a dielectric, then the ratio C/C_0 is called the dielectric constant. It is a dimensionless quantity whose magnitude depends upon the molecular structure and state of the dielectric. Since all materials are polarized by an electric field, all materials have dielectric constants greater than unity. Table 7-1 lists typical values of the dielectric constant.

Most gases at ordinary temperatures and pressures have dielectric constants that are very close to unity. This is a density effect—gases are predominantly empty space. With nonpolar liquids and solids, such as benzene and paraffin hydrocarbons, only distortion polarization is possible and the dielectric constants are low, being in the range of 2 to 3. The much higher values observed with polar liquids and solids are a result of the additional polarization arising from the orientation of the permanent dipoles of these molecules.

The main cause of distortion polarization is displacement of electrons; a secondary cause is the displacement of atomic nuclei. The first is called electronic polarization and the second, atomic polarization. Electronic and atomic polarization are practically independent of temperature.

Therefore the dielectric constants of nonpolar substances are nearly temperature independent.

Orientation polarization, on the other hand, is strongly temperature dependent. As the temperature of the dielectric is increased, the random

Table 7-1

Dielectric Constants of Typical Materials

Substance	Dielectric Constant
Hydrogen (20° C)	1.00025
Benzene (25° C)	2.3
Methanol (25° C)	32.6
Water (25° C)	78.5
Paraffin wax (20°)	2.1
Diamond (20°)	16.5

thermal motion of the dipoles increases. This is a randomizing effect that opposes the orienting effect of the electric field and thereby reduces the effective orientation polarization. Consequently, the dielectric constants of polar substances decrease as the temperature increases.

7-2 Dielectric Behavior in an Alternating Field

Consider a nonpolar substance located between a pair of electrodes to which an alternating potential v is applied. If the potential varies sinusoidally with time, then

$$v = V_m \sin(2\pi f t) \tag{7-3}$$

where V_m is the maximum potential and f the frequency.

Because of the extreme mobility of electrons, electronic polarization occurs almost instantaneously and is in phase with the field. This is also true of atomic polarization at frequencies up to that of infrared radiation. Therefore, with nonpolar materials, the total polarization is in phase with the applied potential.

Since the potential varies sinusoidally with time, its maximum rate of change takes place when its value is zero. Hence, the maximum rate of charge displacement occurs at the zero of potential and the current (the rate of charge displacement) has its maximum values at the zeros of potential. Thus the current is exactly 90° out of phase with the potential. This characteristic of a purely capacitive circuit is obtained with any dielectric in which the polarization keeps up with changes in the potential.

The current is the rate of change of charge on the electrodes. It is given by the equation

$$i = \frac{dq}{dt} = C\frac{dv}{dt} \tag{7-4}$$

Introducing equation (7-3) into (7-4) gives

$$i = 2\pi f V_m C \cos(2\pi f t) \tag{7-5}$$

and since $C = \epsilon C_0$, where ϵ is the dielectric constant, equation (7-5) can be written

$$i = 2\pi f V_m C_0 \epsilon \cos(2\pi f t) \tag{7-6}$$

If equation (7-6) is written in terms of the rms current I instead of the instantaneous current, it becomes

$$I = 2\pi f V C_0 \epsilon \tag{7-7}$$

where V is the rms potential. Equation (7-7) shows that the dielectric behavior of a material in which the polarization remains exactly in phase with the potential depends upon a single material property, the dielectric constant ϵ.

The dielectric behavior of polar materials is more complicated. At low frequencies the orientation polarization is in phase with the potential. However, as the frequency is increased, a point is reached where dipole rotation no longer "keeps up" with changes in the potential. Under these conditions the orientation polarization is no longer in phase with the potential. Therefore the current I_0 associated with orientation polarization is out of phase with the current I_D associated with the displacement polarization. These currents are represented as vectors in Figure 7-1*a*.

The sum of I_D and I_0 is represented by the vector I. Since I_D is exactly 90° out of phase with the potential, the potential is represented by a vector acting along the abscissa. The vector representing the resultant current I is inclined at the angle ϕ, the phase angle, to the potential vector.

The vector I is now resolved into a component I' acting along the abscissa and a component I'' acting along the ordinate, as shown in Figure 7-1*b*. Because both the magnitude and the direction of a vector quantity, such as I, must be specified before it is uniquely determined, it is clear that the dielectric behavior of a polar material in an alternating field cannot be described by a single material constant.

It is customary to consider the current vector I as a complex quantity and to define a complex dielectric constant ϵ^* by the equation

$$\epsilon^* = \epsilon' - i\epsilon'' = -\frac{i}{2\pi f C_0}\frac{I^*}{V^*} \tag{7-8}$$

where

$$I^* = I' + iI''$$

$$V^* = V' + iV'' = V$$

The single primes refer to the real components and the double primes to the imaginary components of the complex quantities.

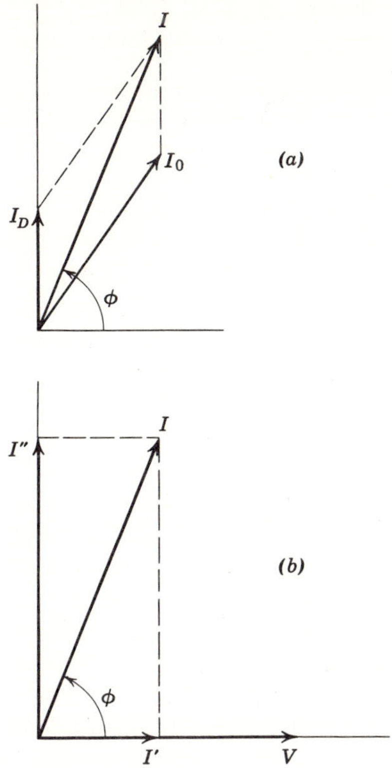

Figure 7-1. Current-voltage relations for a polar material in an alternating field. (*a*) Current I is the vector sum of the orientation and displacement currents and is resolved in (*b*) into an in-phase I' and an out-of-phase I'' component.

By equating real and imaginary components of equation (7-8), one obtains

$$I' = 2\pi f V C_0 \epsilon'' \tag{7-9}$$

$$I'' = 2\pi f V C_0 \epsilon' \tag{7-10}$$

Furthermore, with reference to the above equations and to Figure 7-1*b*, it can be seen that

$$\frac{I''}{I'} = \frac{\epsilon'}{\epsilon''} = \tan(\phi) \tag{7-11}$$

so that the dielectric behavior of a polar substance can be described by any two of the three quantities, ϵ', ϵ'', or ϕ.

Energy dissipation in a dielectric occurs only when the current has a component I' that is in phase with the voltage. The rate of energy dissipation is given by the equation

$$P = I'V = 2\pi C_0 f V^2 \epsilon'' \tag{7-12}$$

If the electrodes are closely spaced parallel plates, the value of C_0 is given by equation (7-2). If this is introduced into (7-12) and the resulting equation divided by the volume of the dielectric, one obtains the power absorption per unit volume p, given by the equation

$$p = 2\pi K\left(\frac{V^2 f}{b^2}\right)\epsilon'' \tag{7-13}$$

Equation (7-13) can be written in the equivalent form

$$p = 2\pi K\left(\frac{V^2 f}{b^2}\right)\epsilon' \cot(\phi) \tag{7-14}$$

where the quantity $[\epsilon' \cot(\phi)] = \epsilon''$ is called the loss factor.

Equation (7-13) shows that the energy dissipation rate is directly proportional to the first power of the frequency and to the second power of the voltage. One would expect power absorption to increase as the frequency and voltage increase, but for polar molecules the dielectric loss factor ϵ'' is, in general, a function of frequency and passes through a maximum as the frequency increases. Therefore the power absorption often passes through a maximum.

At low frequencies, orientation polarization is in phase with the potential and ϵ' has its maximum value ϵ_0, the static dielectric constant. Furthermore, at low frequencies $I' = 0$, so that the loss factor is zero. As the frequency increases, the orientation polarization becomes partially out of phase with the field. This causes ϵ' to decrease and ϵ'' to increase. At very high frequencies, orientation polarization becomes negligible and ϵ' has its minimum value ϵ_∞. Furthermore, since $I' = 0$, the loss factor is also zero.

The dielectric behavior of polar molecules depends upon the forces tending to restrict dipole rotation—the stronger these forces, the longer the time required for the dipoles to change their position. The *relaxation time* λ gives an indication of the strength of the restricting forces. It represents the time for 63% of the dipole polarization to disappear when the field is removed. The theoretical equations of Debye[1] relate ϵ' and ϵ'' to the

[1] See, for example, C. P. Smyth, *Dielectric Behavior and Structure*, McGraw-Hill, p. 55.

frequency f of the field and the relaxation time λ of the dipoles. They can be written in the form

$$f(x) = \frac{\epsilon' - \epsilon_\infty}{\epsilon_0 - \epsilon_\infty} = \frac{1}{1 + x^2} \tag{7-15}$$

$$g(x) = \frac{\epsilon''}{\epsilon_0 - \epsilon_\infty} = \frac{x}{1 + x^2} \tag{7-16}$$

where the dimensionless variable x is defined by the equation

$$x = 2\pi f\lambda \tag{7-17}$$

The limiting values of the function $f(x)$ are as follows:

$$\lim_{x\to 0} f(x) = 1 \qquad \lim_{x\to \infty} f(x) = 0$$

The function $g(x)$ approaches zero as a limit at both the upper and lower extremes of frequency and passes through its maximum at $x = 1$. The frequency f^* at which $g(x)$, and consequently ϵ'', is a maximum is called the critical frequency. If the critical frequency is known, the relaxation time can be calculated with equation (7-17). Note, also, that the maximum value of the function $g(x)$ is 0.5, so that the maximum value of ϵ'' is given by the formula

$$\epsilon''_{\max} = (\tfrac{1}{2})(\epsilon_0 - \epsilon_\infty) \tag{7-18}$$

Plots of the functions $f(x)$ and $g(x)$ versus x, as shown in Figure 7-2, are universal plots applying to all substances that obey the Debye theory. Note that if λ is constant (which would be the case at constant temperature), the theory requires practically all of the change in $f(x)$, or ϵ', to occur over about 2 decades of frequency. Similarly, the function $g(x)$, or ϵ'', rises from a very low value, passes through its maximum, and then decreases in about 2 decades of frequency.

The temperature dependence of the dielectric properties arises primarily through the temperature dependence of the relaxation time λ. Assuming the dipoles to be spheres of radius R in a viscous medium having viscosity μ, the relaxation time is given approximately by the formula[1]

$$\lambda = \frac{4\pi R^3 \mu}{kT} \tag{7-19}$$

where T is the absolute temperature and k is Boltzmann's constant. Since the viscosity of a liquid always decreases with an increase in temperature, equation (7-19) shows that λ must decrease as the temperature increases. The temperature dependence of ϵ_0 and ϵ_∞ is another reason for ϵ' and ϵ'' being temperature dependent.

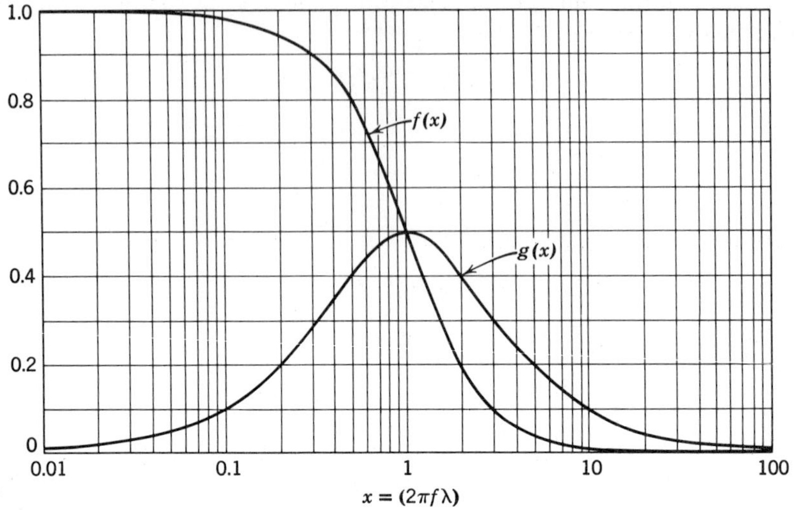

Figure 7-2. The frequency dependence of the functions $f(x)$ and $g(x)$ according to the Debye theory.

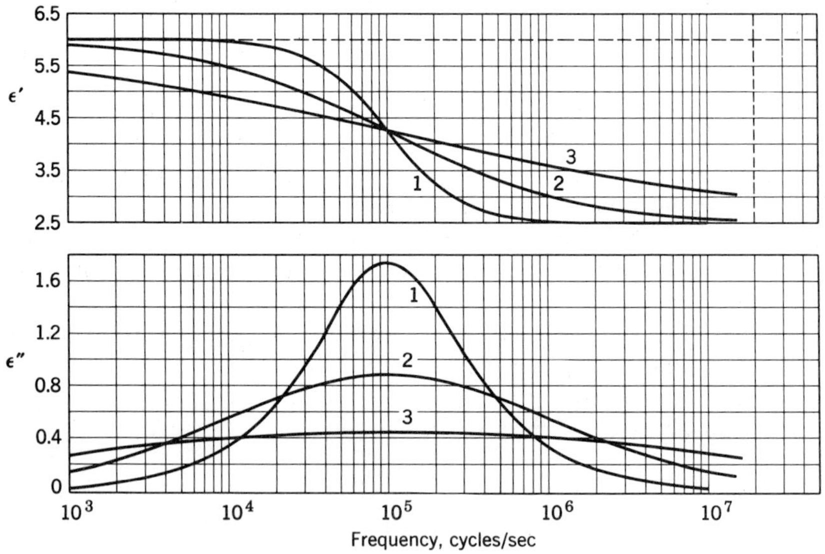

Figure 7-3. Frequency dependence of ϵ' and ϵ'': (1) single relaxation time, (2) moderate distribution of relaxation times, and (3) broad distribution of relaxation times. (By permission, from *High Polymers* by A. X. Schmidt and C. A. Marlies, McGraw-Hill, 1948, p. 442.)

7-3 Dielectric Properties of Polymers

Fuoss[2] measured the critical frequency f^* of a plasticized polyvinyl chloride as a function of temperature. The following table shows the relaxation time calculated from his data with equation (7-17):

t (°C)	λ (milliseconds)
45	1.680
50	0.361
55	0.100
60	0.028
65	0.007

Note the sharp decrease in relaxation time as the temperature increases. This is a characteristic of polar high polymers.

The Debye theory, derived for polar liquids, gives results that agree qualitatively with the measured dielectric behavior of polar high polymers. However, certain significant differences should be noted. For example, a distribution, rather than a single relaxation time, is needed to characterize the dielectric behavior of polymers. With a distribution of relaxation times, the ϵ'-f and the ϵ''-f curves are much broader than the theoretical curves, as shown in Figure 7-3.

Figure 7-4 shows the frequency dependence of the dielectric properties of polyvinyl chloracetate at various temperatures.[3] Note the general similarity to the curves of Figure 7-2. The critical frequency is seen to be a function of temperature, increasing as the temperature increases. Since ϵ''_{max} decreases as the temperature increases, equation (7-18) shows that $(\epsilon_0 - \epsilon_\infty)$ is a function of temperature, decreasing as the temperature increases. The increase in critical frequency with temperature is due to the reduction in viscosity, which enables dipole rotation to remain in phase with the field at frequencies that would give rise to a large out-of-phase component at lower temperatures.

Figure 7-5 shows the temperature dependence of the dielectric properties of polyvinyl acetate at various frequencies.[3] The dielectric constant rises with increasing temperature (at a given frequency) and approaches a limiting curve that represents the static dielectric constant as a function of temperature. For any given frequency, the value of ϵ' will approach the static value if the temperature is increased to a sufficiently high value. This is attributed to the reduction in relaxation time as the temperature increases.

[2] R. M. Fuoss, ch. VI of *The Chemistry of Large Molecules*, edited by R. E. Burk and O. Grummitt, Interscience Publishers, 1943.

[3] D. J. Mead and R. M. Fuoss, *J. Am. Chem. Soc.*, **63**, 2832 (1941).

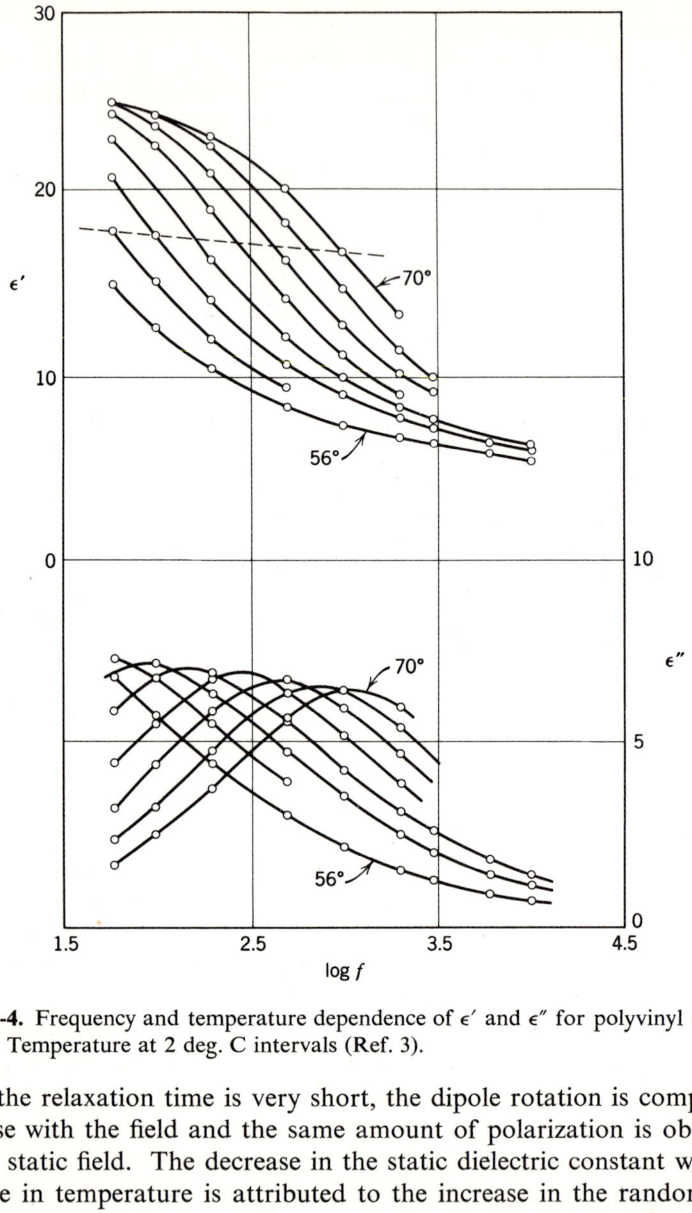

Figure 7-4. Frequency and temperature dependence of ϵ' and ϵ'' for polyvinyl chloroacetate. Temperature at 2 deg. C intervals (Ref. 3).

When the relaxation time is very short, the dipole rotation is completely in phase with the field and the same amount of polarization is obtained as in a static field. The decrease in the static dielectric constant with an increase in temperature is attributed to the increase in the randomizing thermal motion with increasing temperature. This tends to overcome the orienting effect of the electric field and thus reduce the polarization.

The effect of adding plasticizer is similar to that of increasing temperature. The addition of a plasticizer increases the mobility of the polymer

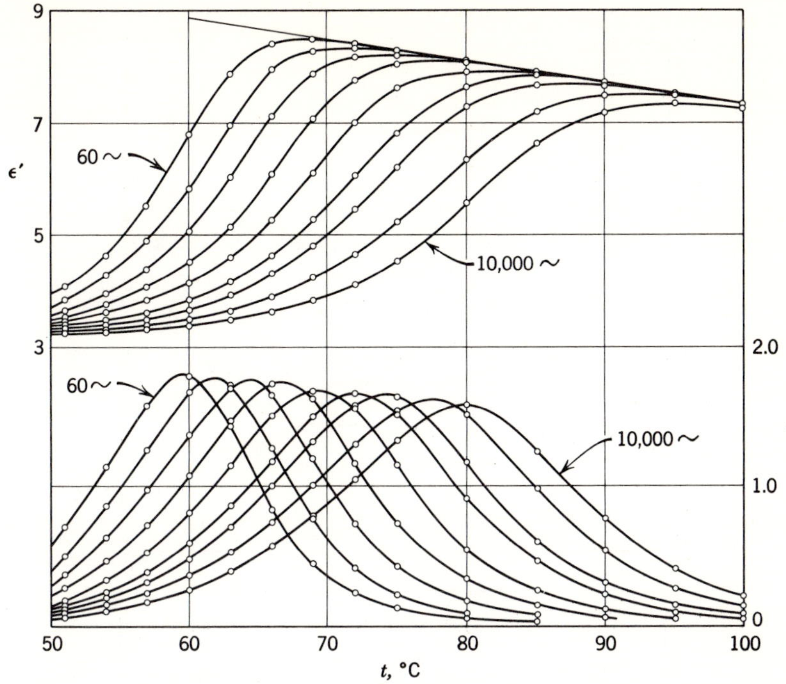

Figure 7-5. Temperature and frequency dependence of ϵ' and ϵ'' for polyvinyl acetate. Frequencies of 60, 120, 240, 500, 1000, 2000, 3000, 6000 and 10,000 cycles/sec (Ref. 3).

chain segments and allows the dipoles to rotate more freely. Hence the relaxation time is reduced. Figure 7-6 shows the effect of different amounts of plasticizer on the dielectric behavior of polyvinyl chloride. The temperature at which the dielectric constant approaches its static value is lowered, as is the critical frequency.

7-4 Dielectric Heating

The dielectric power loss per unit of volume of a slab of material clamped between parallel plate electrodes to which an alternating potential is applied is given by the equation (7-13). By introducing suitable conversion factors, this equation can be written

$$G = 13.3 \times 10^{-14}\left(\frac{V^2 f}{4W^2}\right)\epsilon'' \tag{7-20}$$

where G = heat generation rate (cal/cm^3-sec)
V = potential (volts)
f = frequency (cycles per second)
W = half thickness of the slab (cm)
ϵ'' = dielectric loss factor (dimensionless)

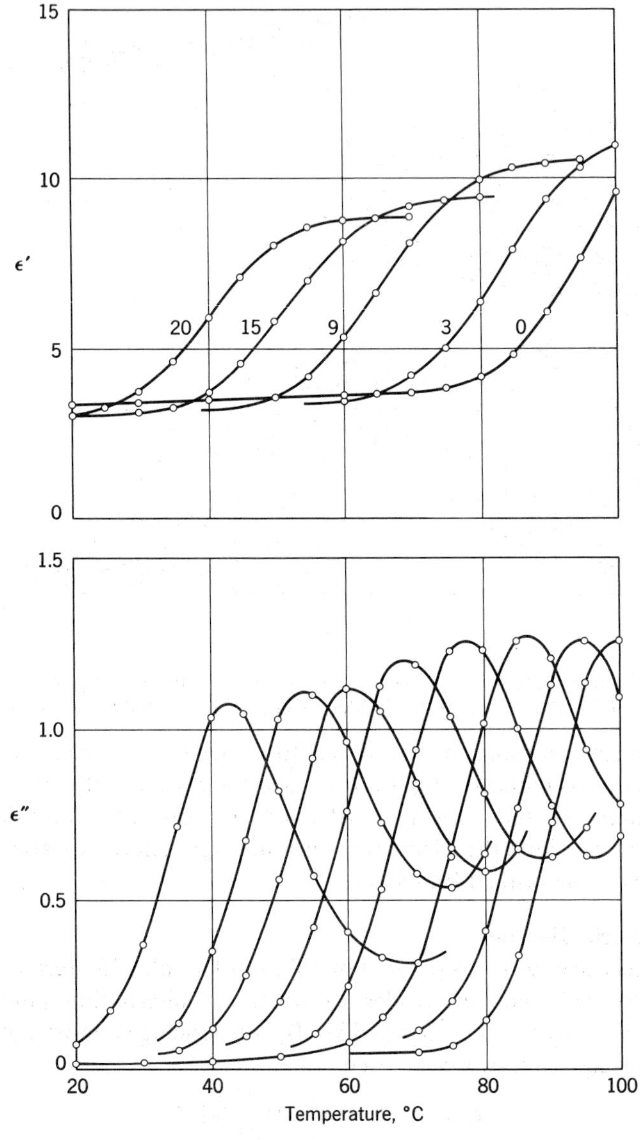

Figure 7-6. Effect on ε′ and ε″ of adding plasticizer to polyvinyl chloride. Measurements made at 60 cycles/sec. Plasticizer concentration ranges from 0 to 20% (Ref. 3).

The dielectric properties of a few thermoplastics at a given frequency of 1 megacycle and room temperature are given in the table below:

Material	Dielectric Constant	Power Factor
Polyethylene	2.2	0.0005
Polystyrene	2.4	0.0001
Nylon	3.5	0.03
"Saran"	2.0	0.003

Because of their extremely low power factors, polyethylene and polystyrene are used as electrical insulators in high frequency applications. They are not readily heated in dielectric heating equipment. Materials that can be heated dielectrically include polyvinyl chloride, "Mylar," nylon, "Kel-F," and "Saran." For convenience, the definitions of some of the terms used for the dielectric properties of materials are listed below:

$$\begin{aligned} \epsilon' &= \text{dielectric constant} \\ \epsilon'' &= \text{dielectric loss factor} \\ \phi &= \text{phase angle} \\ \cos\phi &= \text{power factor} \\ \cot\phi &= \text{dissipation factor} \end{aligned}$$

Because of equation (7-11), which can be written

$$\epsilon'' = \epsilon' \cot\phi$$

it is clear that any two of the above properties determine the dielectric behavior of a material.

In calculating heat generation rates, using equation (7-20), it is important to consider the change in the dielectric properties of the material as the temperature increases. Therefore the dielectric properties at the working frequency and over the range of temperatures to be encountered must be known. Furthermore, often it is important to consider heat conduction from the material into the metal of the electrodes.

The maximum rate of temperature rise would occur if the electrodes were thermally insulated from the material. Then all of the heat generated within the material would be utilized in raising its temperature. The calculation of the temperature rise in a plastic slab under this limiting condition is illustrated in the following example.

Example 7-1.

If T represents the temperature of the material at any time t, then

$$\rho C \frac{dT}{dt} = G \tag{7-21}$$

provided heat generation occurs uniformly and there are no losses. In the above equation ρ represents the density of the material and C its specific heat. Introducing equation (7-20) into (7-21) gives the following relationship between the rate of temperature rise and the operating variables:

$$\left(\frac{dT}{dt}\right) = 13.3 \times 10^{-14}\left(\frac{V^2 f}{4\rho C W^2}\right)\epsilon'' \qquad (7\text{-}22)$$

In order to solve (7-22) and obtain temperature as a function of time, it is necessary to express ϵ'' as a function of temperature. Usually an analytical expression relating ϵ'' to T will not be available, and numerical or graphical methods must be used.

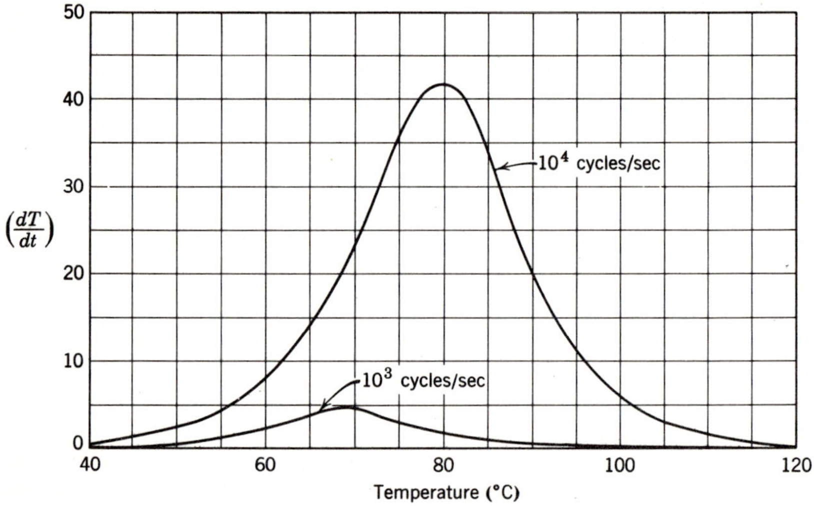

Figure 7-7. Calculated rate of adiabatic temperature rise (deg. C/sec) of a 2-mil sheet of PVA as a function of temperature.

As a specific example, the dielectric heating of a 4-mil thick sheet of polyvinyl acetate (PVA) from 40° C to 120° C using power at 10^4 cycles/sec and 500 volts is considered. It is assumed that the electrodes are thermally insulated from the material. The density of PVA is 1.2 grams/cm^3, and its specific heat is 0.4 cal/gram-deg C. The dielectric loss factor for PVA as a function of temperature at various frequencies is given by Figure 7-5. For example, at a frequency of 10^4 sec^{-1} and a temperature of 80° C, the value of ϵ'' is 1.55. Calculation of the rate of temperature rise at this temperature with equation (7-22) gives the following result:

$$\frac{dT}{dt} = \frac{(13.3 \times 10^{-14})(500)^2(10^4)(1.55)}{(4)(1.2)(0.40)(0.005)^2} = 41.8 \text{ deg C/sec}$$

The results of similar calculations at other temperatures are plotted in Figure 7-7. Curves for frequencies of 1 and 10 kilocycles are shown. Note that at the higher frequency, the curve rises to a much higher peak.

The temperature-time curve is established through graphical integration. The reciprocal of (dT/dt) is plotted against T over the range from 40° C to 120° C. The area under the curve from 40° C to any temperature T represents the time required to heat the sample from 40° C to T. Results are shown in Figure 7-8 for

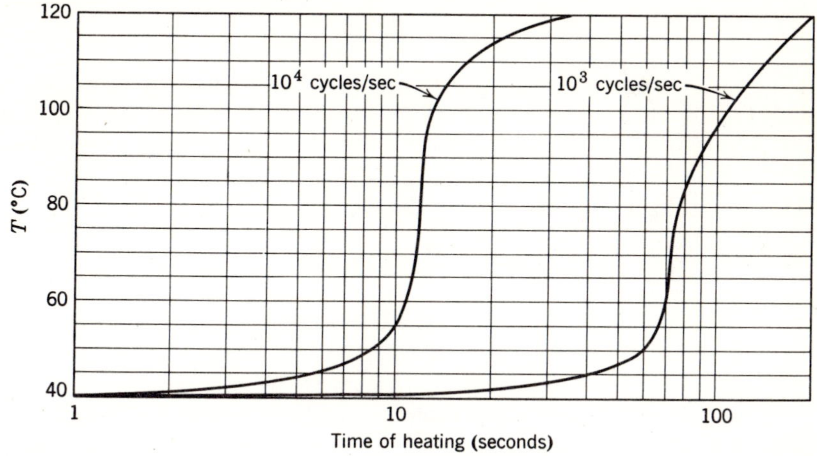

Figure 7-8. Time-temperature curve for the dielectric heating of a 2-mil sheet of PVA.

both frequencies. Note that at the higher frequency, the sample reaches 120° C in about 33 sec, while a heating period of about 190 sec is required at the lower frequency.

Problems

7-1 Equation (7-19) gives the temperature dependence of the relaxation time λ. Show that a plot of log (λT) versus T should be a straight line. Test the data given in Section 7-3 for plasticized polyvinyl chloride.

7-2 Verify equation (7-20).

7-3 Consider the dielectric heating of a flat slab of polyvinyl acetate, 2 mils thick. Determine the steady state temperature distribution in the material when the electrodes are maintained at 40° C and power at 1 kilocycle and 500 volts is applied to the electrodes.

chapter 8

Heat and Energy Requirements in Polymer Processing

Almost all operations of polymer processing involve the heating and cooling of solid or liquid polymers. Two factors must be considered: the total energy required to change the system from the initial to the final state and the rate at which this change can be made to occur.

The calculation of the total energy requirements for a process is based upon the first law of thermodynamics—the principle of conservation of energy. Section 8-1 discusses the application of the first law to steady flow processes and illustrates thermodynamic calculations based upon it.

The basic problem in heat transfer is to calculate temperatures at various times and positions within the system. The temperature distributions depend upon the rate of heat transfer from the surroundings to the system and upon the rate of heat transfer within the system. One or more of the three modes of heat transfer—conduction, convection, and radiation—will be involved.

Sections 8-2 through 8-5 are a brief introduction to the principles of conduction, convection, and radiation heat transfer and present the solution of some heating and cooling problems of particular interest in polymer processing. Although much of this material is available in standard texts on heat transfer, it is presented here for review and reference and will be drawn upon in later sections where specific polymer conversion operations are analyzed.

8-1 Thermodynamics of Steady Flow Processes

The first law of thermodynamics is a statement of the principle of conservation of energy. It asserts that a thermodynamic function E, called the internal energy and defined (partially) by the equation

$$dE = dQ - dW \tag{8-1}$$

is a state function. In equation (8-1) Q represents heat transferred between the system and the surroundings and W represents work done on or by the system.

Certain conventions concerning (8-1) and other thermodynamic equations are followed. Heat transferred to the system from the surroundings and work done by the system on the surroundings are considered positive quantities. Although Q is usually expressed in thermal units and W in mechanical units, equation (8-1) is written without the inclusion of a conversion factor. It is assumed that all quantities will be expressed in consistent units before numerical computations are made. Also, in this textbook, heat and work and all extensive energy functions are expressed on the basis of unit mass.

Because the internal energy E is a state function, it has a unique value for each equilibrium state of a system. Therefore when a system undergoes a process from one equilibrium state to another, the change in E is independent of the nature of the process. On the other hand, Q and W are path functions—their values depend upon the path of the process.

The concept of the reversible process, in which a system passes through an infinite number of equilibrium states, is extremely useful. Thermodynamic efficiencies of actual processes are calculated by comparing their performance with those of reversible processes. The work of a reversible expansion or compression is given by the equation

$$dW = P\,dv \tag{8-2}$$

where P = pressure

v = specific volume

Another function H, called the enthalpy, is defined by the equation

$$H = E + Pv$$

which, upon differentiation, becomes

$$dH = dE + P\,dv + v\,dP \tag{8-3}$$

By combining equations (8-1), (8-2), and (8-3), it can be shown that

$$dE = dQ - P\,dv \tag{8-4}$$

$$dH = dQ + v\,dP \tag{8-5}$$

from which it can be seen that for a *constant volume* process, the transferred heat is equal to the change in internal energy, while for a *constant pressure* process, the transferred heat is equal to the change in enthalpy. Specific heats at constant pressure and constant volume are, therefore, defined by the following equations:

$$c_p = \left(\frac{\partial H}{\partial T}\right)_P \tag{8-6}$$

$$c_v = \left(\frac{\partial E}{\partial T}\right)_v \tag{8-7}$$

The first law of thermodynamics is now applied to a steady flow process. Consider the input and output of mass and energy from the system shown in Figure 8-1. Because the derivation is limited to steady flow, there is no

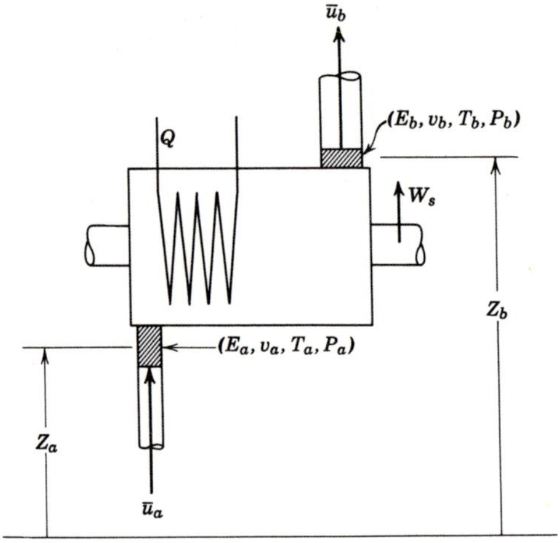

Figure 8-1. Schematic diagram used in the thermodynamic analysis of a steady flow process.

accumulation of mass or energy within the system. The principle of conservation of mass requires that

$$\Delta(\rho A\bar{u}) = 0 \tag{8-8}$$

where $\bar{u}$ is the average fluid velocity, defined as the volumetric flow rate divided by the cross-sectional area A of the pipe.

The principle of conservation of energy requires that the change in energy of the fluid as it passes through the system be equal to the difference between the heat added Q and the net work ΣW done by the fluid. This is expressed mathematically by the equation

$$\Delta E + \Delta KE + \Delta PE = Q - \sum W \tag{8-9}$$

where KE and PE represent the kinetic and potential energy of the fluid, respectively. Note the similarity between equations (8-9) and (8-1).

The change in potential energy, when the elevation of unit mass of the fluid increases by an amount ΔZ, is given by the equation

$$\Delta PE = g\,\Delta Z \tag{8-10}$$

where $\Delta Z = (Z_b - Z_a)$

and g is the gravitational acceleration, 980 cm/sec^2 at sea level.

The change in kinetic energy when the average velocity of the fluid increases from $\bar{u}_a$ to $\bar{u}_b$ is given by the equation

$$\Delta KE = k\,\Delta\bar{u}^2 = k(\bar{u}_b^{\,2} - \bar{u}_a^{\,2}) \tag{8-11}$$

where k is a coefficient that depends upon the shape of the velocity profile. For the parabolic profile resulting from laminar Newtonian flow, $k = 1$. For a flat profile (plug flow), $k = 0.5$. A formula giving k as a function of the flow index n of power law fluids is derived below:

From Table 3-1 we see that the velocity u at any radius r for the steady laminar flow of a power law fluid through a tube of radius R is given by an equation of the form

$$u = \bar{u}\left(\frac{3n+1}{n+1}\right)\left[1 - \left(\frac{r}{R}\right)^{(n+1)/n}\right] \tag{8-12}$$

where the mean velocity $\bar{u}$ is related to the volumetric flow rate Q by the equation

$$\bar{u} = \frac{Q}{\pi R^2}$$

The mass rate of flow through any cross section perpendicular to the tube axis is ρQ. Each fluid element passing through the section carries with it a certain amount of kinetic energy, defined as one-half the mass of the element multiplied by the square of its velocity. Therefore the rate at which kinetic energy passes through an annular ring of radius r and thickness dr is

$$\tfrac{1}{2}[2\pi r\,dr\,\rho u]u^2$$

and the average rate of transport of kinetic energy through the cross section is

$$KE = \frac{\pi}{Q}\int_0^R u^3 r\,dr \tag{8-13}$$

Upon introducing (8-12) into (8-13) and defining the dimensionless variable x as (r/R), we have

$$KE = \bar{u}^2\left(\frac{3n+1}{n+1}\right)^3\int_0^1 x[1 - x^{(n+1)/n}]^3\,dx \tag{8-14}$$

Upon integration, equation (8-14) can be written

$$KE = k\bar{u}^2 \tag{8-15}$$

where the coefficient k, which depends only upon the flow index n, is given by the equation

$$k = \left(\frac{3n+1}{n+1}\right)^3 \left[\frac{1}{2} - \frac{3n}{3n+1} + \frac{3n}{4n+2} - \frac{n}{5n+3}\right]$$

For Newtonian flow ($n = 1$), the above formula reduces to $k = 1$ and for plug flow ($n = 0$), it reduces to $k = 0.5$.

The ΣW term is now considered. A pump can transfer mechanical energy to the fluid, while a turbine can extract mechanical energy from the fluid. In either case, work of this type is usually called shaft work and is designated by the symbol W_s. Flow work must also be considered. The work done on the system in forcing unit mass of the fluid into the system at a is $-(P_a v_a)$, while the system does $+(P_b v_b)$ work in ejecting unit mass of the fluid at b.

Therefore

$$\sum W = W_s + (P_b v_b) - (P_a v_a) \tag{8-16}$$

if all other forms of work are neglected.

By introducing (8-10), (8-11), and (8-16) into equation (8-9), one obtains the general energy balance:

$$\Delta E + g\,\Delta Z + k\,\Delta \bar{u}^2 + \Delta(Pv) = Q - W_s \tag{8-17}$$

In polymer processing applications, the kinetic and potential energy terms are usually negligible, and the energy balance can be written

$$\Delta E + \Delta(Pv) = Q - W_s \tag{8-18}$$

The change in potential energy when there is an increase of 1 meter in elevation of a 1 gram mass is about 10^5 cm²/sec², and the change in kinetic energy when the velocity is increased from rest to 50 cm/sec (about 100 ft/min) is 2.5×10^3 cm²/sec².

If the temperature of a liquid having a specific heat of about 0.5 cal/gram-deg C is increased 1 deg C, the increase in enthalpy of the liquid is 0.5 cal/gram (about 20×10^6 cm²/sec²), which shows clearly that the kinetic and potential energy terms can be neglected in polymer processing applications.

An alternative form of the energy balance that includes the enthalpy function, obtained by introducing (8-3) into (8-18), is

$$\Delta H = Q - W_s \tag{8-19}$$

Each term in the energy balance represents energy per unit mass. Since ΔH, ΔE, and Q are usually expressed in thermal units and the other terms in mechanical units, care must be taken to convert all terms to one system or the other before numerical computations are made.

Example 8-1.

This example presents an analysis of heat and energy requirements for the steady flow process shown schematically in Figure 8-2. In the process, a screw pump forces a molten polymer through a die to form film. At the feed port of the pump, the temperature of the polymer is T_a and its pressure is P_a. At the entrance to the die, the temperature and pressure are T_b and P_b, respectively, and after passing through the die they become T_c and P_c. Heat, which is generated within

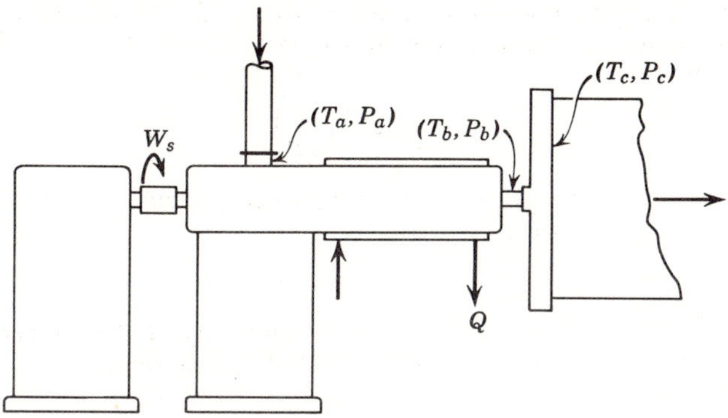

Figure 8-2. Schematic diagram of a continuous steady flow process.

the polymer by viscous shear, is removed with cooling water that circulates through the jacket surrounding the pump. The water flow is regulated so that $T_a = T_b$.

The operating data for the process are as follows:

Mechanical power input to pump	2.0 hp
Pumping rate	100 lb/hr
Die pressure	1000 psi
Pressure at feed port of pump	100 psi
Temperature at feed port of pump	200° C

The physical properties of the polymer melt are shown below. The values given for β, κ, and C_p are assumed to be constant over the range of conditions involved in the process.

Compressibility:	$\beta = 10^{-3}$ atm^{-1}
Coefficient of thermal expansion:	$\kappa = 10^{-4}$ deg C^{-1}
Specific heat at constant pressure:	$C_p = 0.5$ cal/gram-deg C
Density at 200° C and 1 atm:	$\rho = 1.1$ gram/cm^3

In part (*a*) the required rate of heat removal from the pump is calculated. Part (*b*) considers the efficiency of the pump, and in part (*c*) the temperature rise of the polymer at the die is calculated.

(*a*) The basis of the analysis is the general energy balance for a steady flow process. Equation (8-19) is the applicable equation. Here ΔH represents the change in enthalpy of the polymer between the feed port of the pump and the die.

Since there is no change in temperature, only the effect of pressure on enthalpy need be considered. This is given by the thermodynamic equation

$$\left(\frac{\partial H}{\partial P}\right)_T = v(1 - \kappa T) \tag{8-20}$$

which, for a constant temperature change, can be written

$$\Delta H = \int v(1 - \kappa T)\, dP \tag{8-21}$$

In order to integrate (8-21) it is necessary to express the specific volume v as a function of pressure. The relationship can be obtained from the definition of compressibility,

$$\beta = -\frac{1}{v}\left(\frac{\partial v}{\partial P}\right)_T \tag{8-22}$$

if β is assumed to be constant over the range of pressures involved. Equation (8-22) can then be integrated to give

$$v = v_0\, e^{-\beta P} \tag{8-23}$$

where v_0 represents the specific volume of the polymer at 200° C and zero pressure.

Combining equations (8-23), (8-21), and (8-19), one obtains

$$Q = W_s + v_0 \int_{P_a}^{P_b} (1 - \kappa T) e^{-\beta P}\, dP \tag{8-24}$$

Equation (8-24) is integrated directly, assuming κ to be independent of pressure. The result is

$$Q = W_s - \frac{v_0(1 - \kappa T)}{\beta}[1 - e^{+\beta \Delta P}] e^{-\beta P_b} \tag{8-25}$$

where $\Delta P = (P_b - P_a)$. When the product $\beta\, \Delta P$ is small, the following approximation can be used

$$1 - e^{\beta \Delta P} \approx -\beta\, \Delta P \qquad (\beta\, \Delta P < 0.1)$$

and equation (8-25) can be written

$$Q = W_s + v_0(1 - \kappa T) \Delta P\, e^{-\beta P_b} \tag{8-26}$$

The pumping rate of 100 lb/hr is equivalent to 12.6 grams/sec, and the power input of 2.0 hp is equivalent to 356 cal/sec. Therefore

$$W_s = -(356/12.6) = -28.30 \text{ cal/gram}$$

The negative sign indicates that work is put into the system. The values of the other quantities of equation (8-26) are as follows:

$$v_0 = 0.91 \text{ cm}^3\text{/gram}$$
$$\kappa = 10^{-4} \text{ deg C}^{-1}$$
$$T = 473° \text{ K}$$
$$\Delta P = (P_b - P_a) = 61.2 \text{ atm}$$

The value of Q, calculated with equation (8-25) is -27.0 cal/gram, and the required heat removal rate is 340 cal/sec.

(*b*) The efficiency of the pump can be defined as the ratio of the useful work done by the pump to the total work input. In the example under consideration, the only useful work done by the pump is the flow work $\Delta(Pv)$ expended in raising the pressure of the fluid from P_a to P_b. Writing equation (8-23) in the approximate form

$$v \approx v_0(1 - \beta P) \tag{8-27}$$

the formula for flow work becomes

$$\Delta(Pv) \approx v_0 \, \Delta P[1 - \beta(P_a + P_b)] \tag{8-28}$$

The value of $\Delta(Pv)$ calculated with equation (8-28) is 1.25 cal/gram. The efficiency ϵ is

$$\epsilon = \frac{\Delta(Pv)}{-W_s} = \frac{1.25}{28.3} = 0.044$$

The analysis of heat and energy requirements for the pump shows that the input of 28.3 cal/gram of mechanical energy is accounted for by the 27.0 cal/gram of heat removed by the cooling water and the 1.25 cal/gram of flow work. The remaining 0.05 cal/gram is the work of compression.

(*c*) As the polymer flows through the lips of the die, the work of expansion and the flow work, both negative quantities, are released and appear as heat. If one assumes that the flow occurs adiabatically (a reasonable assumption because the residence time in the die lips is very small), the average temperature rise of the polymer can be calculated. For this case, equation (8-19) reduces to

$$\Delta H = 0 \tag{8-29}$$

and the temperature rise is evaluated in the following manner:

Enthalpy, a property of the fluid, is expressed as a function of temperature and pressure by the functional equation

$$H = f(T, P) \tag{8-30}$$

The total differential of H is

$$dH = \left(\frac{\partial H}{\partial T}\right)_P dT + \left(\frac{\partial H}{\partial P}\right)_T dP \tag{8-31}$$

and for an integral change, equation (8-31) can be written

$$\Delta H = \int_{T_b}^{T_c} \left(\frac{\partial H}{\partial T}\right)_P dT + \int_{P_b}^{P_c} \left(\frac{\partial H}{\partial P}\right)_T dP \tag{8-32}$$

where the subscript b refers to the fluid at the inlet to the die and the subscript c to the outlet of the die.

The first integral of equation (8-32) is to be evaluated at the constant pressure P_c and the second integral at the constant temperature T_b. Alternatively, they could be evaluated P_b and T_c, for ΔH is independent of the path of the process.

In view of equations (8-20) and (8-6), equation (8-32) can be written

$$\Delta H = c_p \, \Delta T + (1 - \kappa T) \int_{P_b}^{P_c} v \, dP \tag{8-33}$$

where $\Delta T = (T_c - T_b)$. If the specific volume is expressed as a function of pressure with equation (8-23), then (8-33) becomes

$$\Delta H = 0 = c_p \, \Delta T - v_0 (1 - \kappa T) \, \Delta P \tag{8-34}$$

where $\Delta P = (P_c - P_b)$.

Noting that P_c is atmospheric pressure, the value of ΔT calculated with equation (8-34) is +2.8 deg C. This represents the average temperature increase of the polymer as it passes through the die. The heat generation occurs nonuniformly so that the temperature gradients are set up within the polymer.

8-2 Conduction Heat Transfer

Heat conduction occurs in all materials whenever a temperature gradient exists. The science of heat conduction can be treated in two ways. In a mechanistic treatment, attention is directed toward the exchange of kinetic energy when atoms and molecules collide and to the drift of free electrons through metals. This approach is primarily of interest to physicists. In a phenomological treatment, attention is focused on directly observable quantities, such as time, temperature, and position. This approach is of primary interest to engineers and is the one used in this textbook.

The fundamental problem of heat conduction is to determine the temperature distribution within an object. The temperature distribution may be independent of time, in which case the problem is one of *steady state* conduction, or the temperature distribution may be a function of time, in which case the problem is one of *unsteady state* conduction. In this chapter the analytical solutions of some unsteady state conduction problems of interest in polymer processing are outlined.

The heat flux vector **q** represents the rate of heat transfer across a surface of unit area normal to **q**. The vector **q** is, in general, a function of time and position. Fourier's law, the basis of all heat conduction calculations, states that for an isotropic material, the heat flux vector is proportional to the temperature gradient. This is expressed by the equation

$$\mathbf{q} = -k \, \nabla T \tag{8-35}$$

where k is the thermal conductivity of the material.

If the energy equation, (2-3), is specialized by placing $\boldsymbol{v} = 0$, it becomes

$$\rho c_v \left(\frac{\partial T}{\partial t}\right) = -(\nabla \cdot q) \tag{8-36}$$

and upon introducing equation (8-35), one obtains

$$\rho c_v \left(\frac{\partial T}{\partial t}\right) = k \, \nabla^2 T \tag{8-37}$$

which is the *heat equation* for conduction in an isotropic solid in which k is constant.

If heat is being generated within the solid, a generation term must be added, and the heat equation becomes

$$\frac{\partial T}{\partial t} = \alpha \nabla^2 T + \left(\frac{G}{\rho c_p}\right) \tag{8-38}$$

where G represents the rate of internal heat generation expressed on a unit volume basis. Its metric units are cal/cm³-sec. Examples of internal heat generation include i^2R heating in an electrical conductor carrying a current, absorption of infrared radiation in a semi transparent object, and exothermic chemical reactions. In equation (8-38) c_v has been replaced with c_p, which is permissible for incompressible solids, and the coefficient α, called the thermal diffusivity, is defined by the equation

$$\alpha = \left(\frac{k}{c_p \rho}\right)$$

The quantity $\nabla^2 T$, the Laplacian of the temperature, is given in Table 8-1 in rectangular, cylindrical, and spherical coordinates. Hence the differential equation for many heat conduction problems can be formulated automatically by referring to equation (8-38) and Table 8-1.

Table 8-1

The Laplacian of Temperature Expressed in Rectangular, Cylindrical, and Spherical Coordinates

Rectangular coordinates:

$$\nabla^2 T = \frac{\partial^2 T}{\partial x^2} + \frac{\partial^2 T}{\partial y^2} + \frac{\partial^2 T}{\partial z^2}$$

Cylindrical coordinates:

$$\nabla^2 T = \frac{1}{r}\left(\frac{\partial T}{\partial r}\right) + \frac{\partial^2 T}{\partial r^2} + \frac{1}{r^2}\left(\frac{\partial^2 T}{\partial \theta^2}\right) + \frac{\partial^2 T}{\partial z^2}$$

Spherical coordinates:

$$\nabla^2 T = \frac{1}{r^2}\frac{\partial}{\partial r}\left[r^2\left(\frac{\partial T}{\partial r}\right)\right] + \frac{1}{r^2 \sin\theta}\frac{\partial}{\partial \theta}\left[\sin\theta\left(\frac{\partial T}{\partial \theta}\right)\right] + \frac{1}{r^2 \sin^2\theta}\left(\frac{\partial^2 T}{\partial \phi^2}\right)$$

An analytical solution of a heat conduction problem consists of a function of time and position that is a solution of the heat equation and also satisfies the initial and boundary conditions of the problem.

Four different types of boundary conditions arising in heat conduction problems are mentioned here. The isothermal boundary is one whose temperature is constant during the entire period of time under consideration. If the position of the boundary can be specified by a single number (for example, a distance R), then the isothermal boundary condition is expressed mathematically by an equation of the form

$$T(t, R) = T_R \tag{8-39}$$

where T_R is the boundary temperature.

An isothermal boundary would occur, for example, in a casting process in which a layer of molten polymer is cast upon a cylinder that is maintained at a constant temperature by circulating water. In this case the surface of the polymer in contact with the cylinder instantaneously assumes the temperature of the cylinder and remains constant as long as it is in contact with the cylinder.

A second type of boundary condition arises when there is a *constant heat flux* across a surface. This condition is expressed mathematically by the equation

$$q = -kT'(t, R) \tag{8-40}$$

where q, the heat flux, is a constant and T' represents the partial derivative of T with respect to distance. Some radiant heating problems have boundary conditions of this type.

A *convective* boundary condition arises when the surface is in contact with a fluid. Heat transfer between the surface and the fluid occurs at a rate that is proportional to the difference in temperature between the surface $T(t, R)$ and the bulk of the fluid T_f. The proportionality constant h is called the convective heat transfer coefficient. This boundary condition is expressed mathematically by the equation

$$h[T(t, R) - T_f] = -kT'(t, R) \tag{8-41}$$

The adiabatic boundary condition is the last type considered. It arises when there is no transfer of heat across a boundary, for example, at a well-insulated surface. It is expressed mathematically by an equation of the form

$$-kT'(t, R) = 0 \tag{8-42}$$

8-3 Heating and Cooling of Slabs

This section presents solutions of the heat equation for unidirectional heat flow in an object bounded by two parallel plane surfaces (the slab) and by one plane surface (the semi-infinite slab), as shown in Figure 8-3. For unidirectional heat flow with no internal heat generation, the heat equation reduces to

$$\left(\frac{\partial T}{\partial t}\right) = \alpha\left(\frac{\partial^2 T}{\partial x^2}\right) \tag{8-43}$$

The boundary condition for the heating (or cooling) of the semi-infinite slab by convective heat transfer from a fluid at temperature T_f is

$$h[T(0, t) - T_f] = -kT'(0, t) \tag{8-44}$$

where
$$T' = \left(\frac{\partial T}{\partial x}\right)$$

We shall consider only that case where the slab is initially at a uniform temperature T_i. Hence the initial condition is written

$$T(x, 0) = T_i \tag{8-45}$$

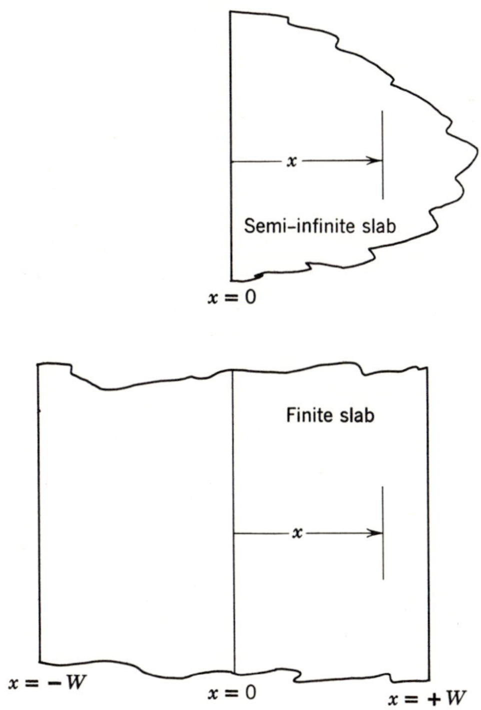

Figure 8-3. Location of coordinates for analysis of heat flow in slabs.

The convective heat transfer coefficient h is assumed constant in this problem, as is the thermal conductivity k.

The partial differential equation (8-43) and the above conditions constitute a boundary value problem. The solution to this problem, obtained through the method of the Laplace transformation, is written

$$Y = \operatorname{erf} \Psi + \exp\left[X + \frac{X^2}{4\Psi^2}\right] \operatorname{erfc}\left[\Psi + \frac{X}{2\Psi}\right] \tag{8-46}$$

where the dimensionless quantities X, Y, and Ψ are defined by the following equations:

$$Y = \frac{T - T_f}{T_i - T_f} \tag{8-47}$$

$$X = \frac{hx}{k} \tag{8-48}$$

$$\Psi = \frac{x}{2\sqrt{\alpha t}} \tag{8-49}$$

In the above equation, erf Ψ denotes the error function of Ψ and erfc the complementary error function. Values of these functions are tabulated

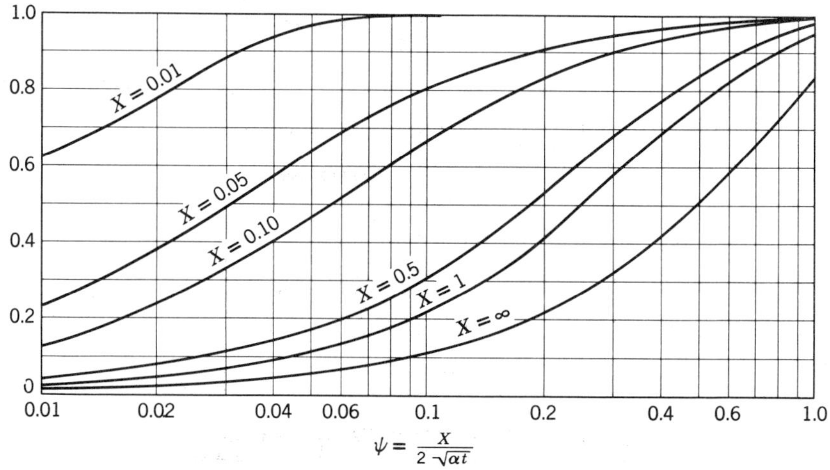

Figure 8-4. Solution of the heat equation for the semi-infinite slab.

in mathematical handbooks. The dimensionless quantity Y depends only on X and Ψ. It is plotted in Figure 8-4 versus Ψ, with X as the parameter.

The solution of the boundary value problem arising when the convective boundary condition is replaced with an isothermal boundary condition can be obtained by specializing equation (8-46). Because an isothermal boundary implies negligible resistance to heat transfer from the fluid to the solid, the convective heat transfer coefficient h (and consequently X) becomes infinitely large, and equation (8-46) reduces to

$$Y = \operatorname{erf} \Psi = \operatorname{erf}\left[\frac{x}{2\sqrt{\alpha t}}\right] \tag{8-50}$$

The graphical solution of equation (8-50) is represented by the curve labeled $X = \infty$ in Figure 8-4.

Heating and cooling a slab by convective heat transfer to both surfaces from a fluid at temperature T_f are now considered. The thickness of the slab is $2W$ and its initial uniform temperature is T_i. Placing the origin of the coordinate system in the mid-plane of the slab (see Figure 8-3), the boundary conditions become

$$\begin{aligned} T(x, 0) &= T_i \\ h[T_f - T(+W, t)] &= kT'(+W, t) \\ h[T_f - T(-W, t)] &= -kT'(-W, t) \end{aligned} \tag{8-51}$$

where $$T' = \left(\frac{\partial T}{\partial x}\right)$$

The solution of the boundary value problem consisting of the above boundary conditions and the differential equation (8-43) is given by Carslaw and Jaeger.[1] It can be written

$$\frac{T_f - T}{T_f - T_i} = \sum_{n=1,2,\cdots}^{\infty} \left[\frac{4 \sin M}{2M + \sin (2M)}\right] e^{-M^2\alpha t/W^2} \cos\left(\frac{Mx}{W}\right) \tag{8-52}$$

where M represents the real positive roots of the equation

$$M \tan M = \frac{hW}{k} \tag{8-53}$$

Equation (8-53) has an infinite number of real positive roots. The first five roots for various values of the parameter $N = hW/k$ are tabulated by Carslaw and Jaeger.[2] Graphical solutions for various special cases are ordinarily used in making numerical computations with equation (8-52).

Equation (8-52) can be specialized to give the surface temperature by replacing x with W. Figure 8-5 gives the graphical solution for this case, with the dimensionless temperature ratio plotted against the quantity $(\alpha t/W^2)$ and N treated as a parameter. Figure 8-6 shows the graphical solution of another special case, that obtained with $x = 0$, which applies to the mid-plane of the slab.

[1] H. S. Carslaw and J. C. Jaeger, *Conduction of Heat in Solids*, 2nd ed., Oxford University Press, 1959, p. 121.

[2] *Ibid.*, p. 491.

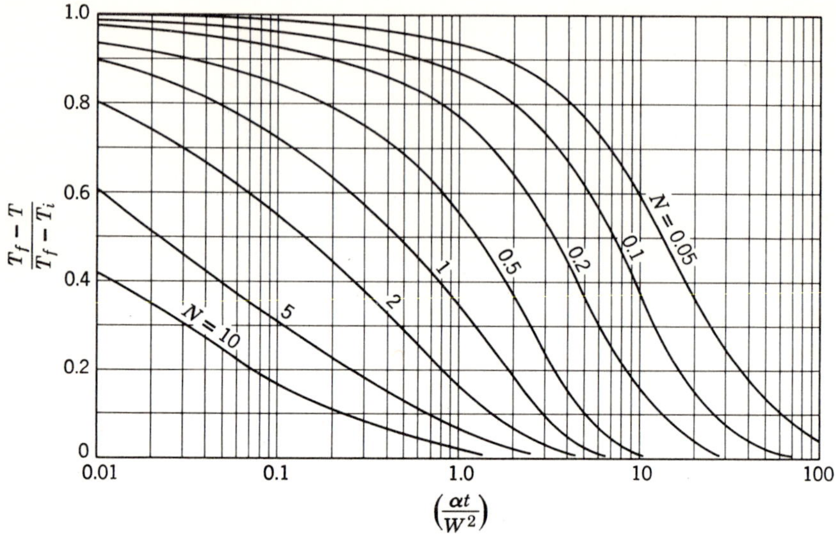

Figure 8-5. Solution of the heat equation for surface of the slab.

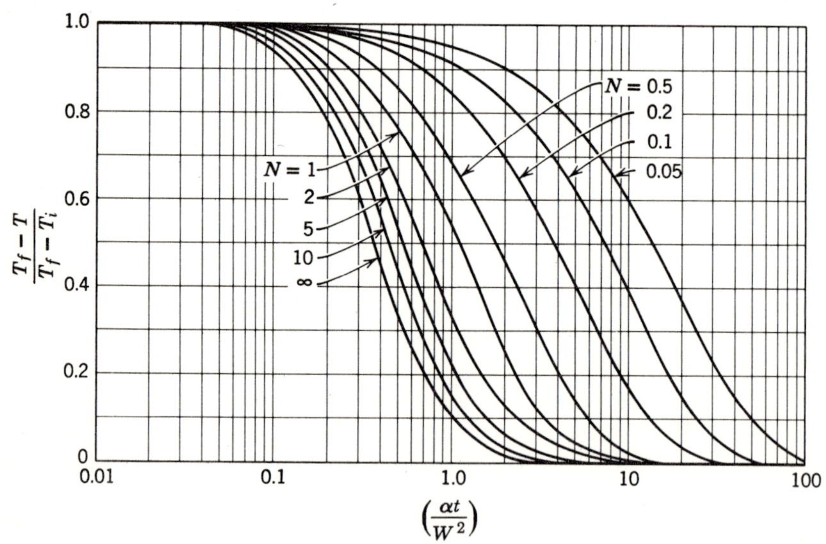

Figure 8-6. Solution of the heat equation for the mid-plane of the slab.

Another special case of interest is obtained by replacing the convective boundary condition with an isothermal condition. With $h = \infty$, the real positive roots of equation (8-53) are $(n\pi/2)$, where $n = 1, 3, \cdots$ and equation (8-52) becomes

$$\frac{T_f - T}{T_f - T_i} = \frac{4}{\pi} \sum_{n=1,3,\cdots}^{\infty} \frac{(-1)}{n} e^{-n^2\pi^2\alpha t/4W^2} \cos\left(\frac{n\pi x}{2W}\right) \tag{8-54}$$

The graphical solution of this equation is represented in Figure 8-6 by the curve labeled $n = \infty$.

8-4 Radiant Heat Transfer

In polymer processing, radiant heating finds its major application in the heating of relatively thin sheets of thermoplastic materials prior to various thermoforming operations. The purpose of this section is to present a brief introduction to the theory of radiant heat transfer and to discuss methods of calculating radiant heat losses in certain extrusion processes.

All substances emit energy in the form of electromagnetic radiation. This energy, requiring no material medium for its transport, travels through space until it encounters an absorbing medium: gas, liquid, or solid. The rate at which radiant energy is emitted is proportional to the fourth power of the absolute temperature of the emitting substance. In the temperature range in which radiant energy becomes a practical means of heating, a substantial portion of it is in the infrared region of the spectrum. For this reason, radiant heating is also called infrared heating.

A hypothetical material that absorbs all incident radiation (absorptivity equal to unity) is known as a *black body*. The intensity E_b of radiation emitted by a black body at absolute temperature T is given by the Stefan-Boltzmann law:

$$E_b = \sigma T^4 \tag{8-55}$$

where σ is the Stefan-Boltzmann constant. Its value is 1.36×10^{-12} cal/cm^2-sec-deg C^4.

The spectral distribution of the emitted energy is a function of the temperature of the emitter. As its temperature increases, the wavelength at which the maximum intensity occurs also increases. This is expressed by the Wein displacement law:

$$T\lambda_m = 0.288 \text{ cm-}^\circ\text{K} \tag{8-56}$$

where λ_m is the wavelength of maximum intensity. The shape of the entire spectral distribution curve for a black body was deduced theoretically by Planck and can be calculated with Planck's law for any temperature.

Real objects do not behave exactly like black bodies. At any given temperature they emit less energy. Their *emissivity* ϵ is defined by the equation

$$\epsilon = \frac{E}{E_b} \tag{8-57}$$

where E is the intensity of radiation emitted by a real body. In general, ϵ is a function of temperature; for most materials it increases as temperature increases. Nonmetalics and metal oxides have high emissivities, usually 0.8 or higher. Highly polished metals have low emissivities, usually 0.1 or lower.

Real objects do not absorb all incident radiation. The *absorptivity* α is defined as the fraction of incident radiation that is absorbed. For a surface in radiation equilibrium, Kirchhoff's law states that $\alpha = \epsilon$.

The calculation of radiation exchange between black bodies is primarily a problem of geometry. Only that part of a body that can be "seen" by the emitting body can be struck by the radiation. Furthermore, the intensity of the emitted radiation is greatest along a normal to the surface and becomes zero along a tangent. For a black body, this is expressed mathematically by the Lambert cosine law.

Hottel[3] has carried out heat exchange calculations for certain simple geometrical configurations. His results are expressed as "viewfactors," representing the fraction of radiation emitted by the one surface that strikes the other surface. For example, the rate at which radiant energy is transferred from black surface 1 to black surface 2 would be $E_1A_1F_{12}$, where A_1 is the area of surface 1 and F_{12} represents the fraction of the radiation emitted by surface 1 that strikes surface 2.

It can be shown that

$$A_1F_{12} = A_2F_{21} \tag{8-58}$$

so that Q_{12}, the net rate of interchange of radiant energy between body 1 and body 2, is given by the equation

$$Q_{12} = E_1A_1F_{12} - E_2A_2F_{12} = A_1F_{12}(E_1 - E_2) \tag{8-59}$$

If the Stefan-Boltzmann equation is introduced into (8-59), one obtains

$$Q_{12} = A_1F_{12}E_1\left[1 - \left(\frac{E_2}{E_1}\right)\right] = \sigma A_1F_{12}T_1^4\left[1 - \left(\frac{T_2}{T_1}\right)^4\right] \tag{8-60}$$

Note that when $(T_2/T_1) \ll 1$, equation (8-60) can be written

$$Q_{12} = \sigma A_1F_{12}T_1^4 \tag{8-61}$$

[3] H. C. Hottel, ch. 3 of *Heat Transmission*, by W. H. McAdams, McGraw-Hill, 1954.

For example, if $T_1 = 800°$ F and $T_2 = 100°$ F, then $(T_2/T_1) = 0.44$ and the rate of radiant energy transfer calculated with (8-61) differs by only 3.9% from that calculated with (8-60).

For nonblack bodies, the calculations become more complicated. One simple case that can be analyzed is that of radiant heat transfer between infinite parallel plates. Since neither plate can see anything but the other plate, the geometrical factors are unity. However, the plates are not black and a part of the emitted radiation from each plate is reflected back and forth before it is finally absorbed. Analysis of this process shows that the net rate of interchange is given by the formula

$$Q_{12} = \sigma F_\epsilon A_1[T_2^4 - T_1^4] \tag{8-62}$$

where F_ϵ is an emissivity factor defined by the equation

$$F_\epsilon = \frac{\epsilon_1 \epsilon_2}{\dfrac{1}{\epsilon_1} + \dfrac{1}{\epsilon_2} - 1}$$

Of course, if one surface is at a predominately higher temperature than the other, the return radiation can be neglected. Then equation (8-61) would be written

$$Q_{12} = \sigma \epsilon_1 A_1 T_1^4 \tag{8-63}$$

The radiation heat transfer coefficient h_r is defined by the equation

$$Q = h_r A_1(T_1 - T_2) \tag{8-64}$$

which is analogous to the equation defining the convection heat transfer coefficient.

Example 8-2.

In this example the radiation heat transfer coefficient for the heat loss from molten polyethylene at 600° F into isothermal surroundings at 80° F is estimated. The emissivity of molten polyethylene at 600° F is taken as 0.90. Since the ratio (T_2/T_1) is about 0.5, the approximate formula (8-63) applies, and since the surface can "see" nothing but the surroundings to which it is radiating its heat the viewfactor F_{12} becomes unity. By combining (8-63) and (8-64), one obtains

$$h_r = \frac{\sigma \epsilon_1 T_1^4}{T_1 - T_2}$$

and the value of h_r is found to be 3.7 Btu/hr-ft²-deg F.

Many solid and liquid polymers are nearly opaque to infrared radiation, so that the incident radiation absorbed by the body is converted to heat at the surface. Some of the heat is immediately lost to the surroundings by convection and radiation. The remainder flows into the interior by conduction. The temperature distribution in a substance being heated by radiant energy, then, depends not only upon the radiant heat flux to it but also upon its thermal conductivity and its convective heat loss at the surface.

8-5 Convection Heat Transfer

A key element in many polymer processing operations is the transfer of heat from the surface of a solid or liquid polymer to a surrounding fluid, such as air or water. In some cases it is desirable to expedite this transfer, in others, to retard it. This section is concerned with methods of estimating values of the convection heat transfer coefficient for two special cases of particular interest in polymer processing.

Heat loss by natural convection from a single horizontal cylinder is considered first. This is important in the extrusion of pipe and other shapes. Next, heat loss through a laminar boundary layer moving parallel to a plane surface is considered. This is important in film extrusion processes.

The convection heat transfer coefficient is usually expressed in dimensionless form by the Nusselt number

$$N_{\mathrm{Nu}} = \frac{hL}{k} \tag{8-65}$$

where L is some characteristic length of the surface and k is the thermal conductivity of the fluid.

The Nusselt number for heat transfer by natural convection from horizontal cylinders can be calculated with the following equation:

$$N_{\mathrm{Nu}} = 0.52\,[N_{\mathrm{Pr}} \cdot N_{\mathrm{Gr}}]^{1/4} \qquad (N_{\mathrm{Pr}} \cdot N_{\mathrm{Gr}} > 10^4) \tag{8-66}$$

where N_{Pr} represents the Prandtl number and N_{Gr} the Grashof number. These dimensionless groups are defined by the equations

$$N_{\mathrm{Pr}} = \left(\frac{c_p\mu}{k}\right) \tag{8-67}$$

$$N_{\mathrm{Gr}} = \frac{g\kappa\rho^2L^3\Delta T}{\mu^2} \tag{8-68}$$

Equation (8-66) is applicable when the product of the Nusselt and Prandtl numbers exceeds about 10^4. Note that the Prandtl number depends only upon the fluid properties, while the Grashof number depends, in addition, upon the temperature difference ΔT between the surface and the bulk of the fluid.

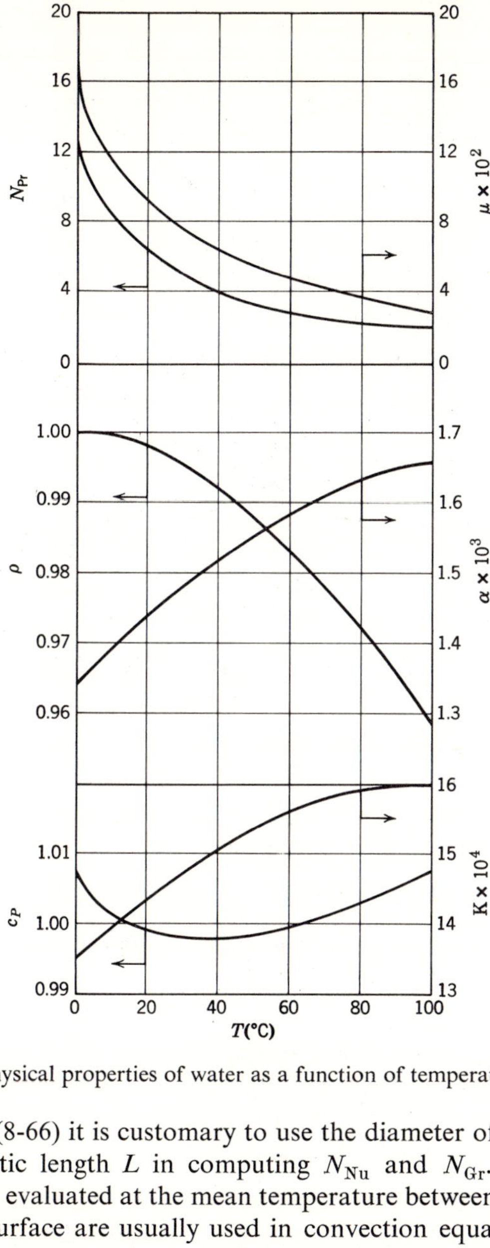

Figure 8-7. Physical properties of water as a function of temperature (cgs units).

In equation (8-66) it is customary to use the diameter of the cylinder as the characteristic length L in computing N_{Nu} and N_{Gr}. Furthermore, fluid properties evaluated at the mean temperature between the bulk of the fluid and the surface are usually used in convection equations. In most polymer processing operations, the fluids involved are either water or air. The physical properties of these fluids are shown in Figures 8-7 and 8-8 as a function of temperature.

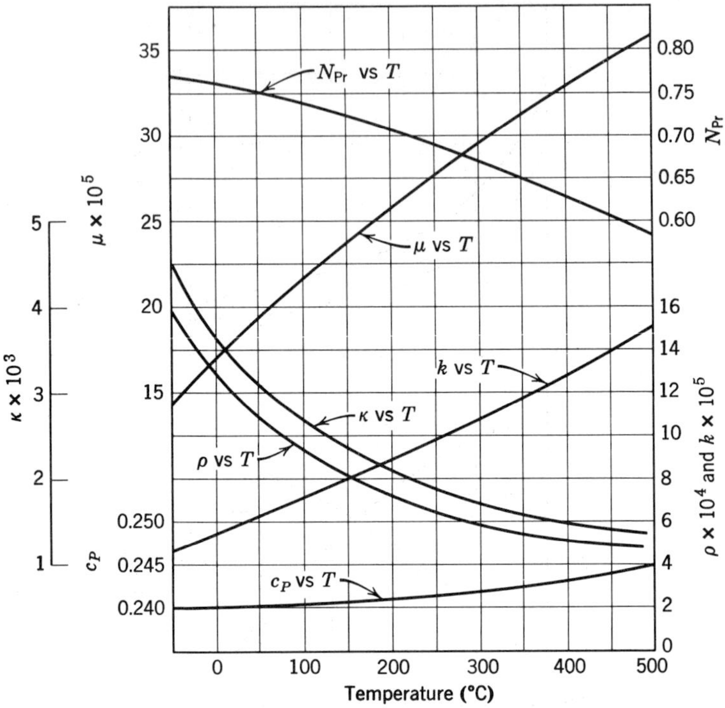

Figure 8-8. Physical properties of air as a function of temperature (cgs units).

Example 8-3.

Calculate the rate of heat loss from a horizontal cylinder having a diameter of 2 cm to air at 30° C when the surface of the cylinder is 170° C.

The value of ΔT is 140° C and the mean temperature is 100° C. At this temperature the physical properties of air, obtained from Figure 8-8, are as follows:

$$N_{Pr} = 0.74$$
$$\mu = 21.5 \times 10^{-5} \text{ poise}$$
$$\rho = 9.4 \times 10^{-4} \text{ grams/cm}^3$$
$$k = 10.8 \times 10^{-5} \text{ cal/cm-sec-deg C}$$
$$c_p = 0.24 \text{ cal/gram-deg C}$$
$$\kappa = 2.68 \times 10^{-3} \text{ deg C}^{-1}$$

The Grashof number, calculated with equation (8-68), is 4.24×10^4. Since the product ($N_{Gr}\, N_{Pr}$) exceeds 10^4, equation (8-66) is applicable. The Nusselt number is found to be 6.93 and the convection heat transfer coefficient h, calculated with (8-65), is 37.5×10^{-5} cal/cm²-sec-deg C (2.8 Btu/hr-ft²-deg F).

The Nusselt number for heat transfer from a plane surface to a fluid in forced laminar flow can be estimated with the equation

$$N_{\mathrm{Nu}} = 0.66\, N_{\mathrm{Re}}^{1/2} N_{\mathrm{Pr}}^{1/3} \tag{8-69}$$

when the Reynolds number N_{Re} is less than about 300,000. The Reynolds number is defined by the equation

$$N_{\mathrm{Re}} = \frac{LV\rho}{\mu} \tag{8-70}$$

where V is the bulk velocity of the fluid relative to the surface and the characteristic length L is taken as the length of the surface parallel to the direction of flow.

Problems

8-1 Using the method of the Laplace transformation, solve the boundary value problem for the heating of a semi-infinite slab, where the slab is initially at the uniform temperature T_i and where the surface of the slab jumps instantaneously to T_f and remains at this temperature. Compare the result with equation (8-49).

8-2 Using the method of separation of variables, solve the boundary value problem for the heating of a slab of thickness $2W$ whose initial temperature is uniform at T_i and which has isothermal boundaries at temperature T_f. Compare the result with equation (8-54).

8-3 A slab having a thickness of 2 cm and an initial uniform temperature of 100° C is immersed in a fluid having a temperature of 20° C. Calculate the temperature of the mid-plane and the surface after 100 seconds for the case where

$$\alpha = 10^{-3}\ \mathrm{cm^2/sec}$$
$$(hW/k) = 5$$

8-4 Derive an equation for the cooling of a long cylinder of radius R by convection heat transfer to a fluid at temperature T_f, where the initial uniform temperature of the cylinder is T_i. Compare the result with that given by Carslaw and Jaeger (p. 202 of *Conduction of Heat in Solids*, Oxford, Clarendon Press, 1959):

$$\frac{T - T_i}{T_f - T_i} = 1 - 2\left(\frac{hR}{k}\right) \sum_{n=1}^{\infty} \exp\left(-\frac{\alpha\beta_n{}^2 t}{R^2}\right) \frac{J_0(\beta_n r/R)}{[\beta_n{}^2 + (hR/k)^2]J_0(\beta_n)}$$

where the β_n are the roots of the equation

$$\beta J_1(\beta) - \frac{hR}{k} J_0(\beta) = 0$$

8-5 Show that for an isothermal surface the above result reduces to

$$\frac{T - T_f}{T_i - T_f} = 2 \sum_{n=1}^{\infty} \frac{e^{-\alpha\beta_n{}^2 t/R^2}}{\beta_n} \frac{J_0(\beta_n r/R)}{J_1(\beta_n)}$$

where the β_n are now the roots of the equation

$$J_0(\beta) = 0$$

8-6 A cylinder having a radius of 1 cm and an initial uniform temperature of 100° C is immersed in a fluid having a temperature of 20° C. Calculate the temperature at the axis of the cylinder and at the surface after 100 seconds for the case where

$$\alpha = 10^{-3}\ \text{cm}^2/\text{sec} \qquad (hR/k) = 5$$

8-7 Derive an equation for the cooling of a sphere of radius R by convective heat transfer to a fluid at temperature T_f, where the initial uniform temperature of the sphere is T_i. Compare the result with that given by Carslaw and Jaeger (p. 238 of *Conduction of Heat in Solids*, Oxford, Clarendon Press, 1959).

$$\frac{T - T_f}{T_i - T_f} = 2\left(\frac{hR}{k}\right)\left(\frac{R}{r}\right)\sum_{n=1}^{\infty}\frac{\beta_n^{\,2} + (N-1)^2}{\beta_n^{\,2}[\beta_n^{\,2} + N(N-1)]}\exp\left(-\frac{\alpha\beta_n^{\,2}t}{R^2}\right)\sin(\beta_n)\sin(\beta r/R)$$

where the β_n are roots of the equation

$$\beta\cot(\beta) + N - 1 = 0$$

and

$$N = \left(\frac{hR}{k}\right)$$

8-8 A sphere having a radius of 1 cm and an initial uniform temperature of 100° C is immersed in a fluid having a temperature of 20° C. Calculate the temperature at the center of the sphere and at the surface 100 seconds after immersion for the case where

$$\alpha = 10^{-3}\ \text{cm}^2/\text{sec} \qquad (hR/k) = 5$$

8-9 Derive the following expression for the emissivity factor F_ϵ which applies to radiant heat transfer between infinite parallel plates:

$$F_\epsilon = \frac{\epsilon_1\epsilon_2}{1/\epsilon_1 + 1/\epsilon_2 - 1}$$

8-10 Consider the temperature distribution in a flat sheet of thickness $2W$ heated from both sides by infrared radiation. Assume that convective heat transfer is negligible, that the sheet is opaque to infrared radiation, and that the heat flux across the surface is independent of the surface temperature. If T_s represents the temperature at the surface of the sheet and T_c the temperature at the center, it has been shown that

$$\frac{T_s - T_i}{T_c - T_i} = f(\phi)$$

where

$$\phi = \frac{\alpha t}{W^2}$$

T_i represents the initial uniform temperature of the sheet and t represents time. Find the form of the function $f(\phi)$. (See p. 104 of *Processing of Thermoplastic Materials*, edited by E. C. Bernhardt, Reinhold Publishing Corp., 1959.)

chapter 9

Hydrodynamic Analysis of Calenders

9-1 Introduction

Calendering is a continuous forming operation utilizing a pair of driven rolls that form a mass of softened material into a sheet of uniform thickness. The operation is shown schematically in Figure 9-1.

Calenders produce film or sheeting from polymeric materials continuously, at high speeds, and with uniform thicknesses controlled to close tolerances. According to Marshall,[1] many thermoplastic materials can be calendered at speeds up to 300 ft/min and with widths up to 6 ft. Films as thin as 2 mils or less can be produced with thickness variations of less than 0.1 mil.

When thermoplastic materials are calendered, the key process occurring

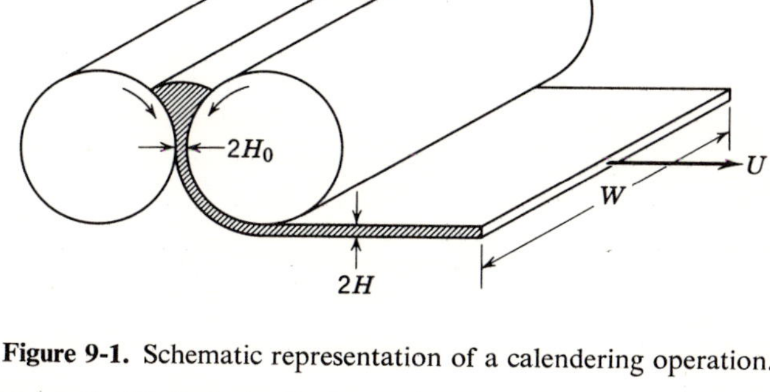

Figure 9-1. Schematic representation of a calendering operation.

[1] D. I. Marshall, "Calendering," ch. 6 of *Processing of Thermoplastic Materials*, edited by E. C. Bernhardt, Reinhold Publishing Corp., 1959.

in the fluid is laminar viscous flow. By making certain simplifying assumptions, it is possible to make a mathematical analysis of the operation. A complete analysis would enable one to solve the following problem: Given the physical properties of the material being calendered, the dimensions of the rolls, their speed, and the gap distance, calculate the rate at which material is drawn through the rolls, the thickness of the film produced by the operation, the temperature distribution in the calendered material, the force tending to separate the rolls, and the power required to operate the calender.

9-2 Simplified Flow Theory

The analysis presented here, which is based upon that of Gaskell,[2] provides a partial solution to the above problem. It ignores the elastic nature of thermoplastic materials and considers the problem to be one of steady isothermal flow of an incompressible Newtonian fluid. A rectangular coordinate system with the axes oriented, as shown in Figure 9-2, is used. The problem is to determine the v_x and v_y components of the velocity vector and the pressure P as a function of the position variables x and y.

Because of incompressibility, the equation of continuity reduces to

$$\frac{\partial v_x}{\partial x} + \frac{\partial v_y}{\partial y} = 0 \tag{9-1}$$

and the x-component of the momentum equation becomes

$$\rho\left[v_x\left(\frac{\partial v_x}{\partial x}\right) + v_y\left(\frac{\partial v_x}{\partial y}\right)\right] = -\left(\frac{\partial P}{\partial x}\right) + \mu\left[\frac{\partial^2 v_x}{\partial x^2} + \frac{\partial^2 v_x}{\partial y^2}\right] \tag{9-2}$$

where the stress components have been replaced with the appropriate values of the velocity gradients given in Table 2-3.

A calender having equal sized rolls rotating at the same speed (the symmetrical calender) is considered here. The radius of the rolls is R,

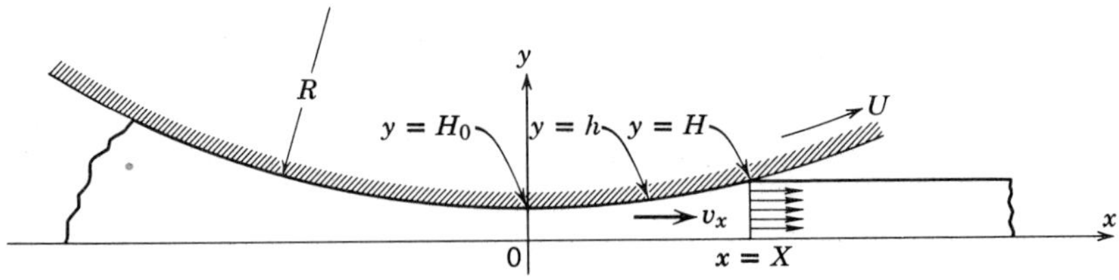

Figure 9-2. Notation for analysis of flow in the symmetrical calender.

[2] R. E. Gaskell, *J. Appl. Mech.*, **17**, 334 (1950).

their surface speed is U, and their separation at the nip is $2H_0$. Because of symmetry about the x-axis, it is necessary to consider only that region where the y-position variable is positive.

Certain simplifying assumptions are now made to reduce equation (9-2) to a manageable form. The left side of (9-2) represents the acceleration term. Because the flow is relatively slow and because viscous forces are large, it can be ignored. Furthermore, it is assumed that the variation in v_x is much less in the x-direction than in the y-direction. Hence the derivatives of v_x with respect to x can be ignored, leaving only derivatives with respect to y. This is reasonable if the boundaries of the fluid approximate parallel plates. Such is the case when H_0 is small in comparison to R and when the equations are applied only to the immediate region of the nip.

With these simplifying assumptions, equation (9-2) reduces to

$$\left(\frac{\partial P}{\partial x}\right) = \mu\left(\frac{d^2 v_x}{dy^2}\right) \tag{9-3}$$

The additional simplifying assumption, that the hydrostatic pressure varies only in the x-direction, is made.

Upon integrating once, equation (9-3) becomes

$$\frac{dv_x}{dy} = \dot{\gamma} = \frac{1}{\mu}\left(\frac{\partial P}{\partial x}\right)y + C_1 \tag{9-4}$$

where C_1 is the constant of integration. The condition $\dot{\gamma}(0) = 0$, arising because of the symmetry of flow about the x-axis, requires that $C_1 = 0$, so that after a second integration, equation (9-4) becomes

$$v_x = \frac{y^2}{2\mu}\left(\frac{\partial P}{\partial x}\right) + C_2 \tag{9-5}$$

The constant of integration C_2 is evaluated from the condition $v_x(h) = U$, and equation (9-5) becomes

$$v_x = U + \frac{y^2 - h^2}{2\mu}\left(\frac{\partial P}{\partial x}\right) \tag{9-6}$$

where h is the distance from the plane of symmetry to the roll surface. Because of the curvature of the rolls, h is a function of the x position variable.

The volumetric rate of flow through the rolls per unit width is given by the integral

$$Q = 2\int_0^h v_x\,dy \tag{9-7}$$

If equation (9-6) is introduced into (9-7) and the integration performed, the following expression for the volumetric flow rate is obtained:

$$Q = 2h\left[U - \frac{h^2}{3\mu}\left(\frac{\partial P}{\partial x}\right)\right] \tag{9-8}$$

It is now convenient to make a change of variable. A dimensionless independent variable ρ, defined by the equation

$$\rho = \frac{x}{\sqrt{2RH_0}} \tag{9-9}$$

is introduced into equation (9-8), which is then rearranged to give the following explicit expression for the gradient:

$$\frac{dP}{d\rho} = \sqrt{2RH_0}\left(\frac{3\mu}{h^2}\right)\left(U - \frac{Q}{2h}\right) \tag{9-10}$$

The quantity h is a function of the x position coordinate, being related to it through the equation

$$h = H_0 + R - \sqrt{R^2 - x^2} \tag{9-11}$$

A useful approximation to equation (9-11) is obtained by expanding the third term on the right in binomial series and then dropping all but the first two terms. One obtains

$$h = H_0 + \left(\frac{x^2}{2R}\right) \tag{9-12}$$

and it can be seen from equations (9-12) and (9-9) that the quantity h is related to the dimensionless variable ρ through the equation

$$\frac{h}{H_0} = 1 + \rho^2 \tag{9-13}$$

Equation (9-10) is now transformed to

$$\frac{dP}{d\rho} = \frac{\mu U}{H_0}\sqrt{\frac{18R}{H_0}}\left[\frac{\rho^2 - \lambda^2}{(1 + \rho^2)^3}\right] \tag{9-14}$$

where the parameter λ is defined by the equation

$$\lambda^2 = \frac{Q}{2UH_0} - 1 \tag{9-15}$$

which can be written in the equivalent form

$$\lambda^2 = \frac{X^2}{2RH_0} \tag{9-16}$$

showing that λ is the value of the variable ρ at the point $x = X$, the point where the sheet loses contact with the rolls.

Upon integration, equation (9-14) becomes

$$P = \frac{\mu U}{H_0}\sqrt{\frac{9R}{32H_0}}\,[g(\rho, \lambda) + C] \tag{9-17}$$

where the function $g(\rho, \lambda)$ is defined by the equation

$$g(\rho, \lambda) = \left[\frac{\rho^2 - 1 - 5\lambda^2 - 3\lambda^2\rho^2}{(1 + \rho^2)^2}\right]\rho + (1 - 3\lambda^2)\tan^{-1}\rho \tag{9-18}$$

The constant of integration C, evaluated from the boundary condition

$$P(\lambda) = 0$$

is given by the equation

$$C = \left(\frac{1 + 3\lambda^2}{1 + \lambda^2}\right)\lambda - (1 - 3\lambda^2)\tan^{-1}\lambda \tag{9-19}$$

for which

$$C = 5\lambda^3$$

is a very good approximation in the region of interest.

Equation (9-14) shows that the slope of the pressure profile is zero at the points

$$\rho = \pm\lambda$$

with the point $\rho = -\lambda$ representing the point of maximum pressure.

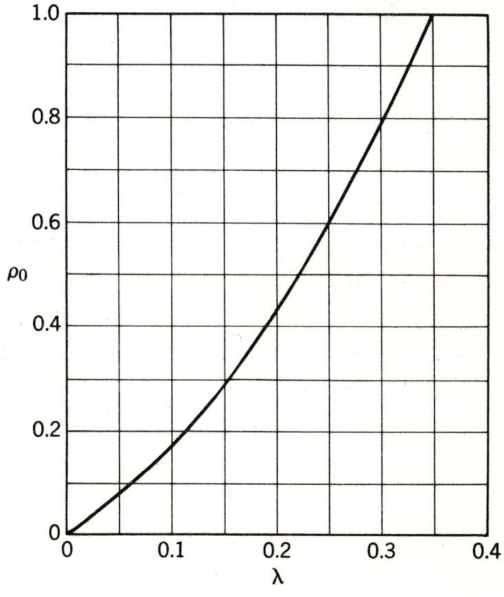

Figure 9-3. Graph showing the relationship between λ and ρ_0.

Because the function $g(\rho, \lambda)$ can be either positive or negative, equation (9-17) has two roots of interest $\rho = -\rho_0$ and $\rho = +\lambda$. At these points, equation (9-17) shows that the following conditions must exist:

$$g(\lambda, \lambda) = -C(\lambda) = -5\lambda^3 \tag{9-20}$$

$$g(-\rho_0, \lambda) = -C(\lambda) = -5\lambda^3 \tag{9-21}$$

from which it can be seen that

$$g(\lambda, \lambda) = g(-\rho_0, \lambda) \tag{9-22}$$

Therefore there exists a unique functional relationship, shown graphically in Figure 9-3, between λ and ρ_0.

9-3 Velocity and Pressure Profiles

An expression for the maximum pressure is obtained from equation (9-17) by making the substitution $\rho = -\lambda$. Because $g(\rho, \lambda)$ is an odd function

$$g(-\lambda, \lambda) = -g(\lambda, \lambda) \tag{9-23}$$

and because of equation (9-20), the expression for maximum pressure becomes

$$P_{max} = P(-\lambda) = \frac{5\mu U\lambda^3}{H_0}\sqrt{\frac{9R}{8H_0}} \tag{9-24}$$

The pressure ratio (P/P_{max}), obtained by dividing (9-17) by (9-24), is

$$\frac{P}{P_{max}} = \frac{P(\rho)}{P(-\lambda)} = \frac{1}{2}\left[1 + \frac{g(\rho, \lambda)}{5\lambda^3}\right] \tag{9-25}$$

For the special case where $\rho = 0$

$$g(0, \lambda) = 0$$

and equation (9-25) reduces to

$$\frac{P}{P_{max}} = \frac{1}{2}$$

which shows that the pressure developed at the nip of the rolls is exactly half of the maximum pressure.

The general nature of the pressure profile obtained when (P/P_{max}) is plotted against ρ is determined by the behavior of the function $g(\rho, \lambda)$. The following key points of the profile are readily established:

at $\rho = -\rho_0$,	$g(-\rho_0, \lambda) = -5\lambda^3$	and	$(P/P_{max}) = 0$
at $\rho = -\lambda$,	$g(-\lambda, \lambda) = +5\lambda^3$	and	$(P/P_{max}) = 1$
at $\rho = 0$,	$g(0, \lambda) = 0$	and	$(P/P_{max}) = 1/2$
at $\rho = +\lambda$,	$g(\lambda, \lambda) = -5\lambda^3$	and	$(P/P_{max}) = 0$

If equation (9-17) is written

$$P = 10\left[\frac{\mu U}{H_0}\sqrt{\frac{9R}{32H_0}}\right]\frac{\lambda^3}{2}\left[1 + \frac{g(\rho, \lambda)}{5\lambda^3}\right] \tag{9-26}$$

then, because of equation (9-25), it reduces to

$$P = 10\frac{\mu U}{H_0}\sqrt{\frac{9R}{32H_0}}\,\lambda^3\left(\frac{P}{P_{max}}\right) \tag{9-27}$$

The effect of the quantity λ on the pressure profile is illustrated by holding quantities μ, U, R, and H_0 constant. Then equation (9-27) can be written

$$P = K\lambda^3\left(\frac{P}{P_{max}}\right) = KP' \tag{9-28}$$

where P' is called the relative pressure.

Figure 9-4 shows a plot of relative pressure P' versus ρ for three values of λ. Note that when λ increases, the profile broadens and the maximum pressure increases. Because of the third power dependence of P_{max} on λ, small variations in λ cause large variations in P_{max}. For example, if λ is doubled, there is an eightfold increase in P_{max}. If λ should increase from 0.28 to 0.32, P_{max} would increase nearly 50%.

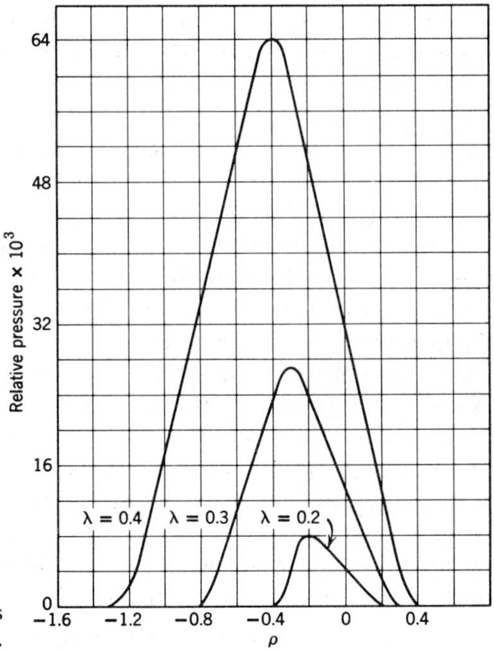

Figure 9-4. Relative pressure profiles for various values of the parameter λ.

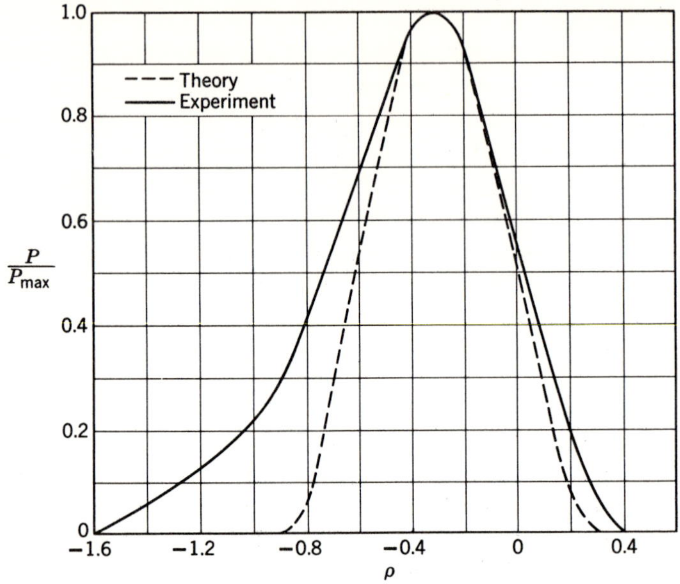

Figure 9-5. Comparison of theoretical and experimental pressure profiles. (Experimental data from Ref. 3).

Although λ has a profound effect on the shape of the pressure profile, the hydrodynamic theory of calendering does not provide an independent means of establishing its value. It must be considered to be an experimentally determined parameter. If a pressure profile is available, then it can be determined from either of the zero pressure points, the point of maximum pressure, or the nip pressure, depending upon which gives the best over-all fit of the data.

Bergen and Scott[3] measured pressure profiles (with a strain gage pressure transducer embedded in the surface of one of the rolls) in polymeric materials being calendered on 10-inch diameter rolls. Figure 9-5 shows the measured pressure ratio (P/P_{max}) as a function of ρ for an experiment with a plasticized thermoplastic resin. The theoretical curve is also shown in the figure. It was constructed by choosing for the value of the parameter λ, the value of the variable ρ at the point of maximum pressure. Hence the theoretical and experimental curves coincide at this point.

In the region ($\rho > -\lambda$), the agreement between theory and experiment is excellent. For ($\rho < -\lambda$), the theoretical curve falls below the experimental. This must be due, at least partially, to the non-Newtonian nature of the fluid used in the experiment. The theory requires the viscosity to be

[3] J. T. Bergen and G. W. Scott, *J. Appl. Mech.*, **18**, 101 (1951).

the same at all points, but the viscosity of the fluid used in the experiment, being pseudoplastic, is a minimum in the nip region. Hence pressure build-up in the fluid must start sooner than required by Newtonian flow theory because its viscosity diminishes as it approaches the nip.

The following expression for the velocity component v_x is obtained by combining equations (9-6) and (9-14), transforming x to ρ and then rearranging:

$$\frac{v_x}{U} = \frac{2 + 3\lambda^2(1 - \eta^2) - \rho^2(1 - 3\eta^2)}{2(1 + \rho^2)} \tag{9-29}$$

The dimensionless variable η is defined by the equation

$$\eta = \frac{y}{h}$$

Equation (9-29) shows that the relative velocity (v_x/U) is a function of the position coordinate (ρ, η) and of the parameter λ.

Figure 9-6 shows velocity profiles calculated for the case where $\lambda^2 = 0.10$. Note that when $\rho = \pm\lambda$ the right side of equation (9-29) reduces to unity and $v_x = U$. The velocity profile is therefore flat at the point of maximum pressure and also at the point where the material loses contact with the roll.

In the region $(-\lambda < \rho < +\lambda)$, the pressure gradient is negative and there is a pressure flow in the forward direction. The velocity profile is therefore convex in this region. In the region $(\rho < -\lambda)$, the positive pressure gradient retards the forward motion of the fluid and the velocity profile is concave. As ρ decreases, a point is eventually reached where the velocity at the mid-plane, where $\eta = 0$, will be zero. This *stagnation point*, designated ρ^*, is found by placing the left side of equation (9-29) equal to zero. One obtains the expression

$$(\rho^*)^2 - 3\lambda^2 - 2 = 0$$

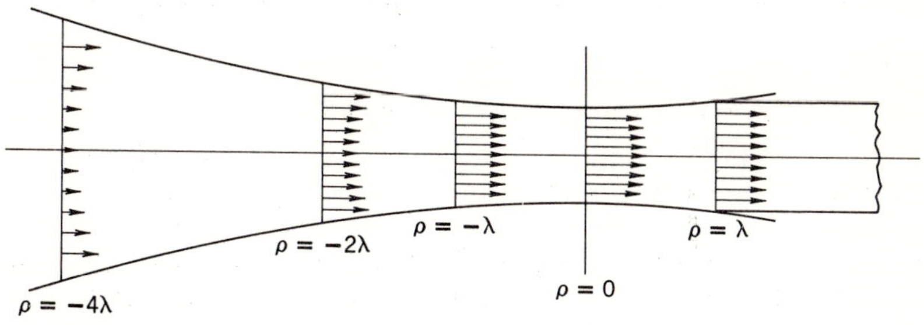

Figure 9-6. Velocity profiles calculated for $\lambda^2 = 0.10$.

showing that ρ^* is a function of the parameter λ. When $\lambda^2 = 0.1$, the stagnation point occurs at $\rho^* = -4.8\lambda$.

In the region ($\rho < \rho^*$), the velocity components have positive and negative values. Toward the mid-plane, fluid moves away from the nip of the rolls because the velocity component is negative. Near the roll surfaces, the velocity component is positive and fluid moves toward the nip. The result is a partial circulation of fluid in the region ($\rho < \rho^*$). In the region ($\rho > \rho^*$), the velocity component has positive values only and there is no circulation.

The shear rate in the fluid, obtained by differentiating (9-29), is given by the equation

$$\dot{\gamma} = \frac{\partial v_x}{\partial y} = \frac{1}{h}\left(\frac{\partial v_x}{\partial \eta}\right) = \frac{3U\eta}{h}\left[\frac{\rho^2 - \lambda^2}{(1 + \rho^2)}\right] \tag{9-30}$$

which, because of equation (9-13), can be written

$$\dot{\gamma} = \frac{3U\eta}{H_0}\left[\frac{\rho^2 - \lambda^2}{(1 + \rho^2)^2}\right] \tag{9-31}$$

At the roll surface $\eta = 1$ and the shear rate $\dot{\gamma}_h$ is

$$\dot{\gamma}_h = \frac{3U}{H_0}\left[\frac{\rho^2 - \lambda^2}{(1 + \rho^2)^2}\right] \tag{9-32}$$

The power p required to drive both rolls is given approximately by

$$p = 2UW\mu \int \dot{\gamma}_h \, dx \tag{9-33}$$

where W is the axial length of the rolls. Equation (9-33) becomes

$$p = 2UW\mu\sqrt{2RH_0}\int_{-\rho_0}^{\lambda} \dot{\gamma}_h \, d\rho \tag{9-34}$$

when transformed to the dimensionless variable ρ.

Introducing equation (9-32) into the above integral and performing the indicated integration gives the following result for the total power:

$$p = 3WU^2\mu\sqrt{2R/H_0}\, f(\lambda) \tag{9-35}$$

where the function $f(\lambda)$ is defined by the equation

$$f(\lambda) = (1 - \lambda^2)[\tan^{-1}\lambda + \tan^{-1}\rho_0] - \left[\frac{(\lambda + \rho_0)(1 + \rho_0\lambda)}{(1 + \rho_0^2)}\right] \tag{9-36}$$

The function $f(\lambda)$, shown graphically in Figure 9-7, depends only upon the parameter λ because the quantity ρ_0 is itself a unique function of λ.

The hydrostatic pressure of the fluid between the rolls exerts a thrust on them that tends to separate them. This roll-separating force is given

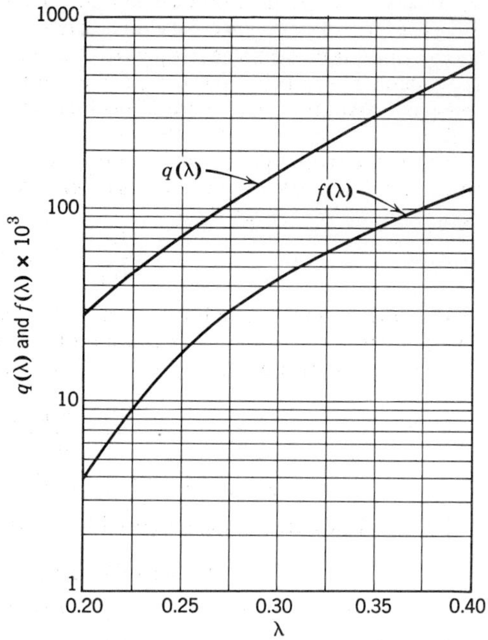

Figure 9-7. Graphs showing the relationship between $q(\lambda)$ and λ. Also the relationship between $f(\lambda)$ and λ.

by the integral

$$F = W\sqrt{2RH_0}\int_{-\rho_0}^{\lambda} P\,d\rho \tag{9-37}$$

Upon introducing the pressure expression, equation (9-17), into the above integral and performing the indicated integration, one obtains

$$F = \left(\frac{3\mu URW}{4H_0}\right)q(\lambda) \tag{9-38}$$

where the function $q(\lambda)$ is defined by the equation

$$q(\lambda) = \left(\frac{\lambda + \rho_0}{1 + \rho_0^2}\right)[\rho_0 - \lambda + 5\lambda^3(1 + \rho_0^2)] + (1 - 3\lambda^2)[\lambda \tan^{-1}\lambda - \rho_0 \tan^{-1}\rho_0]$$

The function $q(\lambda)$ is also plotted in Figure 9-7.

Example 9-1.

A calender having the dimensions

$$R = 10 \text{ cm (3.94 in.)}$$
$$W = 100 \text{ cm (39.4 in.)}$$
$$H_0 = 0.01 \text{ cm (3.9 mils)}$$

operating at a speed of 40 cm/sec (78.8 ft/min) on a material having a viscosity 10^4 poise, produces a sheet having a thickness of 0.0218 cm (8.6 mils). Calculate the maximum pressure developed in the material, the power required to operate the calender, the roll-separating force, and the average temperature rise of the material, assuming that there is no heat transfer between the material and the rolls.

The value of the parameter λ can be determined in this example because the thickness of the calendered sheet is given. At the point of separation, $h = H$ and equation (9-13) reduces to

$$\frac{H}{H_0} = 1 + \lambda^2$$

where H represents the half thickness of the calendered sheet and H_0 is half of the roll separation at the nip. The value of λ is found to be 0.30.

The maximum pressure (at $\rho = -\lambda$) calculated with (9-24) is

$$\begin{aligned} P_{\max} &= \frac{(5)(10^4)(40)(0.3)^3}{(0.01)} \sqrt{\frac{(9)(10)}{(8)(0.01)}} \\ &= 181 \times 10^6 \text{ dynes/cm}^2 \text{ (2660 psi)} \end{aligned}$$

The total power, as calculated with (9-35), is

$$\begin{aligned} p &= (3)(100)(1600)(10^4)\sqrt{(2)(10)/(0.01)}\,(0.043) \\ &= 922 \times 10^7 \text{ ergs/sec (0.92 kw)} \end{aligned}$$

The roll-separating force, as calculated with (9-37), is

$$\begin{aligned} F &= \frac{(3)(10^4)(40)(10)(100)(0.16)}{(4)(0.01)} \\ &= 0.48 \times 10^{10} \text{ dynes (1080 pounds force)} \end{aligned}$$

The volumetric calendering rate Q is

$$\begin{aligned} Q &= 2UHW = (2)(40)(0.0109)(100) \\ &= 87.2 \text{ cm}^3\text{/sec} \end{aligned}$$

and if the material has a density of 1 gram/cm^3, the mass rate is 87.2 grams/sec (690 lb/hr). Taking the specific heat of the material as 0.5 cal/gram-deg C, the average temperature rise is

$$\overline{\Delta T} = \frac{(0.92)(239)}{(87.2)(0.5)} = 5 \text{ deg C}$$

9-4 Non-Newtonian Flow Theory

The hydrodynamic analysis of calendering developed in the previous sections is now extended, following the method of Gaskell,[2] to include non-Newtonian flow. We start with the x-component of the momentum equation, which reduces to

$$\frac{\partial P}{\partial x} = \frac{\partial \tau_{yx}}{\partial y} = \frac{\partial}{\partial y}\left[\eta \frac{\partial v_x}{\partial y}\right] \tag{9-39}$$

after making the same simplifying assumptions as before, except that now the non-Newtonian viscosity η is considered to be a function of the state of shear. Equation (9-39) is the non-Newtonian analog of (9-9).

Upon integration with respect to y, equation (9-39) becomes

$$py = \tau = \eta\left(\frac{\partial v_x}{\partial y}\right) \tag{9-40}$$

where

$$p = \left(\frac{\partial P}{\partial x}\right) \tag{9-41}$$

The constant of integration in equation (9-40) is zero because the flow is symmetrical about the x-axis and therefore $\tau(0) = 0$. The subscript on the stress component has been dropped to simplify the notation.

Eliminating the variable y between the equations (9-40) gives

$$dv_x = \frac{1}{p}\left(\frac{\tau}{\eta}\right) d\tau \tag{9-42}$$

which, upon integration, becomes

$$v_x = U - \frac{1}{p}\int_{\tau}^{\tau_w}\left(\frac{\tau}{\eta}\right) d\tau \tag{9-43}$$

The volumetric flow rate Q is given by (9-7). Upon transformation of the variable y to τ through (9-40), it becomes

$$Q = \frac{2}{p}\int_0^{\tau_w} v_x \, d\tau \tag{9-44}$$

which is a general relationship independent of the nature of the fluid. If equation (9-43) is introduced into (9-44), one obtains

$$Q = \frac{2U_x}{p}\tau_w - \frac{2}{p^2}\int_0^{\tau_w}\int_{\tau}^{\tau_w}\left(\frac{\tau}{\eta}\right) d\tau \, d\tau \tag{9-45}$$

From equation (9-40) it can be seen that $\tau_w = ph$ and, because of (9-13), we have

$$\tau_w = pH_0(1 + \rho^2) \tag{9-46}$$

so that equation (9-45) can be written

$$Q = 2UH_0(1 + \rho^2) - \frac{2}{p^2}\int_0^{\tau_w}\int_{\tau}^{\tau_w} \frac{\tau}{\eta} \, d\tau \, d\tau \tag{9-47}$$

The volumetric flow rate can also be expressed in terms of H_0 by the equation

$$Q = 2UH_0(1 + \lambda^2) \tag{9-48}$$

so that Q can be eliminated between the above equations to yield

$$\rho^2 - \lambda^2 = \frac{1}{UH_0p^2}\int_0^{\tau_w}\int_\tau^{\tau_w}\frac{\tau}{\eta}\,d\tau\,d\tau \tag{9-49}$$

It is shown below that

$$\int_0^{\tau_w}\int_\tau^{\tau_w} f(\tau)\,d\tau\,d\tau = \int_0^{\tau_w}\tau f(\tau)\,d\tau \tag{9-50}$$

Hence equation (9-49) reduces to

$$\rho^2 - \lambda^2 = \frac{1}{UH_0p^2}\int_0^{\tau_w}\frac{\tau^2}{\eta}\,d\tau \tag{9-51}$$

To prove equation (9-50), note that

$$I = \int_0^{\tau_w}\int_\tau^{\tau_w} f(\tau)\,d\tau\,d\tau = \int_0^{\tau_w}\left[\int_0^{\tau_w} f(\tau)\,d\tau - \int_0^{\tau} f(\tau)\,d\tau\right]d\tau$$

$$= \tau_w\int_0^{\tau_w} f(\tau)\,d\tau - \int_0^{\tau_w}\int_0^{\tau} f(\tau)\,d\tau\,d\tau \tag{9-52}$$

Now transform the double integral of equation (9-52) into a single integral, so that

$$I = \tau_w\int_0^{\tau_w} f(\tau)\,d\tau - \int_0^{\tau_w}(\tau_w - \tau)f(\tau)\,d\tau = \int_0^{\tau_w}\tau f(\tau)\,d\tau$$

which is equation (9-50).

The validity of the transformation made on the double integral can be demonstrated with the Laplace transformation. Define $g(\tau_w)$ with the equation

$$g(\tau_w) = \int_0^{\tau_w}(\tau_w - \tau)f(\tau)\,d\tau$$

Applying the convolution theorem,[4] the Laplace transform of the above equation is

$$\mathscr{L}g(\tau_w) = \frac{1}{s^2}\mathscr{L}f(\tau) = \frac{F(s)}{s^2}$$

and because the inverse transform of the function $F(s)/s^2$ can be written

$$\mathscr{L}^{-1}\frac{F(s)}{s^2} = g(\tau_w) = \int_0^{\tau_w}\int_0^{\tau} f(\tau)\,d\tau\,d\tau$$

we see that

$$\int_0^{\tau_w}\int_0^{\tau} f(\tau)\,d\tau\,d\tau = \int_0^{\tau_w}(\tau_w - \tau)f(\tau)\,d\tau$$

---

[4] R. S. Churchill, *Modern Operational Mathematics in Engineering*, McGraw-Hill, 1944, p. 37.

We now specialize the development for power law fluids. Table 2-4 shows that the viscosity of a power law fluid is given by the equation

$$\eta = \eta^\circ \left| \frac{\tau}{\tau^\circ} \right|^{(n-1)/n} \tag{9-53}$$

Introducing equation (9-53) into (9-51) and performing the indicated integration yields

$$\rho^2 - \lambda^2 = \pm \frac{n(\tau^\circ)^{(n-1)/n}}{(2n + 1)UH_0 p^2 \eta^\circ} |\tau_w|^{(2n+1)/n} \tag{9-54}$$

where the negative sign applies to the region ($\lambda^2 > \rho^2$).

Introducing equation (9-46) into (9-54) and solving for the pressure gradient p yields

$$p = \frac{dP}{d\rho} = K \frac{(\rho^2 - \lambda^2)^n}{(1 + \rho^2)^{2n+1}} \qquad (\rho^2 > \lambda^2) \tag{9-55}$$

$$= \frac{dP}{d\rho} = -K \frac{(\lambda^2 - \rho^2)^n}{(1 + \rho^2)^{2n+1}} \qquad (\lambda^2 > \rho^2) \tag{9-56}$$

where the coefficient K is defined by the equation

$$K = \left(\frac{2n + 1}{n}\right)^n \left(\frac{U\eta^\circ}{H_0\tau^\circ}\right)^n \left(\frac{\tau^\circ}{H_0}\right)$$

Example 9-2.

For the calendering process described in Example 9-1, compute the maximum pressure that would be developed in the calender if the fluid, instead of being a Newtonian fluid, were a power law fluid having a flow index of 0.50 and a viscosity η° of 10^4 poise at the standard shear rate of 1 sec^{-1}.

For $n = 0.5$, equation (9-56) reduces to

$$\frac{dP}{d\zeta} = -K\lambda^2 \frac{\sqrt{1 - \zeta^2}}{(1 + \lambda^2 \zeta^2)^2} \tag{9-57}$$

where the variable ζ is defined by the equation

$$\zeta = \frac{\rho}{\lambda}$$

and the coefficient K is given by

$$K = \frac{2\tau^\circ}{H_0} \sqrt{\frac{U\eta^\circ}{H_0\tau^\circ}} \tag{9-58}$$

Equation (9-58) can be written in the equivalent form

$$K = 2\eta^\circ \sqrt{\frac{U\dot{\gamma}^\circ}{H_0^3}}$$

by noting that in the standard state $\tau^\circ = \eta^\circ \dot{\gamma}^\circ$.

Equation (9-57) shows that the slope of the P versus ζ curve is zero at $\zeta = \pm 1$, with the maximum pressure occurring at $\zeta = -1$. This is the same result obtained for Newtonian fluids.

Upon integration of equation (9-57), one obtains

$$P = -K\lambda^2 \int_1^{\zeta} \frac{\sqrt{1 - \zeta^2}}{(1 + \lambda^2\zeta^2)^2}\, d\zeta = -K\lambda^2 \int_1^{\zeta} \phi\, d\zeta \tag{9-59}$$

where the function ϕ is defined by the equation

$$\phi = \frac{\sqrt{1 - \zeta^2}}{(1 + \lambda^2\zeta^2)^2}$$

The pressure at the nip (where $\zeta = 0$) is given by the expression

$$P_{\text{nip}} = -K\lambda^2 \int_1^0 \phi\, d\zeta = K\lambda^2 \int_0^1 \phi\, d\zeta \tag{9-60}$$

and the maximum pressure by

$$P_{\max} = -K\lambda^2 \int_1^{-1} \phi\, d\zeta = K\lambda^2 \int_{-1}^1 \phi\, d\zeta \tag{9-61}$$

Because the function ϕ is an even function (symmetrical about the y-axis), equation (9-61) can be written in the equivalent form

$$P_{\max} = 2K\lambda^2 \int_0^1 \phi\, d\zeta \tag{9-62}$$

and we see that

$$P_{\max} = 2P_{\text{nip}}$$

When $\lambda = 0.30$, the value of the integral

$$\int_0^1 \phi\, d\zeta$$

is about 0.71. This result can be obtained quickly by graphical integration. The coefficient K is given by the following calculation:

$$K = 2\eta^\circ \sqrt{\frac{U\dot{\gamma}^\circ}{H_0^3}} = 2 \times 10^4 \sqrt{\frac{(40)(1)}{10^{-6}}} = 12.6 \times 10^7 \text{ dynes/cm}^2$$

Hence the pressure at the nip, calculated with equation (9-60), is 8.08×10^6 dynes/cm^2 (118 psi). The maximum pressure developed by the calender is 236 psi, which, we see, is of the order of 10% of that developed with a Newtonian fluid having a viscosity of 10^4 poise.

Problems

9-1 By introducing equation (9-29) into the equation of continuity, find an expression for the v_y velocity component in the symmetrical calender.

9-2 The stream function ψ is defined by the equations

$$v_x = -\left(\frac{\partial \psi}{\partial y}\right) \qquad v_y = +\left(\frac{\partial \psi}{\partial x}\right)$$

Use equation (9-29) to find an expression for ψ. Streamlines are lines for which

$$\psi = \text{constant}$$

with the line $\psi = 0$ being a line of stagnation. Find the equations of any such lines in the symmetrical calender.

9-3 Show that an exact formulation of the flow problem for the calender (assuming constant ρ and μ) leads to the following pair of equations:

$$\rho\left[v_x\left(\frac{\partial v_x}{\partial x}\right) + v_y\left(\frac{\partial v_x}{\partial y}\right)\right] = -\left(\frac{\partial P}{\partial x}\right) + \mu\left[\frac{\partial^2 v_x}{\partial x^2} + \frac{\partial^2 v_x}{\partial y^2}\right]$$

$$\rho\left[v_x\left(\frac{\partial v_y}{\partial x}\right) + v_y\left(\frac{\partial v_y}{\partial y}\right)\right] = -\left(\frac{\partial P}{\partial y}\right) + \mu\left[\frac{\partial^2 v_y}{\partial x^2} + \frac{\partial^2 v_y}{\partial y^2}\right]$$

Then show that this pair of equations can be reduced to the following single equation in ψ:

$$\left(\frac{\partial \psi}{\partial x}\right)\left(\frac{\partial \nabla^2 \psi}{\partial y}\right) - \left(\frac{\partial \psi}{\partial y}\right)\left(\frac{\partial \nabla^2 \psi}{\partial x}\right) = \left(\frac{\mu}{\rho}\right)\left[\frac{\partial^4 \psi}{\partial x^4} + \frac{\partial^4 \psi}{\partial x^2\, \partial y^2} + \frac{\partial^4 \psi}{\partial y^4}\right]$$

9-4 Determine the shear stress at the roll surface for the symmetrical two roll calender as a function of the position coordinate ρ.

9-5 Analyze isothermal Newtonian flow in a calender with equal size rolls that turn at different speeds. Determine the stream function for this flow. (See J. T. Bergen, "Mixing and Dispersing Processes," p. 417, ch. 7, of *Processing of Thermoplastic Materials*, edited by E. C. Bernhardt, Reinhold Publishing Corp., 1959.)

chapter 10

Hydrodynamic Analysis of Screw Pumps

Extrusion is one of the most important forming techniques of polymer processing. More thermoplastics are converted into useful products by extrusion than by any other method. Plastic film, pipe, rod and other shapes, and plastic-coated wire and paper are a few examples of commercially important products manufactured by extrusion processes.

In an extrusion process, the polymeric material is either melted or dissolved and then pumped through a die to form a product of the required shape. In some extrusion processes, melting or dissolving occurs separately from pumping, so that the feed to the pump is in the fluid state. *Screw pumps* are often used in processes of this type. Other extrusion processes combine the melting and pumping steps in a *plasticating extruder*, which is a modified screw pump.

If one makes the simplifying assumption that the polymeric fluid is Newtonian, then the screw pump becomes amenable to mathematical analysis. The processes occurring in plasticating extruders are, however, inherently more complex, and a sound mathematical analysis of plasticating extruders has yet to be made. The operation of plasticating extruders is discussed in a qualitative and descriptive manner in Chapter 11.

10-1 Geometrical Considerations

Most screw pumps consist of a single screw that rotates inside a close fitting barrel. A few multiple screw machines have been built. In some of these the screw flights intermesh; in others they do not. If they do intermesh, the action is somewhat similar to that of a gear pump, that is, the forward motion of the fluid is due to positive displacement. Single screw machines and nonintermeshing multiple screw machines lack this positive action, and the flow mechanism is more complicated. The analysis presented in this chapter applies only to single screw machines.

Figure 10-1 is a schematic diagram of a screw pump. The fluid enters at the *feed port*. The rotation of the *screw* moves the fluid forward and increases its hydrostatic pressure so that it will flow through the *die* (or restriction of any kind, such as a pipe line) at the required rate. The input of mechanical energy to the pump occurs through the *shaft* of the screw. Provision is usually made for the transfer of heat either to or from the fluid through the walls of the barrel or through the hollow core of the screw.

In its most general sense, a theory of the screw pump must provide the means for solving the following problem: Given the physical properties of the fluid, the dimensions of the screw and the die, and the operating conditions, calculate the flow rate and the temperature at which the fluid is discharged from the die and the required power input to the pump. The physical properties of the fluid important in the theory of screw pumps include viscosity, thermal conductivity, and specific heat. The speed of rotation of the screw, the inlet temperature and pressure of the fluid, and the type of temperature control imposed upon the barrel and screw constitute the operating conditions.

The theory of screw pumps has been developed to the point where the above problems can be solved, provided certain restrictions are imposed. It should be noted that while the restrictions limit the generality of the theory, many cases of practical interest can be handled with confidence.

The fluid in a screw pump is contained by the bottom and sides of the screw channel and by the barrel surface. Because of the rotation of the screw, the fluid is contained within a system that has both a moving boundary (the bottom and sides of the channel) and a stationary boundary (the barrel surface). Hence a drag flow is established in the fluid. It is

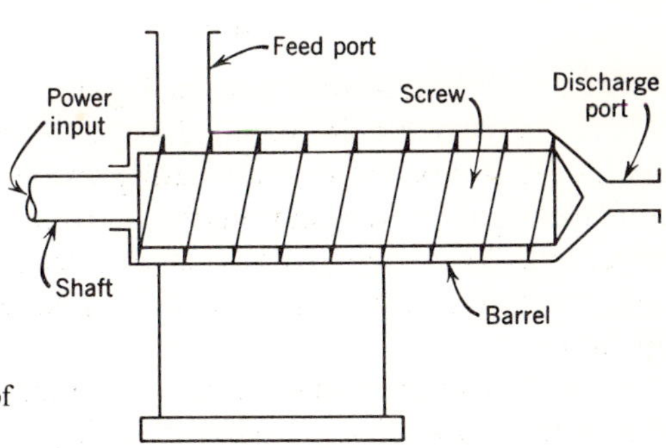

Figure 10-1. Schematic diagram of screw pump.

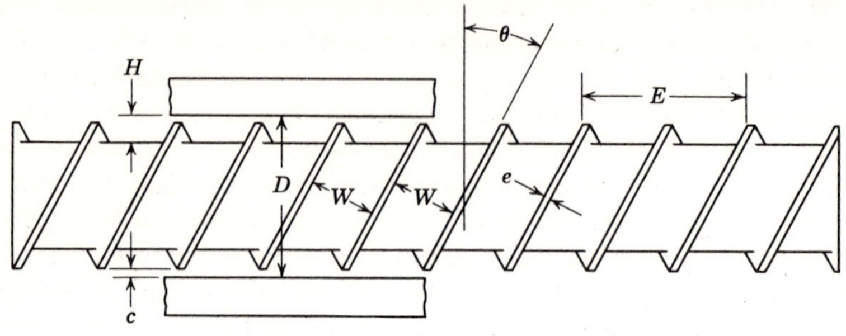

Figure 10-2. Geometry of extruder screw.

only because of the existence of this drag flow that a screw pump is able to "pump." A necessary condition for the existence of a drag flow is that the fluid have a finite viscosity. A perfect fluid (a hypothetical fluid of zero viscosity) could not be pumped with a screw pump because such a fluid could not support a shear stress and therefore could not undergo drag flow.

Figure 10-2 shows the dimensions that establish the geometry of the extruder screw and the symbols used to represent them. In order to make the treatment as general as possible, a multiple flight screw having two parallel channels is shown. The inside diameter of the barrel is D and the distance from the root of the screw to the barrel surface is H. The radial clearance between the top of the screw flight and the barrel is c. The pitch of the screw flight is related to the helix angle θ, measured at the top of the screw flight, by the equation

$$E = \pi(D - 2c) \tan \theta \tag{10-1}$$

where E represents the pitch.

The width of a channel is W, and for a screw having m channels in parallel, the following relation can be deduced:

$$m(W + e) = E \cos \theta \tag{10-2}$$

where e is the width of the screw flight.

Screw pumps can be operated either by rotating the screw and holding the barrel stationary or by rotating the barrel in the opposite direction and holding the screw stationary. In both cases the relative motion is the same, and the same operating characteristics are obtained. Because it is easier to build machines with rotating screws, this is the usual mode of operation.

An hydrodynamic analysis of screw pumps would involve the specification of the path of a fluid particle, as it passed through the pump, relative to a set of reference coordinates. The choice of the reference coordinates is purely arbitrary. Two obvious possibilities would be either to specify the position of a fluid particle relative to the barrel or relative to the screw. It is clear that the better choice is to specify the position of the fluid particle relative to the screw, in which case the fluid motion is described as it would appear to an observer located on the screw. To such an observer, the screw would appear to be stationary and the barrel rotating. The considerations involved in this choice of reference axes are analogous to those that an observer in outer space would be faced with if he wished to describe events happening on the face of the earth, in which case his best choice would be a system of coordinates that rotated with the earth. Of course, one could adopt the alternate point of view that the screw is stationary and the barrel rotating. Exactly the same result would be obtained.

The parallel plate representation of the screw channel is the most useful simplification made in the hydrodynamic analysis of screw pumps. In it the screw channel is figuratively "unwound" from the root of the screw and laid on a flat plate. A second flat plate located in a parallel plane forms the barrel surface. It is the relative motion of the two planes that causes the pump to operate.

The hydrodynamic analysis of screw pumps developed in this chapter uses the parallel plate representation of the screw channel and a left-handed system of rectangular coordinates, as shown in Figure 10-3. The z-axis is the so-called helical axis. Its positive direction is toward the discharge end of the channel and its length is Z. A cross section of the channel perpendicular to the z-axis has width W and height H. The λ-axis, which has a total length L, is parallel to the axis of the screw. Note that it is inclined at the angle θ, the helix angle of the screw, to the x-axis. The shaded cross section shown in Figure 10-3 is in a plane perpendicular to the λ-axis. Hence it represents a plane perpendicular to the axis of the screw. For the fluid to make a positive advance along the screw axis (that is, to advance in a forward direction in the pump) it must move in the $+\lambda$-direction.

At any point in the screw channel, the fluid velocity vector $\boldsymbol{v}$ can be resolved into the components v_x, v_y, and v_z, which act parallel to the reference axes. The top of the channel represents the barrel surface. It moves at a constant velocity in a plane parallel to the x-z plane and at a distance H above it. Its motion is described by the velocity vector $\boldsymbol{V}$, where $\boldsymbol{V}$ is directed perpendicular to the λ-axis. The magnitude of this vector, the surface speed of the barrel, is designated V, and we see that

$$V = \pi DN \tag{10-3}$$

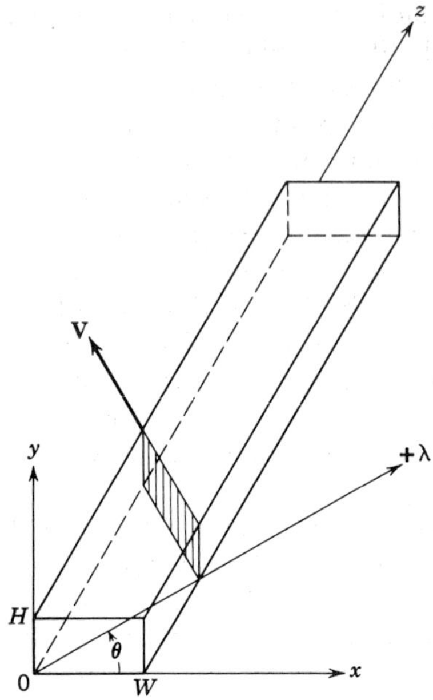

Figure 10-3. Coordinate axes used in derivation of the differential equations for flow and power.

where N is the frequency of rotation of the screw. The vector V, acting in a plane, has components V_x and V_z only.

The parallel plate representation of the screw channel is used in almost all analyses of screw pumps, but a satisfactory analysis of the error introduced by it has never been made. Intuitively, it would appear that if (H/D) is small, the error introduced will be small, and experimental studies of the performance of screw pumps tend to confirm this.

It is possible, however, to analyze a somewhat similar problem—that of approximating the annular drag flow of a fluid by plane parallel flow. The conclusions drawn from this analysis can be applied tentatively, and with certain restrictions, to the screw pump problem.

The problem, as shown in Figure 10-4, is to determine the error involved in using a parallel plate model for annular flow. We consider the case where distance H between the parallel plates is the same as that between the cylinders and the upper plate has the same linear speed V_0 as that of the inner surface of the outer cylinder. The volumetric flow rate Q_{cc} through an area of unit width of the coaxial cylinders is given by the integral

$$Q_{cc} = \int_{R_i}^{R_0} v_\theta \, dr \tag{10-4}$$

The volumetric flow rate Q_{pp} through an area of unit width of the parallel plates is given by the integral

$$Q_{pp} = \int_0^H v_z \, dy \tag{10-5}$$

The quantity ϵ, defined by the equation

$$\epsilon = 1 - \frac{Q_{cc}}{Q_{pp}} \tag{10-6}$$

is a measure of the error introduced by using the parallel plate representation for the annular drag flow.

The integral of (10-4) is evaluated by introducing the velocity distribution for flow in the annulus, equation (3-141). The integral of (10-5) is evaluated by introducing the velocity distribution for flow between parallel plates,

$$v_z = y\left(\frac{V_z}{H}\right) \tag{10-7}$$

Upon evaluating the integrals and after considerable rearranging, equation (10-6) can be written

$$\epsilon + 1 = \left[\frac{1}{\beta^2 - \beta^{2/n} - \beta^{(2-2n)/n} + 1}\right]\left[\beta^2 + \left(\frac{n}{n-1}\right)\beta^{(2-2n)/n} - \left(\frac{1}{1-n}\right)\right] \tag{10-8}$$

where

$$\beta = \frac{R_0}{R_i}$$

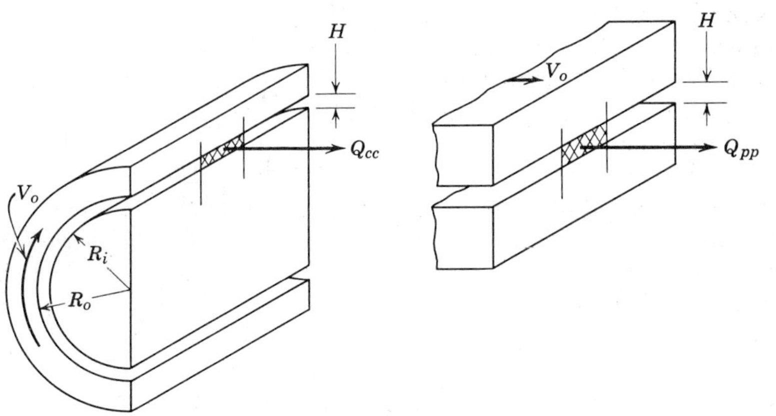

Figure 10-4. Parallel plate model for annular flow.

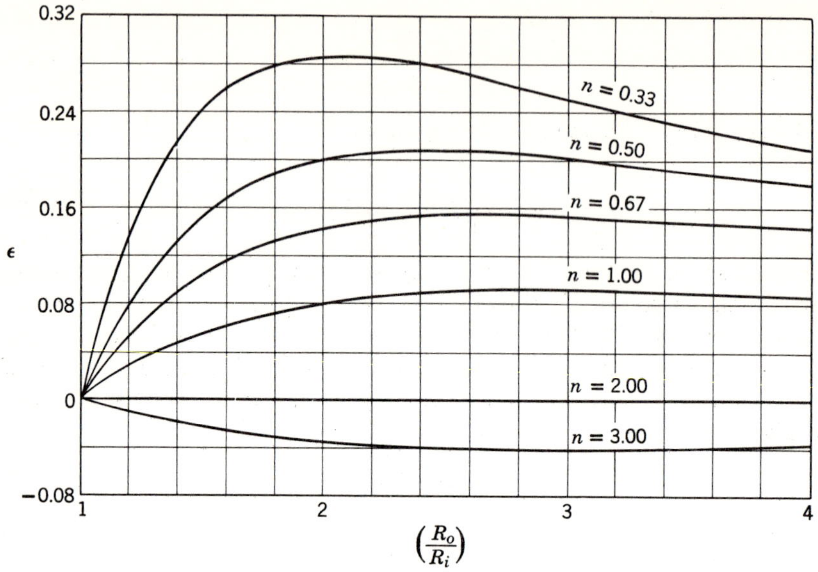

Figure 10-5. Error due to representation of annular flow by plane parallel flow.

Equation (10-8) is valid for all values of n except $n = 1$. For this special case, ϵ is obtained by setting $n = 1$ in equation (3-141) and then performing the integration of equation (10-4). The result obtained is

$$\epsilon + 1 = \frac{\beta}{(\beta - 1)(\beta^2 - 1)} [\beta^2 - 1 - 2 \ln \beta] \tag{10-9}$$

Numerical values of ϵ are plotted against (R_0/R_i) in Figure 10-5. The parameter is the flow index n. ϵ is positive when the volumetric annular flow rate is greater than that calculated for plane parallel flow and negative when it is less. Note that ϵ is positive for $(n < 2)$, negative for $(n > 2)$, and equal to zero for $(n = 2)$.

Many fluids handled in polymer processing operations are pseudoplastic with a minimum flow index of about 1/4, while most screw pumps and extruders have a maximum value of (R_0/R_i) in the range of 1.3. Under these conditions, the maximum ϵ is about 9%. If the fluid were Newtonian, it would be about 4%.

In a screw pump, the annular flow is usually opposed by a pressure gradient, and the above analysis cannot be applied rigorously. Under many conditions, however, the pressure gradient is relatively small and its effect can be neglected. Under these circumstances, the use of the parallel plate model of the screw channel is justified for most engineering purposes.

It should be noted that when unusual circumstances arise so that the effect of the pressure gradient cannot be neglected, there is no theoretical justification for the use of design equations based on a parallel plate model of the screw channel. Then it may be advisable to base designs on the scale-up of models.

10-2 Differential Equations for Flow and Power

A fluid particle starting at the feed end ($z = 0$) of the screw channel describes a complex path as it advances through the channel. It would be extremely difficult to give a complete mathematical description of the motion. However, this is not necessary in order to calculate the volumetric flow rate (pumping capacity) of a screw pump.

The volumetric flow rate Q is obtained by integrating the z-component of the fluid velocity vector over the cross section of the channel perpendicular to the z-axis. This is expressed mathematically by the equation

$$Q = \int_0^H \int_0^W v_z \, dx \, dy \tag{10-10}$$

The z-component of the momentum equation for steady isothermal flow of an incompressible Newtonian fluid (from Tables 2-2 and 2-3) is written

$$\rho\left[v_x\left(\frac{\partial v_z}{\partial x}\right) + v_y\left(\frac{\partial v_z}{\partial y}\right) + v_z\left(\frac{\partial v_z}{\partial z}\right)\right] = -\left(\frac{\partial P}{\partial z}\right) + \mu\left[\frac{\partial^2 v_z}{\partial x^2} + \frac{\partial^2 v_z}{\partial y^2} + \frac{\partial^2 v_z}{\partial z^2}\right] \tag{10-11}$$

The left side of (10-11) represents the acceleration terms and can be neglected for the slow flow of highly viscous materials. Furthermore, if the cross section of the channel is not a function of the z-position coordinate, then v_z does not change with z, and (10-11) reduces to

$$\frac{1}{\mu}\left(\frac{\partial P}{\partial z}\right) = \left[\frac{\partial^2 v_z}{\partial x^2} + \frac{\partial^2 v_z}{\partial y^2}\right] \tag{10-12}$$

If it is assumed that there is no clearance between the top of the screw flight and the barrel surface, and if we are dealing with a single channel screw ($m = 1$), the boundary conditions are written

$$\begin{aligned} v_z(0, y) &= 0 \qquad & v_z(x, 0) &= 0 \\ v_z(W, y) &= 0 \qquad & v_z(x, H) &= V_z \end{aligned} \tag{10-13}$$

where V_z is the z-component of the velocity vector that describes the motion of the top of the channel. Most screw pumps have a small radial clearance through which "leakage" or loss of pumping capacity can occur. Methods of correcting for this are described in Section 10-4.

If one assumes that the pressure depends only upon the z-position coordinate, then the left side of (10-12) does not depend upon x or y and a solution of it that satisfies the above boundary conditions can be found by the method of separation of variables. It is written in the following form:

$$v_z = \frac{4V_z}{\pi} \sum_{g=1,3,\ldots}^{\infty} \frac{\sinh(g\pi y/W)}{g \sinh(g\pi H/W)} \sin\left(\frac{g\pi x}{W}\right) - \frac{1}{\mu}\left(\frac{\partial P}{\partial z}\right) \times \left[\frac{y^2}{2} - \frac{yH}{2} + \frac{4H^2}{\pi^3} \sum_{g=1,3,\ldots}^{\infty} \frac{\cosh\left(\frac{g\pi}{2H}\right)(2x - W)}{g^3 \cosh\left(\frac{g\pi W}{H}\right)} \sin\left(\frac{g\pi y}{H}\right)\right] \tag{10-14}$$

The volumetric flow rate is found by introducing (10-14) into (10-10) and performing the indicated integrations. One obtains

$$Q = \left(\frac{V_z WH}{2}\right) F_D - \frac{WH^3}{12\mu}\left(\frac{\partial P}{\partial z}\right) F_P \tag{10-15}$$

where the first term represents the *drag flow* and the second term the *pressure flow*. Note that if the pressure gradient is positive, the pressure flow opposes the drag flow; and if the gradient is negative, the converse is true.

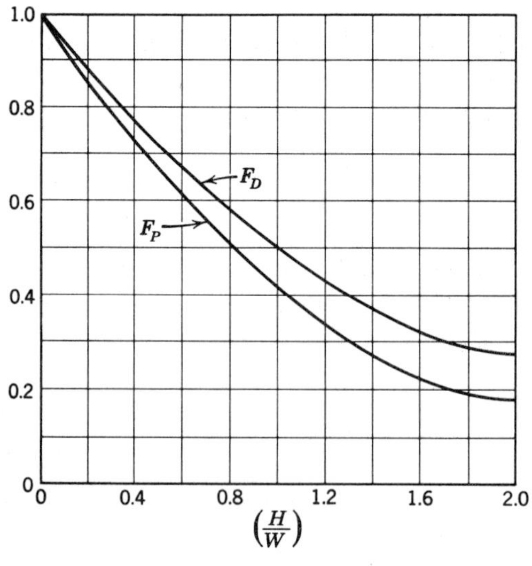

Figure 10-6. Shape factors for drag and pressure flow.

The quantities F_D and F_P are shape factors defined by the equations

$$F_D = \frac{16W}{\pi^3 H} \sum_{g=1,3,\cdots}^{\infty} \frac{1}{g^3} \tanh(g\pi H/2W) \tag{10-16}$$

$$F_P = 1 - \frac{192H}{\pi^5 W} \sum_{g=1,3,\cdots}^{\infty} \left(\frac{1}{g^5}\right) \tanh(g\pi H/2W) \tag{10-17}$$

Note that the shape factors depend only on the (H/W) ratio for the screw. The shape factors are shown graphically in Figure 10-6.

The solution of the boundary value problem consisting of the linear nonhomogeneous partial differential equation, (10-12), and the boundary conditions, (10-13), can be obtained through the method of separation of variables. The variables are separated by expressing v_z as the sum of two functions f and g

$$v_z = f + g \tag{10-18}$$

where both f and g depend upon the position variables x and y. Introducing (10-18) into (10-12) gives

$$\frac{\partial^2 f}{\partial x^2} + \frac{\partial^2 f}{\partial y^2} + \frac{\partial^2 g}{\partial x^2} + \frac{\partial^2 g}{\partial y^2} = \frac{1}{\mu}\left(\frac{\partial P}{\partial z}\right) \tag{10-19}$$

and if the function f is defined by the equation

$$\frac{\partial^2 f}{\partial x^2} + \frac{\partial^2 f}{\partial y^2} = \frac{1}{\mu}\left(\frac{\partial P}{\partial z}\right) \tag{10-20}$$

then the function g is a solution of the equation

$$\frac{\partial^2 g}{\partial x^2} + \frac{\partial^2 g}{\partial y^2} = 0 \tag{10-21}$$

If the following boundary conditions:

$$\begin{aligned} f(0, y) &= 0 & g(0, y) &= 0 \\ f(W, y) &= 0 & g(W, y) &= 0 \\ f(x, 0) &= 0 & g(x, 0) &= 0 \\ f(x, H) &= 0 & g(x, H) &= V_z \end{aligned} \tag{10-22}$$

are assigned to the functions f and g, then the boundary conditions for the function v_z, equations (10-13), are satisfied.

The differential equation (10-20) for the function f and the boundary conditions on f constitute a boundary value problem identical to the one solved previously (see Section 3-5) for pressure flow in a rectangular channel. Similarly, the differential equation (10-21) for the function g and the boundary conditions on g constitute a boundary value problem identical to the one solved previously for drag flow in a rectangular channel with a moving boundary. Hence equation (10-14) is obtained simply by adding the two previously obtained solutions.

Many of the design equations for screw pumps are based on the *simplified flow theory*.[1] This theory is developed in exactly the same manner as above except that the additional simplifying assumption that the variation of v_z with respect to x is small enough so that the derivative $\frac{\partial^2 v_z}{\partial x^2}$ can be neglected. In this case, equation (10-12) reduces to

$$\frac{1}{\mu}\left(\frac{\partial P}{\partial z}\right) = \frac{d^2 v_z}{dy^2} \tag{10-23}$$

and the boundary conditions for the problem become

$$\begin{aligned} v_z(0) &= 0 \\ v_z(H) &= V_z \end{aligned} \tag{10-24}$$

The solution of (10-23) satisfying the above boundary conditions can be written

$$v_z = y\left(\frac{V_z}{H}\right) - \frac{y(H-y)}{2\mu}\left(\frac{\partial P}{\partial z}\right) \tag{10-25}$$

For the simplified flow theory, the integral of equation (10-10) becomes

$$Q = W\int_0^H v_z\,dy \tag{10-26}$$

and the volumetric flow rate, obtained by introducing (10-25) into (10-26), is

$$Q = \frac{V_z W H}{2} - \frac{W H^3}{12\mu}\left(\frac{\partial P}{\partial z}\right) \tag{10-27}$$

Note that (10-27) is the special case of (10-15) applicable to screws having small values of (H/W), where the shape factors are nearly equal to unity.

In the analysis of power requirements, heat transfer, and mixing in screw pumps, it is necessary to consider the motion of the fluid in the x-y plane; hence both the v_x- and v_y-components of the velocity vector must be considered. The x- and y-components of the momentum equation reduce to

$$\frac{1}{\mu}\left(\frac{\partial P}{\partial x}\right) = \frac{\partial^2 v_x}{\partial x^2} + \frac{\partial^2 v_x}{\partial y^2} \tag{10-28}$$

$$\frac{1}{\mu}\left(\frac{\partial P}{\partial y}\right) = \frac{\partial^2 v_y}{\partial x^2} + \frac{\partial^2 v_y}{\partial y^2} \tag{10-29}$$

provided the usual assumptions of incompressibility, steady flow, and constant viscosity are made and the acceleration terms are neglected.

[1] J. F. Carley, R. S. Mallouk, and J. M. McKelvey, *Ind. Eng. Chem.*, **45**, 974 (1953).

A solution of the problem is obtained by making the simplifying assumption that

$$\frac{\partial^2 v_x}{\partial x^2} \ll \frac{\partial^2 v_x}{\partial y^2} \tag{10-30}$$

a condition that will be nearly satisfied by screws having shallow, wide channels. With this simplification (10-28) reduces to

$$\frac{1}{\mu}\left(\frac{\partial P}{\partial x}\right) = \frac{d^2 v_x}{dy^2} \tag{10-31}$$

and the solution of (10-31) satisfying the boundary conditions

$$\begin{aligned} v_x(0) &= 0 \\ v_x(H) &= -V_x \end{aligned} \tag{10-32}$$

is written

$$v_x = -y\left(\frac{V_x}{H}\right) - \frac{y(H-y)}{2\mu}\left(\frac{\partial P}{\partial x}\right) \tag{10-33}$$

The v_x-component of the fluid velocity causes only circulation. Hence

$$\int_0^H v_x \, dy = 0 \tag{10-34}$$

The pressure gradient in the x-direction is found by introducing (10-33) into (10-34) and integrating. One obtains

$$\frac{\partial P}{\partial x} = -\left(\frac{6\mu V_x}{H^2}\right) \tag{10-35}$$

The following expression for v_x is found by introducing (10-35) into (10-33) to give

$$v_x = y\left(\frac{V_x}{H}\right)\left[2 - \left(\frac{3y}{H}\right)\right] \tag{10-36}$$

Note that the velocity distribution is independent of the viscosity of the fluid and the pressure gradient.

The power required for the operation of a screw pump, calculated on the basis of the parallel plate model, is given by the equation

$$p = \int_0^Z \int_0^W (\tau_{yx} V_x + \tau_{yz} V_z) \, dx \, dz \tag{10-37}$$

where the stress components are to be evaluated at the plane $y = H$. The first term of the integral represents the power consumed in circulating the fluid in the x-y plane and the second term the power used to transport the fluid in the z-direction.

Equation (10-37) is easily derived, using vector methods. Consider a small area dA of the moving plate on which the force $\boldsymbol{dF}$ acts. If the velocity of the plate is $\boldsymbol{V}$, then the power dp consumed in the motion of the element dA is

$$dp = \boldsymbol{V} \cdot \boldsymbol{dF} \tag{10-38}$$

where $\boldsymbol{V}$ and $\boldsymbol{F}$ are vectors operating in the plane of the plate. The scalar product of the right side of (10-38) can be written

$$dp = -V_x\, dF_x + V_z\, dF_z \tag{10-39}$$

where the force components are related to stress components by the equations

$$dF_x = \tau_{yx}\, dA \qquad dF_z = \tau_{yz}\, dA \tag{10-40}$$

The total power, found by integrating over the entire area of the plate, is

$$p = \int_0^A \left(-V_x \tau_{yx} + V_z \tau_{yz} \right) dA \tag{10-41}$$

At the upper plate, the shear stresses, obtained from Table 2-3, are

$$\tau_{yx} = \mu \left(\frac{\partial v_x}{\partial y} \right)_{y=H} \qquad \tau_{yz} = \mu \left(\frac{\partial v_z}{\partial y} \right)_{y=H} \tag{10-42}$$

From equation (10-36) we have

$$\left(\frac{\partial v_x}{\partial y} \right)_{y=H} = -\frac{4V_x}{H} \tag{10-43}$$

and from equation (10-25) we have

$$\left(\frac{\partial v_z}{\partial y} \right)_{y=H} = \left(\frac{V_z}{H} \right) + \frac{H}{2\mu} \left(\frac{\partial P}{\partial z} \right) \tag{10-44}$$

Introducing the above equations into (10-37) gives the following result:

$$p = \int_0^Z \int_0^W \left[\frac{4\mu V_x^{\,2}}{H} + \frac{\mu V_z^{\,2}}{H} + \frac{V_z H}{2} \left(\frac{\partial P}{\partial z} \right) \right] dx\, dz \tag{10-45}$$

which is the basic power formula for the screw pump.

10-3 The Fluid Velocity Vector

The motion of a fluid particle in the screw channel is extremely complex, and a complete analysis of it has yet to be made. In this section various components of the fluid velocity vector are examined, and from this examination conclusions concerning the motion of fluid particles in the channel are drawn.

The following treatment is based upon the simplified flow equations developed in Section 10-2. The z-component of the fluid velocity vector is considered first. It is given by equation (10-25), which can be written in the equivalent form

$$\frac{v_z}{V_z} = a(1 - 3\phi + 3a\phi) \tag{10-46}$$

where $a = y/H$ and the function ϕ is defined by the equation

$$\phi = \frac{H^2}{6\mu V_z}\left(\frac{\partial P}{\partial z}\right) \tag{10-47}$$

It is shown below that ϕ represents the ratio of pressure flow to drag flow.

If equation (10-27) is written in the form

$$\frac{Q}{Q_D} = 1 - \frac{H^2}{6\mu V_z}\left(\frac{\partial P}{\partial z}\right) = 1 - \phi \tag{10-48}$$

and one notes that

$$Q = Q_D - Q_P \tag{10-49}$$

it can be seen that

$$\phi = \frac{Q_P}{Q_D} \tag{10-50}$$

Furthermore, the maximum value of ϕ occurs when the pressure gradient reaches its maximum value. This happens when the end of the channel is blocked (closed discharge). Because there is then no net advancement of the fluid down the channel, the condition

$$\int_0^1 v_z \, da = 0 \tag{10-51}$$

is applicable. If one introduces equation (10-46) into the above integral, one obtains

$$\frac{1}{2} - \frac{3\phi}{2} + \phi = 0 \tag{10-52}$$

which shows that the maximum value of ϕ is unity.

Considering positive pressure gradients only, the minimum value of the function ϕ is zero, because μ, H, and V_z are positive quantities.

Now consider the x-component of the fluid velocity vector, given by equation (10-36) which can be written in the equivalent form

$$\frac{v_x}{V_x} = a(2 - 3a) \tag{10-53}$$

Note that v_x depends only upon the dimensionless variable a and is independent of ϕ.

The v_λ-component of the fluid velocity vector can be expressed in terms of the v_x- and v_z-components through the equation

$$v_\lambda = v_x \cos\theta + v_z \sin\theta \tag{10-54}$$

Upon introducing (10-53) and (10-46) into (10-54) and noting that

$$V_z = |V| \cos \theta = V \cos \theta \tag{10-55}$$

$$V_x = |V| \sin \theta = V \sin \theta \tag{10-56}$$

one obtains

$$\frac{v_\lambda}{V} = 3a(1 - a)(1 - \phi) \sin \theta \cos \theta \tag{10-57}$$

where V represents the linear speed of the barrel surface.

Note that the relative value of v_λ depends upon the helix angle θ in addition to a and ϕ.

Example 10-1.

In this example the fluid velocity vector components are examined for a screw having a helix angle of 20° and a linear barrel speed of 10 cm/sec. This corresponds to a 2-inch diameter screw rotating at about 35 rpm.

$$V = \pi DN = 10 \text{ cm/sec}$$
$$V_z = V \cos (\theta) = 9.40 \text{ cm/sec}$$
$$V_x = V \sin (\theta) = 3.42 \text{ cm/sec}$$

Values of v_x, v_z, and v_λ are calculated with equations (10-46), (10-53), and (10-57). Note that v_λ has zero values under the following conditions:

$$a = 0 \quad a = 1 \quad \phi = 1 \quad \theta = 0° \quad \theta = 90°$$

The results of the calculations are plotted in Figure 10-7.

Figure 10-7 shows clearly that, with a screw having a 20° helix, shear rates in the y-z plane are considerably greater than those in the x-y or λ-y planes. As the helix angle is increased, shear in the x-y plane increases and reaches a maximum when $\theta = 90°$. As the helix angle is decreased, shear in the y-z plane increases and reaches a maximum when $\theta = 0°$. Note, also, that the shear pattern in the x-y plane is independent of the parameter ϕ, but that in the y-z plane it depends upon ϕ. The z-component of the velocity vector has positive values only for $\phi \leqq (1/3)$; but for $\phi > (1/3)$, it has both positive and negative values.

In the λ-y plane, the λ-component is positive for all values of ϕ. Therefore all fluid particles in the screw advance in the axial direction of the screw with no "backflow." Some particles, of course, advance at a faster rate than others.

From these shear patterns it is possible to make certain deductions concerning the motion of fluid particles in the channel. For closed discharge ($\phi = 1$), the fluid, making no advance in the λ-direction, simply advances and retreats in the x- and z-directions, and in one plane, the fluid particles describe closed paths, as is shown in Figure 10-8. The solid line indicates the path of a particle near the top of the channel. As the particle nears the screw flight, it is "turned under," and because of the pressure gradient then flows back to the other side of the channel, where the flight turns it up. Note that if a particle had started at a point $\frac{2}{3}$ of the distance up from the bottom of the channel, it would have remained stationary. Under conditions of closed discharge, then, mixing in the λ-direction

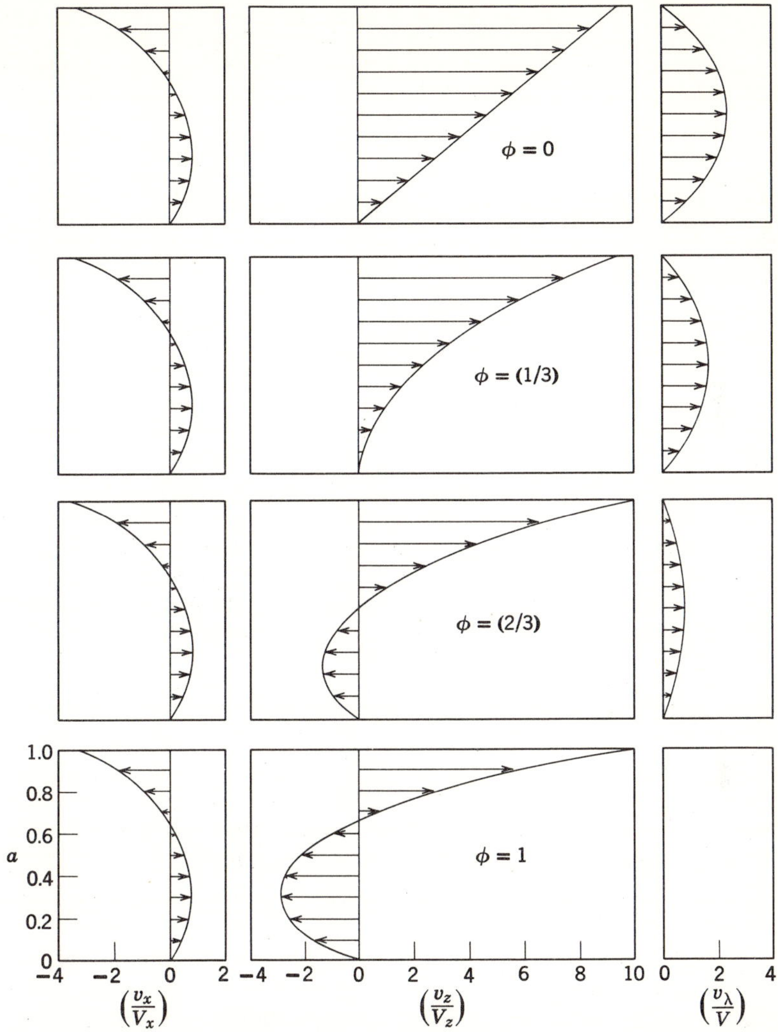

Figure 10-7. Profiles of velocity components for various values of ϕ.

would be nil and mixing in general would be poor because the fluid particles trace closed streamlines.

At the other extreme $\phi = 0$, open discharge, a fluid particle would describe a sort of flattened helix as it advanced toward the discharge port. The projection of a particle would circulate in the x-y plane but would always advance in the positive z- and λ-directions. Its motion would be somewhat as shown in Figure 10-8.

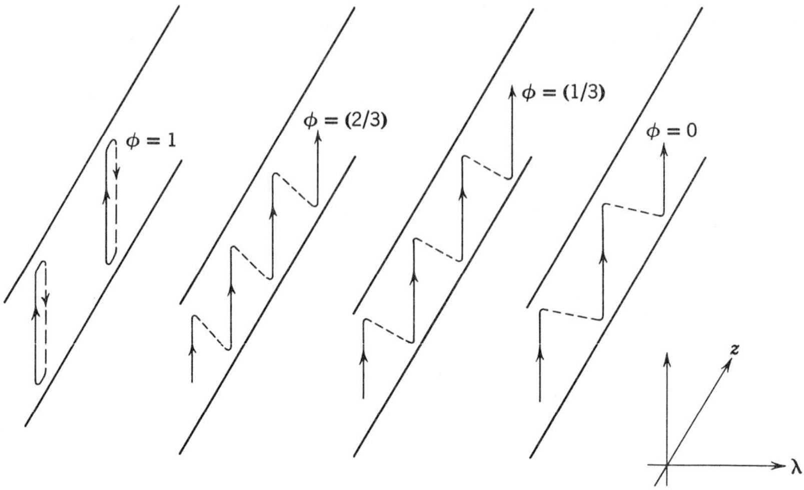

Figure 10-8. Fluid particle paths at different values of ϕ.

Figure 10-8 also shows particle paths for intermediate values of ϕ. Toward the bottom of the channel, for $\phi = 1/3$, the y-components are small but positive. For $\phi = 2/3$, the y-component is negative in the bottom part of the channel, indicating that the projection of a fluid particle would circulate in both the x-y and y-z planes.

In a sense, the effect of increasing ϕ is to slow the advance of fluid particles to the discharge port. This occurs because a particle must make more turns on its helix as ϕ increases. In the limiting case when $\phi = 1$, the number of turns becomes infinite and a particle never reaches the discharge port.

Although the shear patterns and particle paths pictured in Figures 10-7 and 10-8 are based upon several simplifying assumptions, their general validity has been confirmed by direct measurements made by Eccher and Valentinotti.[2] They used a specially constructed extruder having a glass barrel that pumped a fluid consisting of polyisobutylene mixed with paraffin oil and containing small particles of aluminum. The motion of the aluminum particles along the streamlines of the fluid was followed with a measuring microscope. Velocity components were found to be in good agreement with calculations based on the simplified equations presented above.

One of the most important simplifying assumptions of the above analysis is that of isothermal Newtonian flow. In practice, non-Newtonian behavior and heat effects can be quite large and will distort the velocity profiles. These effects are taken into consideration in the following analysis:

Consider the case where the velocity vector V (see Figure 10-3) is

[2] S. Eccher and W. Valentinotti, *Ind. Eng. Chem.*, **50**, 829 (1958).

directed along the z-axis so that the flow is unidirectional. Assume that the (W/H) ratio is large so that the fluid velocity $\boldsymbol{v}$ is a function only of the position variable y, and the parallel surfaces defined by $y = 0$ and $y = H$ are maintained at a constant temperature. Furthermore, assume the channel to be of sufficient length so that *steady state conditions* of flow and temperature have been reached. We now have the steady state problem analyzed by Colwell and Nickolls.[3]

The z-component of the momentum equation and the energy equation, obtained from Table 2-2, are written

$$\left(\frac{\partial P}{\partial z}\right) = \frac{\partial \tau_{yz}}{\partial y} \tag{10-58}$$

$$k\frac{\partial^2 T}{\partial y^2} + \tau_{yz}\left(\frac{\partial v_z}{\partial y}\right) = 0 \tag{10-59}$$

if the usual simplifying assumptions are made. For a non-Newtonian fluid of viscosity, η, the stress components obtained from Table 2-3 reduce the above equations to the following:

$$\frac{\partial P}{\partial z} = \frac{\partial}{\partial y}\left[\eta\left(\frac{\partial v_z}{\partial y}\right)\right] \tag{10-60}$$

$$k\left(\frac{\partial^2 T}{\partial y^2}\right) + \eta\left(\frac{\partial v_z}{\partial y}\right)^2 = 0 \tag{10-61}$$

where the non-Newtonian viscosity η is a function of temperature and shear rate (or shear stress). If the functional relationship is known, then equations (10-60) and (10-61) form a pair of simultaneous ordinary differential equations that can, in principle, be solved for the independent variables T and v_z in terms of the dependent variable y. The boundary conditions for the problem are written

$$v_z(0) = 0 \qquad T(0) = T_0$$

$$v_z(H) = V_z \qquad T(H) = T_H$$

where T_0 and T_H are the temperatures of the upper and lower plate respectively and V_z is the speed of the upper plate in the z-direction.

Colwell and Nickols[3] have obtained numerical solutions for this problem, using digital computing equipment. They expressed η as a function of temperature and shear stress with the equation

$$\eta = \eta^\circ\left(\frac{\tau^\circ}{\tau}\right)^m e^{-a(T-T_0)} \tag{10-62}$$

where η° is the viscosity in the reference state defined by T_0 and τ°, and a and m are constants.

[3] R. E. Colwell and K. R. Nickolls, *Ind. Eng. Chem.*, **51**, 841 (1959).

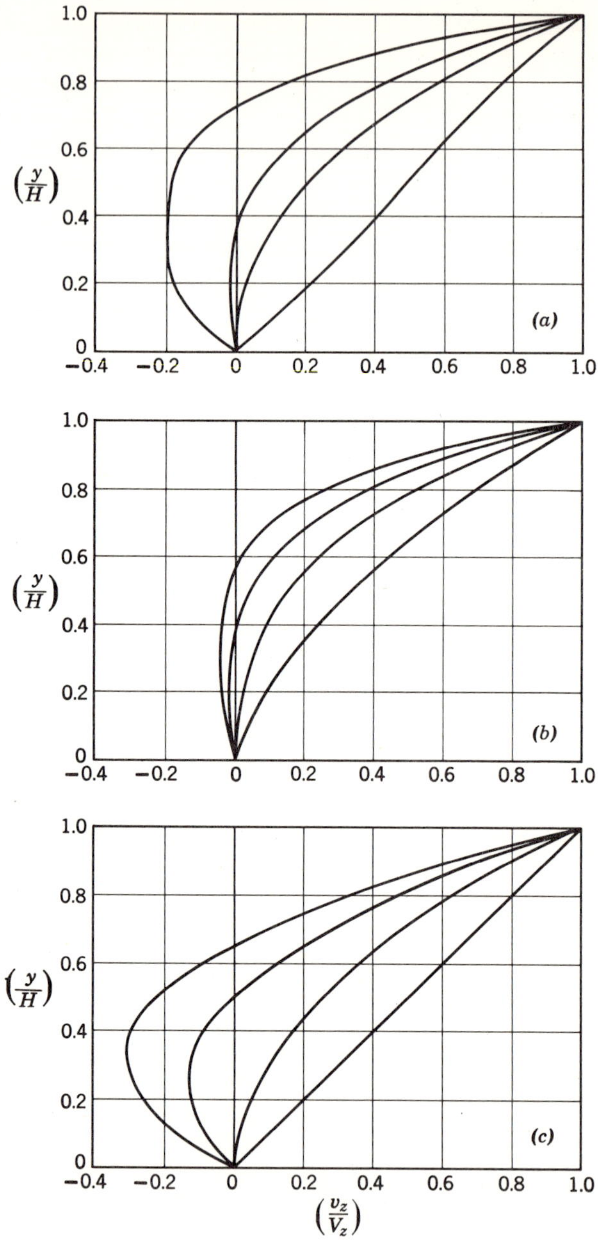

Figure 10-9. Velocity profiles for (*a*) power law fluid with both surfaces at the same temperature, (*b*) power law fluid with upper surface at higher temperature than lower. and (*c*) isothermal Newtonian fluid (data from Ref. 3).

The physical constants used in the computations are as follows:

$$\eta^\circ = 0.09 \text{ lb}(f)/\text{sec/in.}^2$$
$$\tau^\circ = 9.00 \text{ lb}(f)/\text{in.}^2$$
$$T_0 = 400^\circ \text{ F}$$
$$m = 1.56$$
$$a = 0.0089 \text{ deg F}^{-1}$$
$$k = 0.0192 \text{ in.-lb}(f)/\text{in.-sec-deg F}$$

The plate separation H was taken as 0.25 in. and calculations were made for various values of V, T_0, and T_H.

The results for two cases are shown in Figure 10-9, where (v_z/V_z) is plotted against (y/H) for various values of the parameter ϕ. The curves of Figure 10-9*a* apply when both plates are maintained at the same temperature, 450° F. The curves of Figure 10-9*b* apply when the upper plate is at 450° F but the lower plate is at 300° F, which is intended to simulate the conditions of screw cooling in an extruder. Note how the pressure flow near the lower plate is reduced, corresponding to a relatively stagnant layer in an extruder screw under conditions of screw cooling. The lower curves, Figure 10-9*c*, are velocity profiles for isothermal Newtonian flow and are shown for comparison.

10-4 Effect of Radial Clearance

Most screws have a small but finite clearance between the top of the screw flight and the surface of the barrel. Therefore the power and flow formulas derived in Section 10-2 must be modified to account for the power consumed in shearing the fluid within the clearance and for the "leakage" of the fluid through the clearance. It should, however, be noted that a finite clearance is not an essential feature of a screw pump, for small pumps having screws machined from solid rods of "Teflon" and then force fitted inside highly polished barrels have been built.

We first derive an approximate formula for the power dissipated in shearing the fluid contained within the clearance space. The derivation is simplified by assuming that only drag flow, characterized by a shear rate (V/c), is operative in the clearance. Then, if dF represents the tangential force operating on an element dA of the barrel surface

$$dF = \mu\left(\frac{V}{c}\right) dA \tag{10-63}$$

Introducing equation (10-63) into (10-38) gives

$$dp_c = \left(\frac{\mu}{c}\right)(V \cdot V)\, dA \tag{10-64}$$

and for a screw having m flights in parallel, each of width e, the above equation becomes

$$p_c = mV^2 \int_0^Z \left(\frac{e\mu}{c}\right) dz \tag{10-65}$$

where p_c represents the clearance power and V is the surface speed of the barrel.

The existence of a finite radial clearance reduces the effectiveness with which the screw can move the fluid forward; it also enables some of the fluid that is moved forward to "leak" back. Both effects reduce the pumping capacity of the screw.

Gore and McKelvey[4] have proposed that the drag flow formula include the factor

$$\left[1 - \left(\frac{c}{H}\right)\right]^2 \tag{10-66}$$

to correct for the loss of drag flow caused by the clearance. This factor is derived on the basis that the effective region of drag flow is from the root of the screw to the point $(H - c)$ rather than to the barrel surface where $y = H$. Under this assumption, the volumetric drag flow rate is given by the integral

$$Q_D = W \int_0^{H-c} v_z \, dy \tag{10-67}$$

If the drag flow velocity gradient

$$v_z = y\left(\frac{V_z}{H}\right) \tag{10-68}$$

is introduced into the above integral, it becomes

$$Q_D = \frac{V_z W H}{2}\left(1 - \frac{c}{H}\right)^2 \tag{10-69}$$

Mohr and Mallouk,[5] following a somewhat different approach to the problem of leakage, concluded on the basis of an approximate analysis that the drag flow formula should contain the factor

$$\left[1 - \frac{c}{H}\right] \tag{10-70}$$

to correct for the loss of drag flow caused by leakage.

The pressure flow through the radial clearance must now be considered. Most treatments of this flow have used the parallel plate approximation

[4] W. L. Gore and J. M. McKelvey, "Theory of Screw Extruders," ch. 16 of *Rheology*, vol. 3, edited by F. R. Eirich, Academic Press, 1960.

[5] W. D. Mohr and R. S. Mallouk, *Ind. Eng. Chem.*, **51**, 765 (1959).

of the screw and have arrived at formulas in which the pressure flow is proportional to the third power of the clearance. A correction factor for the pressure flow term, based on the derivation of Gore and McKelvey[4], can be written

$$\left[1 + \left(\frac{c}{H}\right)^3 \left(\frac{W}{e}\right) \left(\frac{1}{\sin\theta\cos\theta}\right)^2\right] \tag{10-71}$$

showing that pressure leakage is minimized when the clearance c is small or the flight width e is large.

10-5 Isothermal Operation

The previous sections of this chapter have presented an analysis of flow and energy dissipation in the channel of an extruder screw. Most of the equations are differential equations and apply only to a single cross section of the screw. The performance characteristics of a screw pump as a whole are obtained by integrating the flow and power equations over the entire length of the screw.

In the analysis of the performance of screw pumps several terms are used. The *design variables* are the dimensions that establish the geometrical configuration of the machine. The *operating variables* establish the condition of the operation. They include the frequency of rotation of the screw, the inlet temperature and pressure of fluid, the temperature distribution maintained on the barrel, etc. The *fluid properties* of interest in analyzing the performance of screw pumps include the density, viscosity, and thermal properties of the fluid. The *performance variables* include the output rate, the output temperature, the pressure at the end of the screw, and the power input.

A general theory of operation would enable the performance variables to be calculated once the design variables, the operating variables, and the fluid properties were specified. Such a general and complete theory of operation has not been developed. Operational theories have been developed only for certain special cases. This section considers the special case of *isothermal operation*; another special case, *adiabatic operation*, is considered in the next section.

Isothermal operation is a hypothetical mode of operation that yields easily to mathematical analysis. By definition, it is the mode of operation in which the fluid temperature is the same at all points in the screw channel. No real fluid could ever be pumped isothermally because heat generation due to viscous shear would set up temperature gradients in any fluid having a finite thermal conductivity. Nevertheless, the equations derived for this case are useful in predicting the performance of screw pumps under certain conditions.

For isothermal operation with a given set of design variables, the extrusion rate Q is a unique function of ΔP, the pressure increase generated by the pump. The relationship between Q and ΔP, called the screw characteristic, is derived below:

Equation (10-15), which applies to a single channel, is the starting point of the deviation. Some of the variables of this equation are transformed with the following relationships:

$$V_z = V \cos\theta = \pi DN \cos\theta$$

$$W = \frac{E\cos\theta}{m} - e = \frac{\pi(D - 2c)\sin\theta}{m} - e$$

Upon introducing the correction factors for the effect of radial clearance, equations (10-66) and (10-71), and generalizing the result to apply to a screw having m channels in parallel, we obtain

$$Q = \alpha N - \frac{\beta}{\mu}\left(\frac{\partial P}{\partial z}\right) \tag{10-72}$$

where

$$\alpha = \frac{\pi^2 D^2 H \sin\theta\cos\theta}{2}\left[1 - \frac{2c}{D} - \frac{em}{\pi D\sin\theta}\right]\left[1 - \left(\frac{c}{H}\right)\right]^2 F_D \tag{10-73}$$

$$\beta = \frac{\pi D H^3 \sin\theta}{12}\left[1 - \frac{2c}{D} - \frac{em}{\pi D\sin\theta}\right]\left[1 + \left(\frac{c}{H}\right)^3\left(\frac{W}{e}\right)\left(\frac{1}{\sin\theta\cos\theta}\right)\right]F_P \tag{10-74}$$

Noting that the parameters α and β are generally functions of the position coordinate z but Q (for an incompressible fluid) is independent of z, equation (10-72) can be written

$$\int_{P_0}^{P_Z} dP = \mu N \int_0^Z \left(\frac{\alpha}{\beta}\right) dz - \mu Q \int_0^Z \left(\frac{1}{\beta}\right) dz \tag{10-75}$$

Upon integrating equation (10-75) and solving for Q, one obtains

$$Q = AN - B\left(\frac{\Delta P}{\mu}\right) \tag{10-76}$$

where

$$A = \frac{\int_0^Z \left(\frac{\alpha}{\beta}\right) dz}{\int_0^Z \left(\frac{1}{\beta}\right) dz} \qquad B = \frac{1}{\int_0^Z \left(\frac{1}{\beta}\right) dz}$$

Equation (10-76) shows that there is a linear relationship between Q and ΔP. On linear graph paper, a plot of Q versus ΔP will be a straight line having a negative slope. The slope and intercept of the line depend upon the values of the integrals of equation (10-76). As soon as the functional relationships between α and β and the position coordinate z are established, equation (10-76) can be integrated. We consider, first, the special case where H is the following linear function of z:

$$H = H_0 - \left(\frac{H_0 - H_Z}{Z}\right)z \tag{10-77}$$

If equation (10-77) is substituted into (10-76) and the integrations are performed, one obtains

$$Q = \alpha N\left[\frac{2}{1 + \left(\dfrac{H_Z}{H_0}\right)}\right] - \frac{\beta}{\mu}\left(\frac{\Delta P}{Z}\right)\left[\frac{2}{\left(\dfrac{H_Z}{H_0}\right) + \left(\dfrac{H_Z}{H_0}\right)^2}\right] \tag{10-78}$$

where α and β are based upon the channel depth at the delivery end of the screw. For the special case where $H_Z = H_0$, that is, where there is no compression in the channel, equation (10-78) reduces to

$$Q = \alpha N - \frac{\beta\,\Delta P}{\mu Z} \tag{10-79}$$

Example 10-2.

This example compares the characteristic of a screw having a uniform channel with one having a channel of decreasing depth, all other design variables for the two screws being the same. The depth of the channel of the uniform screw is designated H_Z. For the nonuniform screw, the channel depth at the feed end is H_0, decreasing linearly to H_Z at its discharge end. The special case where $(H_0/H_Z) = 2$ is considered.

The characteristic for the uniform screw is given by (10-79) and for the nonuniform screw by (10-78). From these equations it is readily shown that

$$\frac{Q_{\max}}{Q'_{\max}} = \frac{1 + (H_Z/H_0)}{2} = \frac{3}{4}$$

and that

$$\frac{\Delta P_{\max}}{\Delta P'_{\max}} = \frac{1 + (H_Z/H_0)}{(H_Z/H_0) + (H_Z/H_0)^2} = 2$$

where the primed quantities refer to the nonuniform screw. The maximum pumping rate $Q_{\max}$ is obtained at free discharge ($\Delta P = 0$) and the maximum pressure rise $\Delta P_{\max}$ is obtained at closed discharge ($Q = 0$).

These calculations show that the maximum pumping rate of the nonuniform screw is $\frac{1}{3}$ greater than that of the uniform screw, but that it is capable of developing only $\frac{1}{2}$ of the maximum pressure developed by the uniform screw.

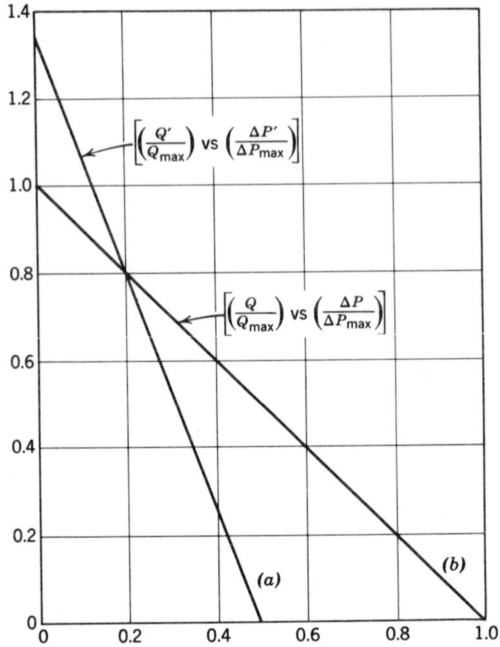

Figure 10-10. Dimensionless screw characteristics showing relative steepness of nonuniform screw (*a*) compared with uniform screw (*b*).

Figure 10-10 shows the relative steepness of the characteristic of the nonuniform screw as compared with the uniform screw.

The power equation is obtained by adding (10-45), which gives the power transferred to the fluid in the channel, to (10-65), which gives the power dissipated to the fluid in the radial clearance. It is written

$$p = mW \int_0^Z \left[\frac{4\mu V_x^{\,2}}{H} + \frac{\mu V_z^{\,2}}{H} + \frac{V_z H}{2} \left(\frac{\partial P}{\partial z} \right) \right] dz + mV^2 \int_0^Z \left(\frac{e\mu}{c} \right) dz \qquad (10\text{-}80)$$

We consider here only the simplest special case, that in which the channel dimensions are constant. In this case (10-80) is readily integrated. One obtains

$$p = \epsilon \mu N^2 Z + \alpha N \Delta P \qquad (10\text{-}81)$$

where

$$\epsilon = \frac{\pi^3 D^3 \sin\theta}{H} \left[1 + 3\sin^2\theta + \left(\frac{eH}{cW} \right) \right] \left[1 - \left(\frac{2c}{D} \right) - \frac{em}{\pi D \sin\theta} \right] \qquad (10\text{-}82)$$

Upon discharge from the pump, the fluid passes through a pipeline or die, or restriction, to flow of some kind. For isothermal laminar Newtonian flow, the flow rate and pressure drop are related by an equation of the type

$$Q = \frac{K}{\mu} \Delta P \tag{10-83}$$

where K is a constant that depends upon the geometry of the flow path and ΔP is the pressure drop. The ΔP in (10-83) is the same as the ΔP in (10-76) for the case where the inlet pressure to the pump and the discharge pressure from the die are the same. This is the most common case. If equation (10-83) is combined with the general screw characteristic equation (10-76) and ΔP eliminated between them, one obtains

$$Q = \frac{AKN}{B + K} \tag{10-84}$$

which shows that the discharge rate is directly proportional to the frequency of rotation of the screw. This result applies, of course, only to the special case of isothermal operation with Newtonian fluids. Note that the result is independent of the viscosity of fluid and depends only upon the design variables of the pump and die. With non-Newtonians, a different result is obtained. This is discussed in Section 10-9.

The extrusion rate Q could also be eliminated between (10-76) and (10-83) to obtain the following expression for ΔP, the pressure at the end of the screw:

$$P = \frac{\mu AN}{(B + K)} \tag{10-85}$$

Equation (10-85) shows that the pressure rise is directly proportional to the viscosity of the fluid and to the screw speed. The proportionality constant depends only upon the design variables.

Many "optimum" designs have been discussed in the literature.[6] An optimum design is one for which, under a given set of conditions, a particular quantity, such as the extrusion rate, the pumping efficiency, or any other quantity of interest, is maximized. The result that one obtains depends upon the quantity selected for maximization and the conditions under which the maximization is carried out.

Example 10-3.

This example considers a screw of fixed length and diameter. The problem is (1) to determine the screw design that will give the maximum extrusion rate at a given frequency of rotation and (2) to investigate pumping efficiencies of screws designed for maximum extrusion rates.

[6] J. F. Carley and R. A. Strub, *Ind. Eng. Chem.*, **45**, 988 (1953).

In order to simplify the analysis, we consider the case where the radial clearance c and the flight width e are small and may be neglected in the flow and power equations and where H and θ are not functions of z. Under these conditions, the constants A and B of the general screw characteristic reduce to

$$A = \alpha = \frac{\pi^2 D^2 H \sin(\theta) \cos(\theta)}{2} \tag{10-86}$$

$$B = \frac{\beta}{Z} = \frac{\pi D H^3 \sin^2(\theta)}{12L} \tag{10-87}$$

The extrusion rate, given by equation (10-84), can be written:

$$Q = \frac{\pi^2 D^2 N H \sin(\theta) \cos(\theta)}{2 + \pi D H^3 \sin^2 \theta / 6KL} \tag{10-88}$$

If there is a channel depth H° that will maximize Q in (10-88), then

$$\left(\frac{\partial Q}{\partial H}\right) = 0$$

when $H = H^\circ$. The value of H°, found by differentiating (10-88), is

$$H^\circ = \left[\frac{6KL}{\pi D \sin^2 \theta}\right]^{1/3} \tag{10-89}$$

The extrusion rate for the screw having the optimum channel depth H° is found by substituting (10-89) into (10-88). One obtains

$$Q^\circ = \frac{\pi^2 D^2 N H^\circ \sin(\theta) \cos(\theta)}{3} \tag{10-90}$$

If there is helix angle θ° that will maximize Q in (10-88), then

$$\left(\frac{\partial Q}{\partial \theta}\right) = 0 \qquad \text{at } \theta = \theta^\circ$$

The value of θ°, found through differentiation of (10-88), is given by the equation

$$\sin^2 \theta^\circ = \frac{1}{2 + \pi D H^3 / 12KL} \tag{10-91}$$

If the screw is to have optimum H and θ, their values, which must satisfy (10-89) and (10-91), are

$$\theta^* = 30^\circ \qquad H^* = \left[\frac{24KL}{\pi D}\right]^{1/3}$$

and the extrusion rate for a screw designed with θ^* and H^* is obtained by substituting the above values into (10-88). One obtains

$$Q^* = \frac{\sqrt{3}\, \pi^2 D^2 N H^*}{12} \tag{10-92}$$

It is interesting to compare the ratio (Q°/Q^*) at various helix angles. Dividing equation (10-90) by (10-92) gives

$$\frac{Q^\circ}{Q^*} = \frac{4 \sin(\theta) \cos(\theta)}{\sqrt{3}\ \sqrt[3]{4 \sin^2 \theta}} \tag{10-93}$$

Note that the value of the ratio depends only on the helix angle θ. Figure 10-11 shows a plot of (Q°/Q^*) versus θ. The curve passes through its maximum at 30°. However, since the slope of the curve in the region between 10° and 50° is not very steep, any helix angle in this range will have nearly the same pumping capacity. Note that this result is independent of the fluid properties and of the dimensions of the die and the pressure rise over the length of the screw.

If the pumping capacity is nearly the same over a wide range of helix angles, would there be any design advantage, then, in selecting one over another? To answer this question one must consider the efficiency of the pump. The pumping efficiency is defined by the equation

$$E = \frac{Q\ \Delta P}{p} \tag{10-94}$$

where p is the total power input to the screw. It is given by (10-81), which, for the case where the channel depth is equal to H°, is written

$$p = \epsilon^\circ \mu N^2 Z + \alpha^\circ N\ \Delta P \tag{10-95}$$

Since the pressure drop and extrusion rate are related by (10-83), the expression for pumping efficiency can be written

$$\frac{1}{E^\circ} = \frac{\epsilon^\circ N^2 K Z}{(Q^\circ)^2} + \frac{\alpha^\circ N}{Q^\circ} \tag{10-96}$$

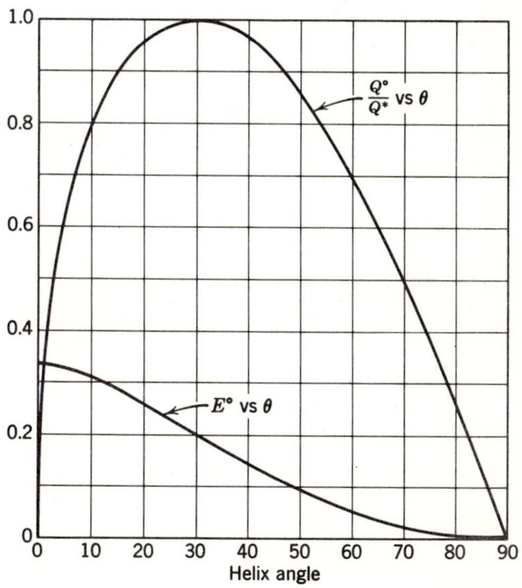

Figure 10-11. Effect of helix angle on efficiency and capacity of a screw pump.

where the quantities ϵ° and α° are given by the equations

$$\epsilon^\circ = \frac{\pi^3 D^3 \sin(\theta)}{H^\circ}[1 + 3\sin^2\theta]$$

$$\alpha^\circ = \frac{\pi^2 D^2 H^\circ \sin(\theta)\cos(\theta)}{2}$$

If these are introduced into (10-96) and the expression rearranged, one obtains

$$E^\circ = \frac{\cos^2\theta}{3(1 + \sin^2\theta)} \tag{10-97}$$

Values of E° as a function of θ are shown also in Figure 10-11.

Note that the maximum value of E°, which is $\frac{1}{3}$, occurs with a zero degree helix angle. Here all of the flow is directed down the channel so that no energy is dissipated in the transverse flow. This is a limit value since the pumping capacity is zero with a zero helix. When $\theta = 30^\circ$, the efficiency is 20%, and because most screw pumps have a helix angle between 10° and 20°, their theoretical pumping efficiencies will be about 25%. When the effect of a finite radial clearance is included, the actual efficiency will be less, for the extrusion rate will be reduced by leakage and the power requirement increased because of the shear force in the clearance.

Considerable experimental data verify the isothermal operating equations for Newtonian fluids. At the present time, experimental verification of the power formulas is lacking. Rowell and Finlayson[7] experimented with a variety of screws using water, soap solutions, and oils. In all cases the data confirmed the isothermal flow equations presented in this chapter. Additional experimental verification was provided by Pigott[8] in 1951. Using oils of Newtonian behavior that had a viscosity of 0.5 poise and 1.1 poise, he determined the screw characteristics of a 1-inch diameter screw. The linear relationship between extrusion rate and pressure, as indicated by equation (10-76), with the slopes of the curves being inversely proportional to the viscosity and the intercepts being directly proportional to the frequency of rotation, was confirmed.

Further experiments were reported by McKelvey[9] in 1953. A 2-inch diameter screw was used to pump corn syrup, a Newtonian liquid. Linear screw characteristics were observed and the linear relationship between extrusion rate and frequency of rotation, as predicted by equation (10-84), was observed.

10-6 Adiabatic Operation

The term *adiabatic operation* refers to the hypothetical case where there is no transfer of heat through the walls of the extruder barrel; thus all of

[7] H. S. Rowell and D. Finlayson, *Engineering*, **126,** 249 (1928).
[8] W. T. Pigott, *Trans. A.S.M.E.*, **73,** 947 (1951).
[9] J. M. McKelvey, *Ind. Eng. Chem.*, **45,** 982 (1953).

the heat and pressure energy in the extrudate come from the mechanical energy dissipated by the screw.

In the following mathematical analysis of adiabatic operation, the simplifying assumption is made that the fluid temperature and viscosity are uniform in the transverse direction (the x-direction). Furthermore, the analysis is based upon the simplified one-dimensional flow and power equations.

Although the conditions under which the adiabatic analysis applies are highly restrictive, the analysis does give some insight into the operation of screw pumps when internal heat generation occurs at a rate fast enough to cause a considerable increase in fluid temperature. Such cases do arise in practice. For example, solid polyethylene can be melted and its temperature raised several hundred degrees in a conventional screw pump (extruder) without the transfer of any external heat into the polymer.

First, the equation for the adiabatic screw characteristic is derived. In order to simplify the analysis, a uniform channel is considered. This is the special case where the dimensions of the screw channel are not functions of z. The flow equation, (10-75), is written in the form

$$\Delta P = \frac{\alpha N - Q}{\beta} \int_0^Z \mu \, dz \tag{10-98}$$

where μ, being a function of temperature, is also a function of z.

The viscosity and temperature are related to the position variable z by making an energy balance over a small element of the screw channel. The length of the element is dz and its cross section is perpendicular to the z-axis. As the fluid passes through this section, its pressure rises an amount dP and its temperature an amount dT. The total increase in energy of the fluid passing through the element is $(CQ\,dT + Q\,dP)$, where the first term is the increase in thermal energy and the second term the increase in pressure energy. For adiabatic operation, this is exactly equal to the rate at which mechanical energy is dissipated by the screw. Therefore, with reference to (10-81), the energy balance is formulated as

$$CQ\,dT + Q\,dP = \epsilon \mu N^2\,dz + \alpha N\,dP \tag{10-99}$$

where C is the heat capacity per unit volume of the fluid.

The second and fourth terms of (10-99) are nearly equal and are an order of magnitude smaller than the first and third terms. Therefore they can be dropped from the energy balance with little error. One then obtains the desired relationship between μ, T, and z, which is written

$$\mu\,dz = \left(\frac{CQ}{\epsilon N^2}\right) dT \tag{10-100}$$

The adiabatic screw characteristic is obtained by introducing (10-100) into (10-98) and integrating. Upon rearranging, one obtains

$$Q = \alpha N - \left(\frac{\epsilon\beta}{C}\right)\frac{N^2\,\Delta P}{Q\,\Delta T} \tag{10-101}$$

which is analogous to the isothermal screw characteristic given by equation (10-79).

Some of the characteristics of adiabatic operation can be deduced from the adiabatic screw characteristic. Note that the maximum extrusion rate, $Q_{\max}$ is obtained at zero pressure and is equal to αN. If the screw characteristic is written in the form

$$\left(\frac{Q}{Q_{\max}}\right)^2 - \left(\frac{Q}{Q_{\max}}\right) + \left(\frac{\epsilon\beta}{C\alpha^2}\right)\frac{\Delta P}{\Delta T} = 0 \tag{10-102}$$

it can be seen that $(\Delta P/\Delta T)$ passes through a maximum when $(Q/Q_{\max})$ is equal to $(\frac{1}{2})$. Therefore

$$\left(\frac{\Delta P}{\Delta T}\right)_{\max} = \frac{\alpha^2 C}{4\beta\epsilon} \tag{10-103}$$

The screw characteristic can be written in dimensionless form by introducing (10-103) into (10-102). One obtains

$$\left(\frac{Q}{Q_{\max}}\right)^2 - \left(\frac{Q}{Q_{\max}}\right) + \frac{(\Delta P/\Delta T)}{4(\Delta P/\Delta T)_{\max}} = 0 \tag{10-104}$$

Figure 10-12 is a plot of this equation.

Let g represent the $(Q/Q_{\max})$ ratio. At open discharge, the value of g is unity. Since ΔP is zero, the $(\Delta P/\Delta T)$ ratio is zero and the initial point on the curve, in the upper left-hand corner, has the coordinate (0, 1). As the discharge is restricted, both ΔP and ΔT increase, but since the $(\Delta P/\Delta T)$ ratio becomes larger, ΔP must increase more rapidly than ΔT. This condition continues until the value of g decreases to 0.50, at which point $(\Delta P/\Delta T)$ reaches its maximum value.

For values of g less than 0.5, Figure 10-12 shows that the $(\Delta P/\Delta T)$ ratio becomes smaller. Therefore in this region

$$\frac{d(\Delta P)}{dg} < \frac{d(\Delta T)}{dg} \qquad \left(g < \frac{1}{2}\right) \tag{10-105}$$

Under most conditions ΔP will pass through a maximum, after which $d(\Delta P)/dg$ will be negative. The value of g at which the maximum occurs, designated as g^*, can be determined approximately in the following manner:

Equation (10-102) is rearranged and written as

$$\Delta P = \frac{C\alpha^2}{\epsilon\beta}(g - g^2)\,\Delta T \tag{10-106}$$

Then the empirical viscosity-temperature relationship

$$\mu = ae^{-bT} \tag{10-107}$$

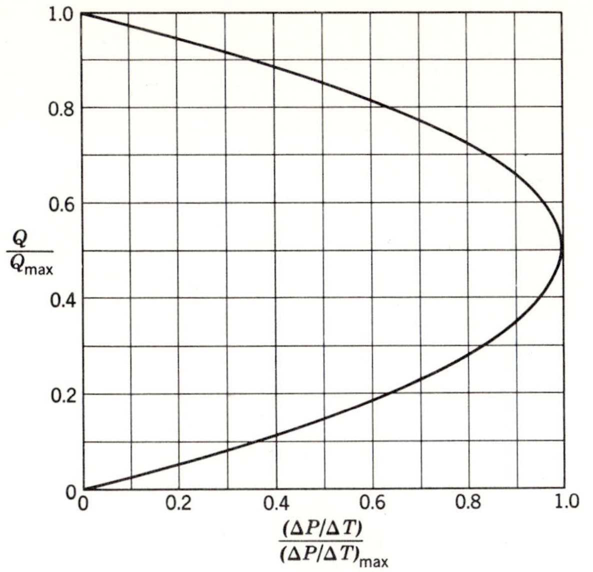

Figure 10-12. Dimensionless adiabatic screw characteristics.

is introduced into equation (10-100). Upon integration, one obtains

$$\exp(b\,\Delta T) - 1 = \frac{A}{g} \tag{10-108}$$

where

$$A = \frac{b\mu_i \epsilon NZ}{\alpha C}$$

Under conditions where ΔP reaches its maximum value, the discharge is quite severely restricted and ΔT will be large if the pump is working on high viscosity polymer melts and solutions. Under these conditions equation (10-108) can be written in the approximate form

$$\exp(b\,\Delta T) \approx \frac{A}{g}$$

If this result is introduced into (10-106), an equation having ΔP and g as its variables is obtained. The value of g at which ΔP passes through its maximum, found in the usual manner, is given by the equation

$$\ln\frac{A}{g^*} = \left(\frac{1 - g^*}{1 - 2g^*}\right) \quad (g \leq 0.5) \tag{10-109}$$

Figure 10-13 shows a plot of g^* versus A.

For large values of A, which would occur with long screws at high speeds and with high viscosity melts, the value of g^* is nearly $\frac{1}{2}$. In other words, the maximum pressure occurs when the discharge of the pump is restricted to slightly

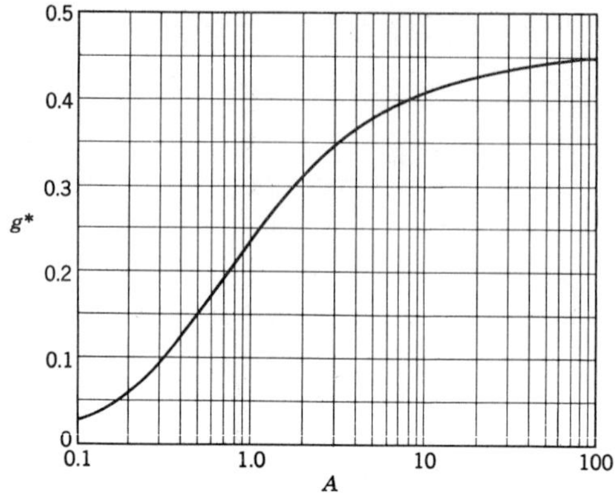

Figure 10-13. Plot of equation (10-109).

less than 50% of its free discharge rate. At the other extreme, that is, with low viscosity fluids, short screws and low speeds (conditions under which near isothermal operation would be expected), the value of g^* is nearly zero. In other words, the maximum pressure is built up at nearly closed discharge, which is similar to the isothermal case.

Just as in isothermal operation, the operating point of the pump is determined by the interaction of the die characteristic with the screw characteristic. In the following development it is convenient to introduce the dimensionless quantity X, which represents the ratio of the viscosity at the inlet to the pump to the viscosity at the outlet of the pump. The viscosity-temperature relation (10-107) can be written

$$X = \exp(b\,\Delta T) = \frac{\mu_i}{\mu_0} \tag{10-110}$$

where $\Delta T = (T_0 - T_i)$.

The extrusion rate Q can be expressed in terms of X by introducing (10-110) into the previous result, equation (10-108), and rearranging. One obtains

$$Q = \frac{\mu_i b \epsilon N^2 Z}{C(X - 1)} \tag{10-111}$$

The flow through the die is given by the equation

$$Q = \frac{K\,\Delta P}{\mu_0} = \frac{KX\,\Delta P}{\mu_i} \tag{10-112}$$

It is now possible to eliminate ΔT and ΔP from the screw characteristic, equation (10-101), by introducing (10-112) and (10-110). One obtains an equation containing Q and X as the variables. This equation can then be combined with (10-111) to eliminate Q, obtaining the single equation in X

$$N_1 = N_2 X \ln(X) - \frac{X \ln(X)}{(X-1)} \tag{10-113}$$

where N_1 and N_2 are dimensionless groups defined by the equations

$$N_1 = \frac{\beta}{KZ} \qquad N_2 = \frac{\alpha C}{NZ\mu_i b\epsilon} \tag{10-114}$$

If equation (10-113) is written in functional form

$$X = f(N_1, N_2)$$

it is readily seen that X is a unique function of the two dimensionless groups N_1 and N_2. Unfortunately, X cannot be written as an explicit function, and usually graphical methods are used to evaluate it. Figure 10-14 shows a typical chart used for the evaluation of X.

Once the value of X has been determined, the value of Q can be calculated with (10-111), the value of ΔP with (10-112), and the value of ΔT

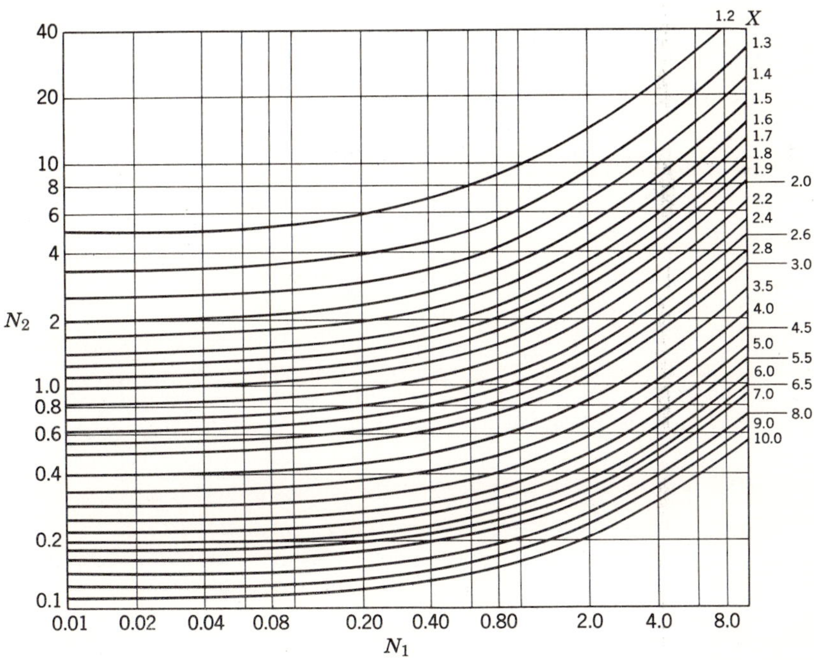

Figure 10-14. Chart to be used in the graphical solution of equation (10-113). (With permission from "Theory of Screw Extruders," ch. 16 of *Rheology*, vol. 3, edited by F. R. Eirich, Academic Press, 1960).

from (10-110). The power required to operate the screw pump is readily calculated as the sum of the increase of thermal energy and pressure energy of the extrudate.

Example 10-4.

This example considers the adiabatic operation of a screw pump at free discharge. The adiabatic temperature rise ΔT is calculated for various frequencies of rotation of the screw, as is the power and pumping rate. In addition, the fluid temperature is calculated as a function of position in the screw channel for a single frequency of rotation.

At free discharge there is no pressure build-up, and the constant K becomes infinite. This reduces N_1 to zero, and (10-113) can be solved directly for X. One obtains

$$X = 1 + \frac{1}{N_2}$$

Since X is related to ΔT by (10-110), the adiabatic temperature rise is given by

$$\Delta T = \left(\frac{1}{b}\right) \ln \left[1 + \frac{1}{N_2}\right] \tag{10-115}$$

The temperature is related to position z in the screw channel with (10-100). Replacing Q with (10-111), one obtains

$$\mu\, dz = \frac{\mu_i b Z}{(X - 1)}\, dT$$

and then by replacing μ with (10-107) and integrating

$$T = T_i + \frac{1}{b} \ln \left[1 + \frac{z(X - 1)}{Z}\right] \tag{10-116}$$

The power and pumping rate are given by the equations

$$Q = \alpha N$$
$$P = CQ\, \Delta T\, Q \Delta P$$

The numerical computations are made using the same physical constants and screw dimensions employed in a previous example by McKelvey.[10] They are as follows:

$D = 2.00$ in. $\quad H = 0.20$ in. $\quad \mu_i = 0.20$ lb(f)-sec/in.^{2}

$L = 15$ in. $\quad e = 0.20$ in. $\quad b = 0.025$ deg K^{-1}

$E = 3.50$ in. $\quad c = 0.005$ in. $\quad C = 300$ in.-lb(f)/in.^{3}-deg K

$\alpha = 1.57$ in.^{3} $\quad \epsilon = 2430$ in.^{2}

Figure 10-15 shows Q, ΔT, and p as a function of frequency of rotation. Figure 10-16 shows the temperature rise in the screw channel as a function of position for the case where the frequency of rotation is 2 rps.

---

[10] J. M. McKelvey, *Ind. Eng. Chem.*, **46**, 660 (1954).

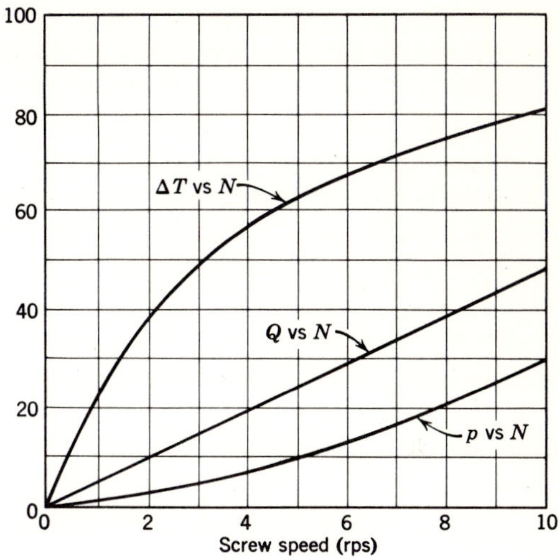

Figure 10-15. Adiabatic performance curves.

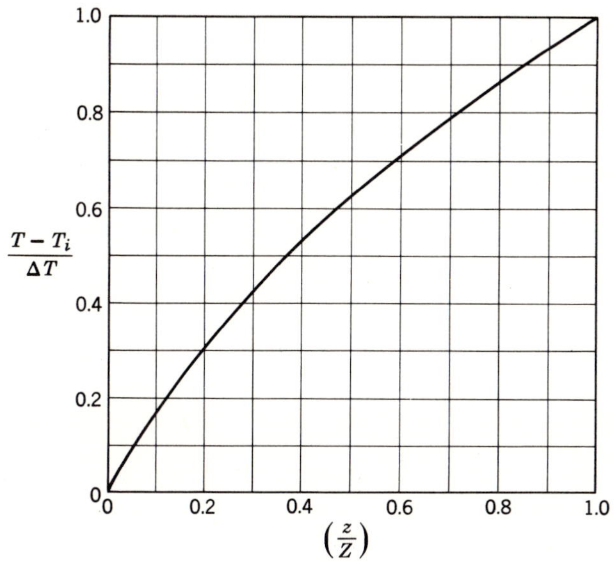

Figure 10-16. Dimensionless temperature as a function of position along screw.

10-7 Scale-Up Procedures

The scale-up of models is a useful basis of design when reliable scale-up procedures can be established. This section presents the derivation of a set of scale-up rules applying to the pumping of Newtonian fluids under adiabatic and under isothermal conditions.

With the rules derived here, the performance of a screw pump can be predicted from the performance of a geometrically similar scale model pumping the same fluid and at the same frequency of rotation. This is the most useful type of scale-up for screw pumps, for it insures that the shear rates at corresponding points in the two pumps are the same. This enables the scale-up rules to be extended under certain circumstances to the pumping of non-Newtonian fluids (discussed in Section 10-8) and to plasticating extruders, as discussed in Section 11-4.

For the case where the screws of the model and the pump are geometrically similar and differ in size by the factor X,

$$X = \frac{e'}{e} = \frac{D'}{D} = \frac{H'}{H} = \frac{L'}{L} = \frac{c'}{c} \qquad \text{and} \qquad \theta' = \theta$$

where the unprimed symbols refer to the model. Since the linear dimensions of the die are increased by the same factor, the following relationship exists between the die constants:

$$K' = X^3 K$$

Furthermore, the operating conditions and the fluid properties are the same in both cases. Therefore

$$\frac{b'}{b} = \frac{C'}{C} = \frac{T_i'}{T_i} = \frac{N'}{N} = 1$$

and the following relationships between the screw constants can be established:

$$\epsilon' = X^2 \epsilon$$

$$\alpha' = X^3 \alpha$$

$$\beta' = X^4 \beta$$

The scale-up of a model operated adiabatically is now considered. Referring to equation (10-114), it can be seen that the dimensionless quantities N_1 and N_2 must have the same value for the model as for the full-size pump. Therefore (10-113) shows that $f' = f$, and one can deduce the following relationship between the performance of the model and the full-size machine:

$$Q' = X^3 Q \qquad \Delta P' = \Delta P \qquad \Delta T' = \Delta T$$

Furthermore, the total power is $(CQ\,\Delta T + Q\,\Delta P)$, so that

$$p' = X^3 p$$

For isothermal operation, the pumping rate and the pressure rise are given by (10-84) and (10-85). For the special case where the channel dimensions are constant (not functions of z), these equations can be written

$$Q = \frac{\alpha KNZ}{\beta + KZ} \qquad \Delta P = \frac{\alpha \mu NZ}{\beta + KZ}$$

from which it follows that

$$Q' = X^3 Q \qquad \text{and} \qquad \Delta P' = \Delta P$$

and from (10-81) the following relation for power consumption can be derived:

$$p' = X^3 p$$

It can be shown that the scale-up rules derived for isothermal operation with screws having constant channel dimensions also apply to the more general case where the channel dimensions are any arbitrary function of z. In order to show that this is true, one must prove that

$$A' = X^3 A \qquad B' = X^3 B$$

The proof of the second of these is shown below. A similar procedure will prove the first relation. From the definition of B, given by (10-76), one can write

$$\frac{B'}{B} = \frac{\displaystyle\int_0^Z \frac{dz}{\beta}}{\displaystyle\int_0^{Z'} \frac{dz}{\beta'}} \tag{10-117}$$

It is noted that the point z on the model corresponds to the point (zX) on the full-size pump. Therefore

$$X^3 \beta(z) = \beta'(Xz)$$

and since $Z' = XZ$, equation (10-117) becomes

$$\frac{B'}{B} = X^3 \frac{\displaystyle\int_0^{Z'/X} \frac{dz}{\beta'(Xz)}}{\displaystyle\int_0^{Z'} \frac{dz}{\beta'(z)}} = X^3 \tag{10-118}$$

A similar examination of the power equation shows that $p' = X^3 p$.

10-8 Pumping of Non-Newtonian Fluids

For Newtonian fluids, the theory of screw pumps is on a sound theoretical basis, and the flow equations have been confirmed by experiment.

Since the power equation rests upon the same assumptions, one can be reasonably sure of its validity even though experimental verification is lacking.

Most applications of screw pumps, however, are with non-Newtonian fluids; a completely satisfactory theory for non-Newtonians has not been developed. In the approach taken here, the frame of the Newtonian theory is retained, and each part of it is modified to correct for non-Newtonian effects. The die characteristic can be handled satisfactorily with the empirical power law, provided elastic effects are not important. For drag flow, non-Newtonian behavior has a relatively small effect and can probably be handled satisfactorily by the method outlined below. The problem of pressure flow in the screw channel is different. There are several ways in which one might proceed, but unfortunately there is little experimental evidence to enable one to decide upon the validity of the methods. One must consider this problem as being unsolved at the present time.

The Newtonian die characteristic is defined by the equation

$$Q = \frac{K\,\Delta P}{\mu} \tag{10-119}$$

where μ is the Newtonian viscosity of the fluid in the die, and the die constant K depends only on the geometry of the die. This characteristic plots as a straight line on linear graph paper. Using the power law, the die characteristic for non-Newtonian flow can be written

$$Q = K'(\Delta P)^{1/n} \tag{10-120}$$

where n is the flow index of the fluid and K' is a constant that includes geometrical factors as well as material constants. For the isothermal flow of a given fluid through a given die, K' should not vary. Reference should be made to Chapter 3 for methods of computing values of K'. Note that the non-Newtonian die characteristic plots as a straight line on logarithmic paper.

The effect of non-Newtonian behavior on drag flow, or the free discharge rate, is now considered. The drag flow rate derived on the basis of the flat-plate Newtonian model is given by αN, where α depends only upon geometrical factors. Therefore, as a first approximation, the drag flow is independent of the nature of the fluid, and one would expect the same free discharge rate with a non-Newtonian as with a Newtonian fluid.

It should be noted, however, that the flat plate model is only an approximation to the actual geometry of the screw channel. It was shown in Section 10-2 that channel curvature introduces an error into the drag flow formula. The quantity $(1 - \epsilon)$ can be considered as a correction factor to be applied to the conventional drag flow formula. The function ϵ depends upon both the channel curvature and the flow index of the fluid. Values of ϵ can be obtained from Figure 10-5. For example, with a Newtonian fluid and a screw in which the channel depth is 10% of the diameter of the screw, the free discharge rate would be 97% of that calculated with the drag flow formula. With the same screw and a power law fluid having a flow index of 0.5, the free discharge rate would be about 92% of that calculated with the drag flow formula. It would appear that with deep-flighted screws and strongly non-Newtonian fluids, the drag flow correction may become significant and should be made.

Pressure flow in the screw channel is now considered. The problem is somewhat similar to that of calculating the axial pressure flow between concentric cylinders when one of the cylinders is rotating. With Newtonian fluids the axial flow will be independent of the speed of rotation of the cylinder. In other words, there is no interaction between the axial flow and the circular flow because the viscosity of a Newtonian fluid is independent of shear stress.

With a non-Newtonian fluid the situation is unclear. If one proceeded on the same basis, it would be equivalent to assuming that the apparent viscosity for flow in the axial direction is completely independent of the fluid stresses existing normal to the flow. Such an assumption seems improbable. It would appear more likely, for example, that axial flow would occur more readily when the fluid was being sheared in a direction normal to the flow and that the effect would be different for different fluids, depending upon their molecular structure. However, experimental data are needed to clarify this point.

Several alternative assumptions that might be made in order to handle the problem are discussed below. If the Newtonian viscosity in the equation for the screw characteristic, equation (10-76), is replaced with the average non-Newtonian viscosity $\bar{\eta}$, one obtains the non-Newtonian screw characteristic

$$Q = AN - \frac{B\,\Delta P}{\bar{\eta}} \tag{10-121}$$

where $\bar{\eta}$ is some unknown function of the shear stresses existing in the fluid.

For power law fluids, the viscosity is directly proportional to $\dot{\gamma}^{n-1}$. Therefore, by restricting the treatment to power law fluids, the screw characteristic can be written

$$Q = AN - C(\bar{\dot{\gamma}})^{1-n}\,\Delta P \tag{10-122}$$

where C is a constant for isothermal operation with a given fluid.

The quantity $\bar{\dot{\gamma}}$ represents the average shear rate giving rise to the average viscosity $\bar{\eta}$. The problem now becomes one of deciding how one should calculate $\bar{\dot{\gamma}}$. We examine, below, three alternate methods of calculating $\dot{\gamma}$.

(1) This method assumes that the pressure flow is completely independent of all other flows. In this case we must calculate the average shear rate resulting from pressure flow between parallel plates. Since the flow is symmetrical about the mid-plane, we have

$$\bar{\dot{\gamma}} = \frac{2}{H}\int_0^{H/2} \dot{\gamma}\, dy = -\frac{2v_0}{H}$$

where v_0 is the velocity of the fluid at the mid-plane, given by equation (3-99). Note that v_0 is proportional to $\Delta P^{1/n}$. Therefore we conclude that $\bar{\dot{\gamma}}$ is proportional to $\Delta P^{1/n}$ and equation (10-122) can be written

$$Q = AN - C'\Delta P^{1/n} \tag{10-123}$$

where C' is a constant for isothermal operation with a given fluid.

Note that the screw characteristic defined by (10-122) is nonlinear. For pseudoplastics, $n < 1$, and the negative slope of the screw characteristic increases as the pressure increases. This offers a possible test of the validity of the assumption that there is no interaction of the pressure flow with the other flows.

If this screw characteristic is combined with the non-Newtonian die characteristic, one can solve for the extrusion rate and the pressure, obtaining

$$Q = \frac{AN}{1 + (C'/K')} \tag{10-124}$$

$$\Delta P = \left[\frac{AN}{C' + K'}\right]^n \tag{10-125}$$

Note that the extrusion rate would be directly proportional to the screw speed but that the pressure developed would be proportional to the nth power of the screw speed. Thus, for pseudoplastic fluids, the increase in pressure for a given increase in screw speed would be less than for Newtonian fluids.

The maximum pressure developed by the screw occurs at closed discharge and can be obtained by solving (10-123). One obtains

$$(\Delta P)_{\max} = \left(\frac{AN}{C'}\right)^n \tag{10-126}$$

Note that another test of the validity of the assumption that there is no interaction between the flows would be to plot log $(\Delta P)_{\max}$ versus log (N).

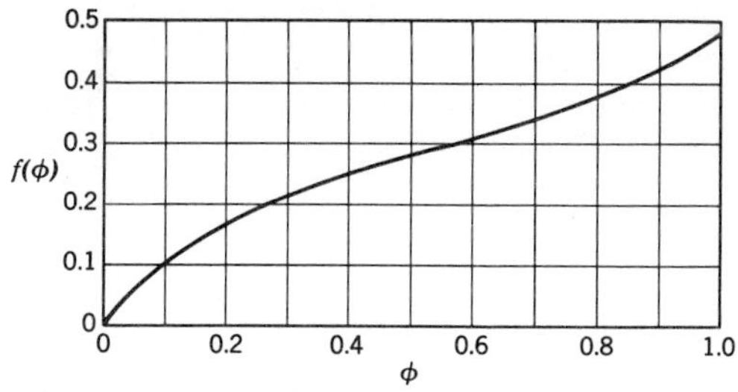

Figure 10-17. Graph for determining the value of the function $f(\phi)$.

The points should fall on a straight line and have a slope of n, the flow index for the fluid.

(2) The second approach is due to Pigott.[8] He used the volume-weighted average shear, defined by the equation

$$\bar{\dot{\gamma}} = \frac{\int_0^H |v\dot{\gamma}|\, dy}{\int_0^H |v|\, dy} \tag{10-127}$$

and computed from the velocity profile in the y-z plane established by the interaction of drag and pressure flows. Note that absolute values of v and $\dot{\gamma}$ are used in evaluating the integrals.

The velocity distribution to be used in (10-127) is given by (10-46) and the shear rate can be derived from it by differentiation. The result obtained upon evaluating (10-127) can be written

$$\bar{\dot{\gamma}} = \frac{V}{H} f(\phi) \tag{10-128}$$

where the function $f(\phi)$ is plotted in Figure 10-17.

The screw characteristic is obtained by introducing (10-128) into (10-122) to obtain

$$Q = AN - C\left[\frac{V}{H} f(\phi)\right]^{1-n} \Delta P \tag{10-129}$$

Since $f(\phi)$ is a function of ΔP, the screw characteristic must be nonlinear. Noting that V is directly proportional to N, it can be seen that the pressure developed at closed discharge is directly proportional to N^n, as was the result in the previous case.

(3) In this third case, the average shear rate is taken as (V/H), where V is the linear speed of the barrel and H the channel depth. The screw characteristic becomes

$$Q = AN - C\left(\frac{V}{H}\right)^{1-n} \Delta P \tag{10-130}$$

In this case the screw characteristic is linear, but as the screw speed increases, the negative slope of the characteristic decreases. The maximum pressure developed by the pump at closed discharge is proportional to N^n, as in the first two cases.

Example 10-5.

This example considers the operation of a screw pump under isothermal conditions with a power law fluid having a flow index of 0.50. At 30 rpm the pump delivers 50 lb/hr (0.30 in.^{3}/sec) against a head of 250 psi. It is required to estimate the pumping rate and the pressure developed upon increasing the speed to 60 rpm. The free discharge pumping rate at 30 rpm is 0.50 in.^{3}/sec.

The following methods of calculation are used in order to illustrate the above discussion:

(*a*) Newtonian flow in the screw and the die is assumed.
(*b*) Newtonian flow in the screw is assumed, but the fluid in the die is treated as a power law fluid.
(*c*) Method I, as outlined above.
(*d*) Method II, as outlined above.
(*e*) Method III, as outlined above.

Under the assumptions of (*a*), both the extrusion rate and the pressure developed are directly proportional to the screw speed. Therefore at the second operating point the pumping rate is 100 lb/hr and the pressure developed is 500 psi. This is shown graphically in Figure 10-18, where the die characteristic is shown as a straight line passing through the origin and the screw characteristics are parallel straight lines.

Under the assumptions of (*b*), the screw characteristics remain straight parallel lines, but the die characteristic shows an increasing slope. The die characteristic is conveniently determined by plotting the known point of 50 lb/hr and 250 psi on logarithmic graph paper and then drawing a straight line through the point

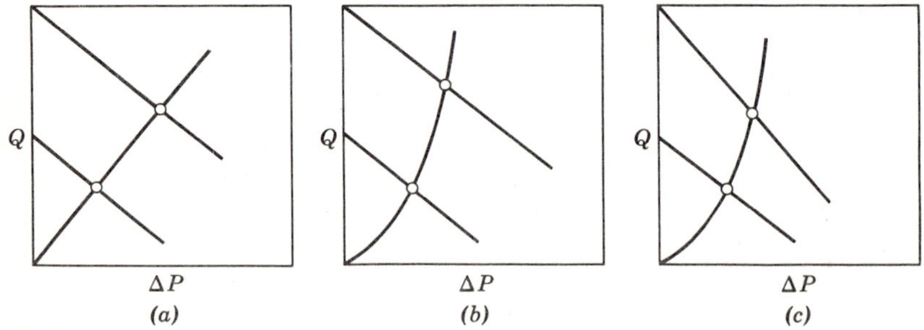

Figure 10-18. Screw and die characteristics assuming (*a*) Newtonian behavior in screw and die, (*b*) Newtonian behavior in screw and power law applicable in die, and (*c*) power law applicable in die and Method III in screw.

having a slope of n. The intersection of this characteristic (replotted on linear paper) with the screw characteristic gives the second operating point, as shown in Figure 10-18*b*. The pumping rate is 117 lb/hr and the pressure is 380 psi.

With Method I, it is noted from equation (10-124) that the extrusion rate is directly proportional to N. Therefore at the second operating point the pumping rate is 100 lb/hr. The increase in pressure, as shown by (10-125), is seen to be proportional to the nth power of the increase in screw speed. Therefore its value at the second operating point is 354 psi.

Using Method II, one can combine (10-130) and the die characteristic equation (10-120) to obtain, after considerable rearranging,

$$\frac{\phi_2}{\phi_1} = \left[\frac{f(\phi_2)}{f(\phi_1)}\right]^{1-n}\left[\frac{1-\phi_2}{1-\phi_1}\right]^n \tag{10-131}$$

Equation (10-131) contains only one unknown, ϕ_2, which is found most conveniently by a trial and error calculation. One obtains the pumping rate as 94 lb/hr and the pressure as 343 psi.

Using Method III, the solution is found in a manner somewhat similar to Method II. The screw characteristic, equation (10-130), is combined with the die characteristic to give

$$\frac{\phi_2}{\phi_1} = \left[\frac{1-\phi_2}{1-\phi_1}\right]^n \tag{10-132}$$

which is analogous to (10-131). The value of ϕ_2 is found to be 0.44, and the pumping rate and pressure are nearly the same as in Method II because the function $f(\phi)$ changes very little between $\phi = 0.40$ and $\phi = 0.44$. Figure 10-18*c* shows this result graphically. The results of all of the calculations are summarized below:

Method of Calculation	*Pumping Rate (lb/hr)*	*Pressure (psi)*
(*a*)	100	500
(*b*)	117	380
(*c*)	100	354
(*d*)	94	340
(*e*)	94	340

Unfortunately, there is little experimental data available to test the various methods of calculating screw characteristics for non-Newtonian fluids. The experiments of McKelvey[9] with polyethylene terephthalate are somewhat inconclusive. In these experiments, which were carried out under isothermal conditions, the molten polymer was pumped through a sand-pack and a spinnerette die. Two different dies were used, and three points were obtained for each die characteristic. These points are plotted in Figure 10-19. Note that on the logarithmic graph paper the points fall on a straight line indicating that the power law is applicable. The slopes of both lines are the same and the flow index is 0.90, indicating that the fluid is mildly non-Newtonian.

Screw characteristics for this material at 0.087 rps and at 0.256 rps were also reported. According to Method III, the negative slope of the screw characteristics should be proportional to the screw speed raised to the (1-n) power. Since the ratio of the two screw speeds is 2.94, one would expect the negative slope of the screw characteristic at the higher speed to be about 11% greater than the slope of the characteristic at the lower speed. However, the measured slopes of the two characteristics are nearly equal. Since the experimental error was about the same order of magnitude, the experiment must be considered to be inconclusive. Also, it should be noted that the measured characteristic covered only the range from $0 < \phi < \frac{1}{3}$, so that its apparent linear nature might have been lost at higher pressures. One can conclude from this experiment that with fluids for which $n > 0.9$ and for which $\phi < \frac{1}{3}$, the screw characteristics can be calculated with the Newtonian formulas with little error, although the power law should be used for the die characteristic.

Corrections to the power formula are also necessary. The recommended procedure is to use equation (10-81), replacing μ with $\bar{\eta}$ evaluated at a shear rate of (V/H), and calculating the constant ϵ with the modified formula

$$\epsilon' = \frac{\pi^3 D^3 \sin\theta}{H}\left[1 + 3\sin^3\theta + \frac{\bar{\eta}_c}{\bar{\eta}}\left(\frac{eH}{cW}\right)\right]\left[1 - \left(\frac{2c}{D}\right) - \frac{em}{\pi D \sin\theta}\right]$$

where $\bar{\eta}_c$ represents the viscosity of the fluid in the clearance. This will

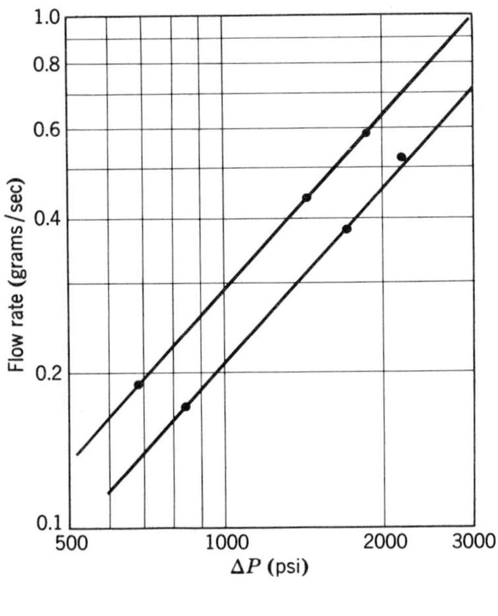

Figure 10-19. Die characteristics for polyethylene terephthalate (data from Ref. 9).

usually be less than in the bulk of the channel because of the higher shear rates that exist in the clearance. Assuming that the power law is applicable,

$$\frac{\bar{\eta}_c}{\bar{\eta}.} = \left(\frac{\dot{\gamma}_c}{\gamma}\right)^{n-1} = \left(\frac{H}{c}\right)^{n-1}$$

so that the formula for ϵ' can be written

$$\epsilon' = \frac{\pi^3 D^3 \sin\theta}{H}\left[1 + 3\sin^2\theta + \frac{e}{W}\left(\frac{H}{c}\right)^n\right]\left[1 - \left(\frac{2c}{D}\right) - \frac{em}{\pi D \sin\theta}\right] \tag{10-133}$$

The scale-up of screw pumps operating isothermally with non-Newtonian fluids is now considered. The treatment is restricted to pumps having geometrically similar screws operating at the same frequency of rotation. For a given value of ϕ, the shear patterns established in the two screws will be geometrically similar. Therefore at corresponding points in the model and the full size pump, the shear rates are the same, and, the average viscosity $\bar{\eta}$ must be the same for the model as for the full size machine. Referring to the generalized screw characteristic, equation (10-121), and noting that the constants A and B for the full size pump are X^3 times larger than for the model, it is concluded that the screw characteristic for the full size machine is the same as for the model, except that it is displaced upward an amount X^3Q.

If the die for the model is geometrically similar to the full size die, differing in all linear dimensions by the factor X, then the two die constants, as defined by (10-120), will differ by a factor X^3. This is easily proved for cylindrical and flat plate dies and power law fluids. Refer to equations (3-9) and (3-101).

The die characteristic is combined with the generalized screw characteristic to eliminate Q. One obtains

$$(\Delta P)^{1/n} + \left(\frac{B}{K'\bar{\eta}}\right)\Delta P - \left(\frac{AN}{K'}\right) = 0$$

and because of the relations

$$A = X^3 A'$$

$$B = X^3 B'$$

$$(K') = X^3 (K')'$$

one concludes that the pressure developed by the model is the same as that developed by the full size pump. Because of the relationship between the die constants, the full size pump must pump at a rate X^3 times greater than that of the model. By similar arguments it can be shown that the power requirement for the full size pump is X^3 times that of the model.

10-9 Summary of Flow and Power Equations for Screw Pumps

For convenient reference, the flow equations, the power equations, and the notation used in the hydrodynamic analysis of screw pumps are summarized in the following tables:

1. Differential Equations for Flow and Power

$$Q = \alpha N - \frac{\beta}{\mu}\left(\frac{\partial P}{\partial z}\right)$$

$$dp = \epsilon\mu N^2\, dz + \alpha N\, dP$$

2. Equations for Isothermal Operation

Screw characteristic:

$$Q = AN - B\left(\frac{\Delta P}{\mu}\right)$$

Die characteristic:

$$Q = K\left(\frac{\Delta P}{\mu}\right)$$

Operating equations:

$$Q = \left(\frac{AK}{B + K}\right)N$$

$$\Delta P = \left(\frac{A}{B + K}\right)\mu N$$

3. Equations for Adiabatic Operation (Screw channel uniform)

Screw characteristic:

$$Q = \alpha N - \left(\frac{\epsilon\beta}{C}\right)\left(\frac{N^2\,\Delta P}{Q\,\Delta T}\right)$$

Operating equations:

$$Q = \frac{\mu_i b\epsilon N^2 Z}{C(X - 1)}$$

$$\Delta T = \left(\frac{1}{b}\right)\ln X$$

$$\Delta P = \frac{\mu_i Q}{KX}$$

$$p = CQ\,\Delta T + Q\,\Delta P$$

4. Equations for Geometric Parameters

$$\alpha = \frac{\pi^2 D^2 H \sin\theta\cos\theta}{2}\left[1 - \frac{2c}{D} - \frac{em}{\pi D \sin\theta}\right]\left[1 - \frac{c}{H}\right]^2 F_D$$

$$\beta = \frac{\pi D H^3 \sin\theta}{12}\left[1 - \frac{2c}{D} - \frac{em}{\pi D \sin\theta}\right]\left[1 + \left(\frac{c}{H}\right)^3\left(\frac{W}{e}\right)\left(\frac{1}{\sin\theta\cos\theta}\right)\right]F_P$$

$$\epsilon = \frac{\pi^3 D^3 \sin\theta}{H}\left[1 + 3\sin^2\theta + \frac{eH}{cW}\right]\left[1 - \frac{2c}{D} - \frac{em}{\pi D \sin\theta}\right]$$

$$A = \frac{\int_0^Z \left(\frac{\alpha}{\beta}\right) dz}{\int_0^Z \left(\frac{1}{\beta}\right) dz}$$

$$B = \frac{1}{\int_0^Z \left(\frac{1}{\beta}\right) dz}$$

5. Notation for Screw Pumps (with cgs units)

C = specific heat of fluid (cal/cm³-deg C)
D = inside diameter of barrel (cm)
E = pitch of screw flight (cm)
F_D = shape factor for drag flow, defined by equation (10-16)
F_P = shape factor for pressure flow, defined by equation (10-17)
H = distance from root of screw to barrel surface (cm)
K = die constant (cm³)
L = axial length of screw (cm)
N = frequency of rotation of screw (sec^{-1})
P = pressure (dynes/cm²)
Q = volumetric flow rate (cm³/sec)
T = temperature (°C)
W = width of screw channel (cm); $m(W + e) = E\cos\theta$
X = parameter defined by equation (10-113)
Z = helical length of screw channel (cm)
b = temperature coefficient of viscosity (deg C^{-1})
c = radial clearance (cm)
e = width of screw flight (cm)
m = number of channels in parallel
p = power (ergs)
z = position coordinate measured along helix (cm)
θ = helix angle
μ = viscosity (dyne-sec/cm²)

Problems

10-1 Show that the general formulas for isothermal operation given in Section 10-9 reduce to the following for the special case where the channel dimensions are independent of z, and where $m = 1$, $e \approx 0$, $c \approx 0$, and $W \gg H$:

$$Q = \frac{\pi^2 D^2 N H \sin\theta \cos\theta}{2} - \frac{\pi D H^3 \sin^2\theta \, \Delta P}{12 \mu L}$$

$$p = \frac{\pi^3 D^3 N^2 L}{H} [1 + 3\sin^2\theta] + \left[\frac{\pi^2 D^2 N H \sin\theta \cos\theta}{2}\right] \Delta P$$

10-2 Using the flow and power equations given in Problem 10-1, show that the efficiency of a screw pump that pumps against a fixed resistance is independent of the viscosity of the fluid and the frequency of rotation of the screw. Calculate the efficiency of a pump whose screw has the following dimensions:

$$D = 10 \text{ cm} \qquad L = 100 \text{ cm}$$
$$H = 1 \text{ cm} \qquad \theta = 20^\circ$$

The resistance against which the screw is pumping is such that with a fluid having a viscosity of 100 poise, a pressure drop of 100 atm is required to force the fluid through at a rate of 500 lb/hr. The density of the fluid is 1 gram/cm^3.

10-3 Design a screw pump that will deliver 1000 lb/hr of a Newtonian fluid having a density of 1 gram/cm^3 against a pressure of 500 psi. The viscosity of the fluid is 100 poise. For your design specify the length, diameter, helix angle, channel depth, and frequency of rotation of the screw. List all assumptions involved in the design.

10-4 Calculate the efficiency of the pump described in problem 2, assuming that it operates adiabatically and that

$$C = 0.20 \text{ cal/cm}^3\text{-deg. C}$$
$$b = 0.025 \text{ deg. C}^{-1}$$
$$\mu_i = 10^4 \text{ poise}$$
$$N = 2 \text{ sec}^{-1}$$

chapter 11

Plasticating Extruders

The plasticating extruder is one of the most versatile machines used in polymer processing. It is the basis of many commercially important thermoplastic molding, extruding, and mixing processes. A plasticating extruder is essentially a screw pump that is fed a solid thermoplastic polymer. In addition to functioning as a pump, it functions as a conveyer and a melter and, in some cases, as a heat exchanger and a mixer.

11-1 Introduction

One can consider a plasticating extruder to be a screw pump to which a solids transport machine and a polymer melting machine have been attached. This concept is illustrated by Figure 11-1. In zone 1 the polymer is a solid; in zone 2 it is a "slush" consisting of a mixture of the molten

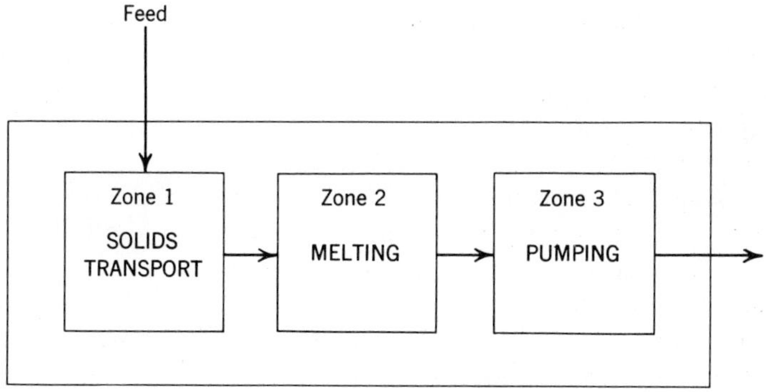

Figure 11-1. Functional representation of a plasticating extruder.

polymer and solid particles; and in zone 3 it is completely molten and thus a fluid. The over-all performance of the machine obviously depends upon how well it performs its three individual functions. In principle, its over-all performance can be determined by analyzing the performance of the three zones and then coupling the results.

The unrestricted solids transport rate is designated r_1^*, the unrestricted melting rate r_2^*, and the unrestricted pumping rate r_3^*. These are the rates at which these processes would occur if they operated *independently* of the others. Under the conditions

$$r_1^* < r_3^* \qquad \text{or} \qquad r_2^* < r_3^* \tag{11-1}$$

the screw is commonly said to be "starved." A starved screw usually leads to erratic operation and an extrudate of poor quality. In order to avoid "starving" the screw it is necessary to design it so that

$$r_1^* \geq r_2^* \geq r_3^* \tag{11-2}$$

When the inequalities of (11-2) apply, zone 3 is said to "control" the operation. When zone 3 controls, the operation is more stable and the extrudate quality is better than when either of the other zones is controlling. However, if the inequalities of equation (11-2) are too large, the control of zone 3 can be "overridden," and again an erratic operation may result.

The performance of zone 3 of the plasticating extruder can be analyzed, using the concepts developed in the previous chapter on screw pumps. Some progress has been made in developing a theory of solids transport, although little experimental evidence is available to justify using the results as the basis of design. With regard to zone 2, few investigations of either a theoretical or experimental nature have been reported, and serious difficulties stand in the way of developing a general solution of the problem. It must be concluded, therefore, that plasticating extrusion has not yielded to mathematical analysis, and that one must be content at this time with a much less precise description of the operation of plasticating extruders than of screw pumps.

When zone 3 controls the operation, and neither of the other two zones has a tendency to override this control, then one can apply the operating equations developed for screw pumps with some success to the operation of a plasticating extruder. This is discussed in Section 11-4. Furthermore, under these conditions one can draw some general conclusions concerning the "fine scale" performance characteristics—primarily the temperature homogeneity of the extrudate and uniformity of extrusion rate. These fine scale characteristics are closely related to the quality of the extrudate.

11-2 Solids Transport

Solids transport occurs only in zone 1 of a plasticating extruder. The function of this zone is to transport the solid particles from the bottom of the feed hopper to zone 2, where melting occurs. From a practical viewpoint, solids transport does not appear to be a particularly serious problem for plasticating extruders whose feed consists of free-flowing particles having a relatively high bulk density. With nonflowing powders of low bulk density, solids transport in zone 1 may present a serious problem.

It appears likely that directly underneath the feed hopper, the screw normally acts as a conveyer. Under ideal conditions the individual particle paths would be straight lines parallel to the axis of the screw. The rate of mass transport under these conditions is readily calculated.

Upon being conveyed into the barrel proper, the mechanism of solids transport probably changes. The particles become somewhat compacted and tend to move as a single mass or plug. The motion of the plug results from frictional forces exerted on it from the metal surfaces of the barrel and screw, and can be analyzed[1] if certain simplifying assumptions are made.

Certain general conclusions concerning solids transport in zone 1 of a plasticating extruder have been drawn. Unfortunately, no comprehensive experimental investigations of solids transport in extruders have been made, so one cannot be sure if the conclusions are based upon a correct theory.

The basic assumption made in the analysis is that the solid particles pack together to form an "elastic" plug that does not undergo shear flow. The plug is acted upon by frictional forces between its surfaces and the surfaces of the barrel and channel. The analysis is facilitated by considering the barrel to be stationary and the screw to be in motion, as shown in Figure 11-2.

The initial position of the screw channel is shown in the upper diagram. Point A is a reference point on the plug. The motion of the screw channel is downward at the speed V. In time t it has moved a distance A-B and is located in the position shown in the lower diagram of the figure. Note that the reference point has now moved to point A'. Therefore the plug has moved distance B-A' in the direction of the channel.

The motion of the reference point relative to the barrel is along the line A-A'. The motion of the reference point relative to the screw is along the line B-A'. The angle ϕ gives the inclination of A-A' from the vertical. Note that the positive direction of ϕ is taken to be counterclockwise.

[1] W. H. Darnell and E. A. J. Mol, *S.P.E. Journal*, **12**, 20 (1956).

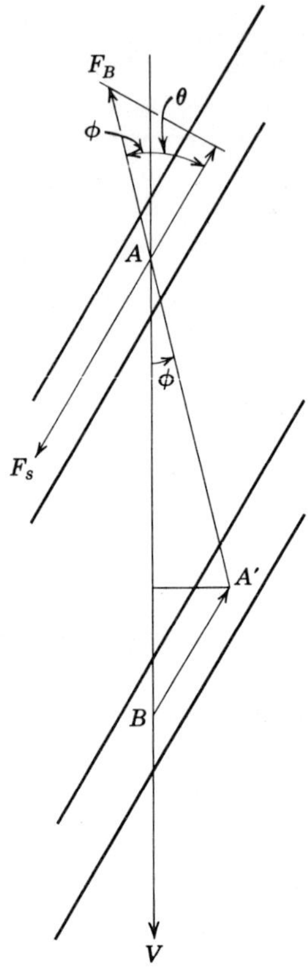

Figure 11-2. Motion of an elastic plug in screw channel.

The speed v at which the plug advances down the channel is readily determined as a function of ϕ:

$$v = \frac{(B - A')}{t} = V\frac{(B - A')}{(A - B)} \tag{11-3}$$

Noting that B-A' is inclined at the angle θ from the vertical, where θ is the helix angle of the screw, equation (11-3) can be transformed to

$$v = V\frac{\sin(\phi)}{\sin(\phi + \theta)} \tag{11-4}$$

and by using standard trigonometric identities, equation (11-4) can be transformed into

$$v = \frac{V}{\cos(\theta) + \sin(\theta)\cot(\phi)} \tag{11-5}$$

We now consider the frictional forces acting upon the plug. For steady motion, these forces must be balanced, and we can write, with reference to Figure 11-2, that

$$F_s = F_B \cos(\phi + \theta) \tag{11-6}$$

Since θ and ϕ must always be positive, the frictional forces acting upon the plug from the barrel must *always exceed* the frictional forces acting upon the plug from the screw surfaces.

Equation (11-6) can be written in the form

$$\cos(\theta + \phi) = R \qquad (R \leq 1) \tag{11-7}$$

where R is the ratio (F_s/F_B).

If the value of R is known, then the value of ϕ can be determined from (11-7) and the solids transport rate calculated from (11-5). Note that up to this point in the analysis, the only assumption made has been that the mass of particles moves as a plug. If one wishes to carry the analysis further, then additional assumptions concerning the nature of the forces F_B and F_S must be made. Before doing this, however, it is possible to deduce the behavior of the plug for certain limiting cases.

Note first that when $R \geq 1$, motion is not possible. In this case the angle ϕ is zero. Equation (11-5) shows that v is also zero. Thus the plug makes no advance down the channel. It simply "rides the screw, round and round." Under these conditions the extrusion operation obviously comes to a halt. Note that this must occur whenever the value of R becomes equal to or greater than unity.

The second limiting case occurs when the screw is so exceedingly "slippery" that it can exert no frictional forces on the plug. If there is no pressure gradient in the channel, the value of F_S becomes zero, so that $R = 0$. It can be seen from equation (6-7) that $(\theta + \phi) = 90°$. Therefore equation (11-5) can be written

$$v = \frac{V}{\cos(\theta) + \sin(\theta)\cot(90° - \theta)} = V\cos(\theta) = V_z = 2\bar{v} \tag{11-8}$$

showing that the speed of the plug is just equal to the component of the barrel velocity directed down the screw channel. Since the average drag velocity for fluids $\bar{v}$ in the direction of the channel is $V_z/2$, the screw is capable of transporting solids at a volumetric rate twice that at which it can pump fluids under drag flow conditions.

A third special case is now considered. If longitudinal grooves were cut into the surface of the barrel to restrict the motion of the plug to the axial direction, then the angle ϕ would be 90° and equation (11-5) would reduce to

$$v = \frac{V}{\cos(\theta)} \tag{11-9}$$

which can be written in the equivalent form

$$v = \frac{2\bar{v}}{\cos^2 \theta}$$

where $\bar{v}$ is the average drag velocity for fluids. These results are summarized below:

Limiting Conditions	Speed of Plug
$R \geq 1$	0
$R = 0$	$2\bar{v}$
$\phi = 90°$	$2\bar{v}/\cos^2 \theta$

Now consider the conditions under which motion of the plug in the screw channel would cease. Equation (11-5) shows that this condition will arise whenever the angle ϕ becomes zero. Designate the force ratio at which this condition occurs as R^*. From equation (11-7) it can be seen that R^* is a function of the helix angle. For large helix angles, a more favorable force ratio is required to keep the operation going than at small helix angles. For example, as the helix angle approaches 90°, steady operation could be obtained only with a screw that has a zero coefficient of friction with regard to the plug, a condition that obviously cannot be obtained. On the other hand, as the helix angle is decreased, a less favorable force ratio can be tolerated. In the limit, as the helix angle decreases to zero, the value of R^* becomes unity.

It seems reasonable to assume that during the operation of a plasticating extruder, the force ratio is not a constant. For a given set of conditions it undoubtedly fluctuates between limits. Furthermore, as screw speed, barrel temperature, die pressure, and all other factors change, it is likely that they will affect the force ratio. With screws having large helix angles in the feed zone, one would have a greater risk of having an intermittent operation, or complete stoppage of feeding, than with a screw having a small helix angle.

While the analysis indicates that small helix angles give a more positive and certain solids transport action, it does not reveal how small the helix angle should be made nor what the solids transport rate should be. To obtain information of this type, a more detailed analysis of the forces acting upon the plug is required.

In order to do this, the simplifying assumption is made that the plug exerts "hydrostatic" pressure equally on all surfaces with which it is in contact. Furthermore, it is assumed that there is no pressure gradient in the direction of the channel. Under these conditions the tangential (frictional) forces exerted on the plug are obtained by multiplying the surface area by the "pressure" and then by the coefficient of friction. The force ratio R, for this special case, becomes

$$R = \frac{F_s}{F_B} = \frac{A_s\mu_s}{A_B\mu_B} \tag{11-10}$$

where A_s is the total surface area of the screw in contact with the plug and A_B is the area of the barrel in contact with the plug. The μ's represent coefficients of friction for the barrel and screw surfaces.

If the width of the screw flight is neglected

$$A_B = \pi DL \tag{11-11}$$

and the value of A_s is given by the equation

$$A_s = \pi(D - 2H)L + 2\int_{R_0}^{R} Z\, dr \tag{11-12}$$

where Z is the length of a helical path. The helical path length is a function of the radial position coordinate r and can be related to the axial length L by the equation

$$Z = \frac{L}{\sin(\zeta)} \tag{11-13}$$

where the helix angle ζ is a function of radial position. Combining equations (11-11), (11-12), and (11-13) gives

$$\frac{A_s}{A_B} = 1 - 2\left(\frac{H}{D}\right) + \frac{2}{\pi D}\int_{R_0}^{R} \frac{dr}{\sin(\theta)} \tag{11-14}$$

The width of the screw flight perpendicular to the z-axis of the channel is given by the equation

$$W = \pi D \sin\theta \qquad \theta = \zeta(R) \tag{11-15}$$

where D and θ are measured at the crest of the flight. Because W does not vary with r, the following relationships are readily established:

$$D \sin\theta = D_0 \sin\zeta_0 = 2r \sin\zeta \tag{11-16}$$

where ζ_0 is the helix angle at the root of the screw and D_0 is the root diameter. Therefore we have

$$\frac{A_s}{A_B} = 1 - 2\left(\frac{H}{D}\right) + \frac{4}{\pi D^2 \sin\theta}\int_{R_0}^{R} r\,dr \tag{11-17}$$

which, upon integrating, becomes

$$\frac{A_s}{A_B} = 1 - 2\left(\frac{H}{D}\right) + \frac{2(R - R_0)(R + R_0)}{\pi D^2 \sin\theta} \tag{11-18}$$

Equation (11–18) can be rearranged to

$$\frac{A_s}{A_B} = 1 - 2\left(\frac{H}{D}\right)\left[1 - \frac{1 - (H/D)}{\pi \sin\theta}\right] \tag{11-19}$$

The use of (11-19) is illustrated with the following numerical example.

Example 11-1.

In this example the solids transport rate for a 2-inch diameter screw having a 36° helix angle and a channel depth of $\frac{1}{2}$ inch is estimated. It is assumed that all surfaces have the same coefficient of friction, so that the force ratio reduces to the area ratio (A_s/A_B).

The area ratio, computed with equation (11-19), is 0.7. From equation (11-7) we find that

$$\theta + \phi = 46^\circ$$

so that the angle ϕ has the value of 10°.

The average speed of the plug v in the direction of the channel, computed with equation (11-5), is $0.24V$. The volumetric solids transport rate is given by

$$Q = vWH$$

and since $V = \pi DN \cos\theta$, the volumetric rate per revolution is

$$\frac{Q}{N} = (0.24)\,\pi DWH \cos\theta$$

If W is expressed in terms of the helix angle with the equation $W = \pi D \sin\theta$, the volumetric rate becomes

$$\left(\frac{Q}{N}\right) = (0.24)\,\pi^2 D^2 H \sin(\theta)\cos(\theta)$$

and for the particular case

$$\frac{Q}{N} = 2.25 \text{ in.}^3/\text{rev}$$

If the particles have a bulk density of 0.613 grams/cm^3, the solids transport on a mass basis becomes

$$\frac{W}{N} = 22 \text{ grams/rev}$$

This can be compared with the value of 27.9 grams/rev calculated by Darnell[2] for this same problem by using somewhat different assumptions in establishing the force balance, and the value of 32.9 grams/rev, which he reported as having been measured in an experiment.

11-3 Melting

It is useful to distinguish between two essentially different mechanisms by which melting can occur in a plasticating extruder. The first mechanism involves the transfer of heat from the barrel of the extruder to the particles of the polymer. The rate at which this process can happen depends upon the surface area of the barrel, the thermal conductivity of the plastic, and the flow patterns within the screw channel. The second mechanism does not involve transferred heat. In it, the heat energy required to melt the polymer comes from the dissipation of mechanical energy into heat through the deformation of the solid plastic.

If the first mechanism is operative, the maximum melting rate (and consequently the maximum extrusion rate) is governed, roughly, by the surface area of the barrel—by the factor πDL. If the second mechanism is operative, the melting rate is limited by the amount of mechanical power that can be transmitted through the drive shaft and the root of the screw. Usually this gives a much higher limit.

In most plasticating extruders, the first mechanism of melting is operative. Therefore the ultimate extrusion rate of any given extruder is limited or controlled by heat transfer, and the extruder should be considered primarily as a heat exchanger. This section presents an analysis (primarily qualitative) of heat transfer within the screw channel. Certain general conclusions, useful in the design of heat-transfer-controlled plasticating extruders, are drawn.

Sometimes it is possible to operate a plasticating extruder so that the second mechanism of melting is operative. Doing so offers certain advantages over the first method. Much higher rates can usually be obtained with machines of the same length and diameter, and the plastic is heated more uniformly and with less danger of local overheating. However, it appears likely that operation in this manner is possible only with plastics that are relatively "soft" at room temperature, as will be discussed later in this section.

In the extrusion of those plastics in which melting occurs through the first mechanism, "metering" type screws are almost always used. Figure 11-3 is a sketch of a typical metering type screw. Many modifications exist, but they all share the common features described below. The (L/D) ratio

[2] W. H. Darnell, in ch. 4 of *Processing of Thermoplastic Materials*, edited by E. C. Bernhardt, Reinhold Publishing Corp., 1959.

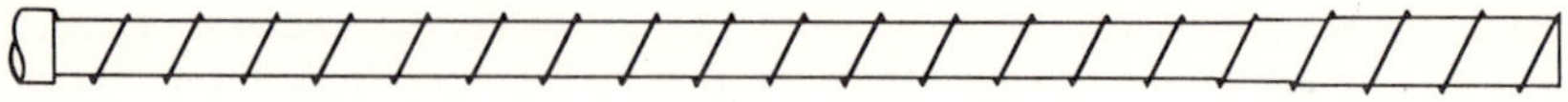

Figure 11-3. Typical "metering" screw. Constant pitch with compression occurring over 2 to 3 turns of the helix.

ranges from about 16 to 24, and the (E/D) ratio is nearly always about unity. Solids transport occurs within the first 2 or 3 turns of the flight and fluid pumping (metering) in the last 3 or 4 flights. The middle (melting) zone accounts for most of the screw length. The transition from the melting zone to the metering zone can occur over as little as half a turn of the flight to as much as 2 or 3 turns. The design of these screws has evolved over a period of 10 years or more, and the screws work reasonably well with almost all types of plastics. Their design illustrates the importance of the melting zone and heat transfer in plasticating extrusion.

The rate at which the polymer can be melted is determined by the rate at which heat can be transferred from the barrel surface to the polymer. As a first approximation one might expect the heat transfer rate for screws of similar design and for the same material to be proportional to the second power of the diameter. On this basis a 4-inch extruder would have 16 times the capacity of a similar 1-inch extruder. This exponential rule is a conservative basis for the scale-up of plasticating extruders in which heat transfer is the controlling factor. Jepson[3] has shown that many commercial extruders are rated and their drive motors supplied on this basis.

A survey of more recent estimates[4] of extruder capacities for nine different plastics shows that the exponential rule is still followed, with the exponent ranging from about 2 to $2\frac{1}{2}$. The larger exponent probably reflects the fact that heat generation due to shear deformation can be a significant factor for certain plastics.

The melting process is now examined in closer detail. The investigation of Maddock[5] reveals some of the details of the process. He extruded polyethylene under various conditions. When steady state was reached, the screw was stopped and the barrel and the screw chilled. After solidification of the plastic in the screw channel, the screw was forced from the barrel with a pneumatic piston under the influence of slight barrel heating. The plastic was then peeled from the channel and sectioned.

Upon examination of the sectioned samples, Maddock concluded that the melting process starts when the particles in contact with the barrel surface partially melt and smear a film of molten polymer over the barrel

[3] C. H. Jepson, *Ind. Eng. Chem.*, **45**, 992 (1953).

[4] *Modern Plastics Encyclopedia* for 1958, Breskin Publ., Inc., New York.

[5] B. H. Maddock, *S.P.E. Journal*, **15**, 383 (1959).

surface. This might be called surface melting and is represented schematically by Figure 11-4*a*. During this stage of the melting process, heat transfer is undoubtedly slow, as the porous solid mass or plug under the surface film offers a poor path for heat flow.

As surface melting continues, the bulk density of the plug in the channel increases, and the particles lose firm contact with the screw surfaces. They are then relatively free to move in the channel, and the motion of the screw relative to the barrel causes the molten material, together with some unmelted particles, to collect against the rear flight of the screw, as shown in Figure 11-4*b*.

As the "slush" in the rear of the channel increases in volume, it begins a circulating flow, somewhat like that shown in Figure 11-4*c*. Unmelted particles are carried into the slush and, being surrounded by the hot fluid, rapidly melt. Heat transfer between the barrel and the fluid is good because of the motion of the fluid. Once the "slush" stage is reached, melting should occur quite rapidly.

The possibility of providing the heat needed to melt the plastic by generating it through shear deformation has been mentioned. Polyethylene is one of the "soft" plastics for which this mode of melting is a possibility. The results of Bernhardt and McKelvey[6] are cited as an example. They extruded polyethylene with three different screws and over a wide range of operating conditions, under conditions where heat

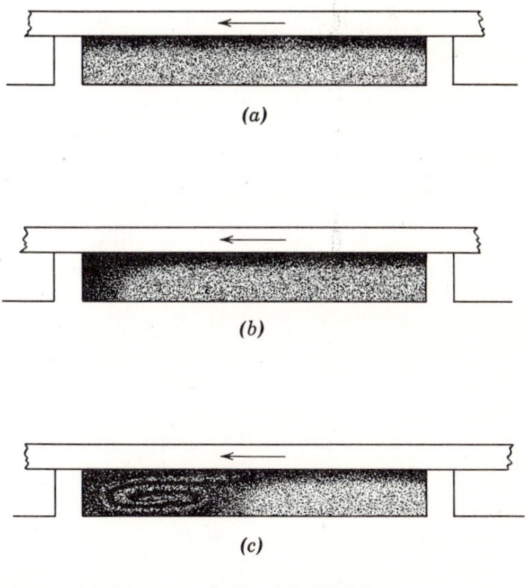

(a) (b) (c)

Figure 11-4. Representation of melting processing in extruder screw channel.

[6] E. C. Bernhardt and J. M. McKelvey, *S.P.E. Journal*, **10**, 419 (1954).

generation through mechanical working predominated. The results of their Run No. 10, which are typical, are shown below:

Operating Variables	
Frequency of rotation of screw:	2.50 rps
Rate of heat input:	0.24 kw
Performance Variables	
Extrusion rate:	71.4 lb/hr
Melt temperature at die:	311° F
Melt pressure at die:	1760 psi
Mechanical power input from motor:	5.62 kw

The total power input is 5.86 kw, of which 0.24 is supplied as thermal energy and 5.62 as mechanical energy. The rate of increase of the enthalpy of the polyethylene in the operation is 5.23 kw. Even if all of the thermal energy were utilized, it could account for only 4.6% of the energy supplied to the polyethylene.

This extrusion, therefore, is almost completely independent of external heat transfer to the plastic. In this case the upper limit to the capacity of the extruder may be the strength of the shaft (or the root of the screw) through which the mechanical power is transmitted. Of course, at extremely high rotational frequencies, it may become difficult to transfer the plastic from the feed hopper to the screw, and the conveying action of the screw may become less efficient.

The possibility of melting by mechanical working in a plasticating extruder has been demonstrated. However, all plasticating extruders will not be equally efficient in mode of operation. For example, the screw shown in Figure 11-3 would generate little frictional heat in the plastic until the point was reached where the channel decreased abruptly in depth. At that point melting would occur. However, it is so close to the end of the screw that there would be insufficient length to raise the temperature of the melt to the desired level.

The screws used by Bernhardt and McKelvey differed radically from the type shown in Figure 11-3. Their screws decreased abruptly in channel depth after only 2 or 3 turns of the flight. The objective of this design was to melt the plastic as quickly as possible in order to provide sufficient length to raise the melt to the desired temperature. Under normal operating conditions, it was found that the solids transport action was insufficient to force the solid particles into the melting zone. Therefore the melting capacity of the screw was far less than its pumping capacity.

Bernhardt and McKelvey found, however, that the screws could be made to work by placing electrical heaters in their core at the point of compression. It is believed that this caused a layer of polyethylene next to the root of the screw to be melted, thus enabling the plug of polyethylene to slip easily over the root of the screw. Thus the solids transport action was intensified and the plug was forced into the melting zone at the required rate. It appears likely that a screw designed somewhat as a compromise between that of Figure 11-3 and the ones described above would be well suited for melting plastics through mechanical working.

11-4 Performance Characteristics

The performance characteristics of plasticating extruders include extrusion rate and temperature, pressure developed by the screw, and power. Their relationship to the design and operating variables is discussed in this section, together with an additional factor—the uniformity of temperature and pressure in the extrudate. This fine-scale character of the extrudate is an extremely important factor in determining the quality of products produced in certain extrusion processes, for example, in the manufacture of plastic film.

The over-all performance of a plasticating extruder depends upon the capabilities of the screw to perform each of its three functions: solids transport, melting, and pumping. This point is illustrated by the experiments of Bernhardt and McKelvey, which were discussed in the last section. It was pointed out that the performance of the melting section of the screw could be drastically changed by changing the temperature of the root of the screw in the melting section. Since the extruder used in the experiments had pressure gages installed in the barrel, it was possible to observe directly the effect of changing the ability of the screw to perform one of its functions.

Figure 11-5 shows typical pressure profiles obtained under three different conditions of operation. Curves of type A are obtained when the melting capacity of the screw is less than its pumping capacity. Under these conditions the screw is "starved" and

$$r_2^* < r_3^*$$

where r_2^* is the unrestricted melting rate and r_3^* the unrestricted pumping rate. Since, for steady operation, the two rates must be equal, only a fraction of the pumping section of the screw is required to match the performance of the melting section. Hence the pumping section operates partially empty for part of its length. The point at which the channel becomes filled is the point at which the pressure rises.

Curves of type B are obtained when

$$r_2^* = r_3^*$$

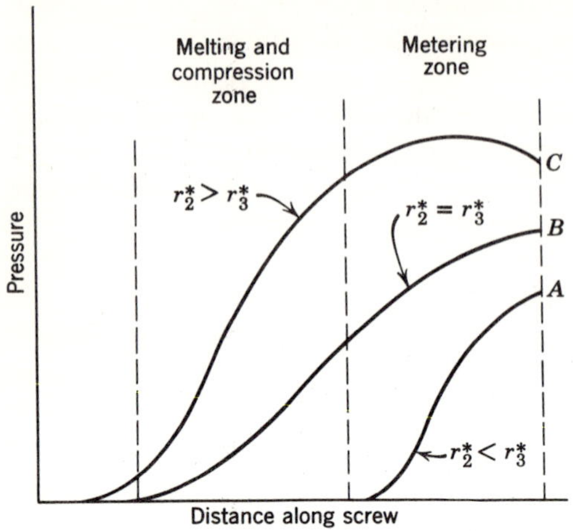

Figure 11-5. Pressure profiles in a plasticating extruder under different conditions of operation.

that is, when the screw melts polymer at the same rate that the full pumping section can force it through the die. Under these conditions the pressure starts to increase at the beginning of the pumping section, and a higher pressure is developed at the die, so that the extrusion rate rises above that of type *A*.

Curves of type *C* are obtained when

$$r_2^* > r_3^*$$

Under these conditions the unrestricted melting capacity of the screw is greater than its pumping capacity. The polymer becomes molten at a point farther back in the melting section, thereby reducing the melting capacity somewhat and increasing the pumping capacity. Note that the pressure rises very sharply in the melting zone due to the sharp compression of the screw channel there. In the extreme case shown in the figure, the pressure at the entrance to the pumping zone is higher than the pressure at the end of the pumping zone. Thus, the pressure flow in the screw channel is in the forward rather than in the reverse direction, and extrusion rates higher than the drag flow rate of the screw are observed.

If a plasticating extruder is operated so that the conditions of curve *B* are attained, then its performance characteristics will be nearly the same as those of a screw pump having the same dimensions as the plasticating extruder. As in the case of screw pumps, it is useful to distinguish between the isothermal and the adiabatic modes of operation.

The experiments of Grant and Walker[7] illustrate the performance of a

[7] D. Grant and W. Walker, in *Plastics Progress*, edited by P. Morgan, Iliffe and Sons, Ltd., 1951.

plasticating extruder under isothermal operating conditions and under conditions where the three functions of the screw are nearly balanced. They extruded polyethylene with a 1.25-inch steam-jacketed extruder and measured the extrusion rate and the pressure developed by the screw as a function of screw speed, temperature, and die resistance.

They found the extrusion rate to be a linear function of screw speed, indicating that a relationship of the form

$$Q = K_1 N \tag{11-20}$$

was applicable. Runs made with different die restrictions showed that K_1 increased as the die restriction was reduced. Runs made with the same die restriction but at different temperatures showed that K_1 was nearly independent of temperature.

They also obtained straight lines, with slope 0.42, by plotting log (ΔP) versus log (N), indicating that a relationship of the form

$$\Delta P = K_2 N^{0.42} \tag{11-21}$$

was also applicable. Runs made at different die restrictions showed that K_2 increased as the die restriction was reduced. Runs made at different temperatures showed that K_2 increased as the temperature was reduced.

These results can be interpreted with the aid of the non-Newtonian pumping theory presented in Section 10-9, where for Case I the following results were obtained for power law fluids:

$$Q = \frac{AN}{1 + (C'/K')} \tag{11-22}$$

and

$$\Delta P = \left[\frac{N}{C' + K'}\right]^n \tag{11-23}$$

where C' is the non-Newtonian screw constant and K' the non-Newtonian die constant. Both constants are temperature dependent, increasing as the temperature increases. The constant A depends only on the geometry of the screw for a screw of constant channel dimensions.

Equation (11-22) indicates that K_1 should be relatively independent of temperature, for both C' and K' are affected equally by a change in temperature. However, equation (11-22) shows that K_1 should decrease as the die restriction is increased, which again is in agreement with the experimental results.

Equation (11-23) predicts a straight line relationship when log (ΔP) is plotted against log (N). The slope of the line should be equal to n, the flow index for the polyethylene. Note that the measured slope of 0.42 falls within the range of n values for polyethylene. It is also noted that (11-23) predicts that K_2 should increase as the temperature decreases and should increase as the die resistance K' decreases, results that again are in agreement with the experiments.

Grant and Walker also reported that the screw characteristics were linear. However, the range of pressures included in their experiments was relatively small, and the scatter of the points is sufficiently large to make this conclusion appear to be somewhat unjustified.

During the operation of an extruder, variations of ambient temperature coupled with the continual adjustment of the barrel heaters by the controllers cause the temperature (and consequently the viscosity) of the molten polymer flowing into the die to fluctuate. This, in turn, causes the pressure at the discharge end of the screw to fluctuate. At the other end of the screw, variations in the rates of solids transport and melting cause the pressure to fluctuate. Thus the pressure developed by the pumping section fluctuates about a mean value and can be represented by an expression of the form

$$\Delta P = \overline{\Delta P} \pm \tfrac{1}{2}\delta(\Delta P) \tag{11-24}$$

where $\overline{\Delta P}$ represents the mean pressure rise and is the value normally measured and used in calculations.

The total variation is $\delta(\Delta P)$, and it is this variation that causes the extrusion rate to fluctuate about its mean value, as shown by the equation

$$Q = \bar{Q} \pm \tfrac{1}{2}\delta Q \tag{11-25}$$

The magnitude of δQ is an indication of the stability of the extrusion operation. Figure 11-6 shows how δQ depends upon the screw characteristic. Curve A represents a steep characteristic giving rise to the large variation δQ, while B represents a flat characteristic giving rise to the small variation $\delta Q'$.

For isothermal operation, the screw characteristic is defined by (10-76), from which we can write

$$\delta Q = -\frac{B}{\mu}\,\delta(\Delta P) \tag{11-26}$$

Equation (11-26) can serve as the basis for the definition of a stability index, defined as the variation in pressure required to cause a 1% change in the extrusion rate. For a 1% change in the extrusion rate, $(\delta Q/Q) = 0.01$ and equation (11-26), with the negative sign dropped, becomes

$$\delta(\Delta P) = \frac{\mu Q}{100B} \tag{11-27}$$

The larger the value of $\delta(\Delta P)$, the more stable the operation. The index is highly arbitrary but does serve as a useful means of comparing the performance of different screws.

Another factor that will be useful in characterizing the performance of plasticating extruders is now introduced. This is the adiabatic heating factor ΔT^*, defined by the equation

$$\Delta T^* = \frac{1}{b} \ln \left(1 + \frac{1}{N_2}\right) \tag{11-28}$$

where

$$N_2 = \frac{\alpha C}{Z N \mu_i b \epsilon}$$

The adiabatic heating factor represents the adiabatic temperature rise at free discharge. Its calculation was illustrated in Example 10-4.

The ratio of ΔT^* to the total temperature rise required is a rough indication of the importance of internal heat generation in the extrusion operation. If the ratio is greater than unity, heat generation is excessive. Note, however, that in some extrusion operations it may be advantageous to operate nearly adiabatically. Under such conditions the polymer is heated internally rather than through conduction. This should result in the reduction of temperature gradients in the polymer, thereby increasing the uniformity of the extrudate temperature.

In some extrusion processes a uniform extrudate temperature is essential. For example, in film extrusion the molten polymer entering the die divides

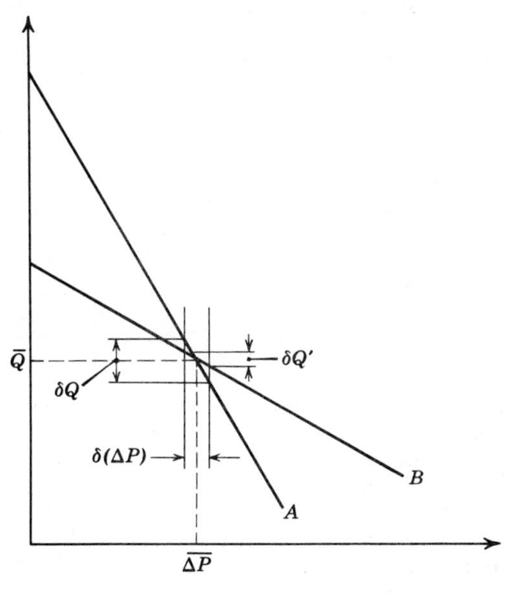

Figure 11-6. Output variation δQ depends upon steepness of the screw characteristic for a given pressure variation.

into many streams. A mass of "cold" melt followed by a mass of "hot" melt may end up at widely separated points along the die. Since the viscosity is a strong function of temperature, the "hot" melt will extrude at a faster rate than the "cold" melt, leading to gage variations across the width of the film.

Another useful factor in characterizing the performance of an extruder is the *residence time*, which we shall define as the total volume of the screw channel divided by the extrusion rate. It represents the average length of time required for a particle of the polymer to pass through the extruder. This is a particularly important factor when one is extruding heat sensitive polymers at relatively high temperatures.

Example 11-2.

A proposed polyethylene extrusion process calls for an extrusion rate of 150 lb/hr (1.55 in.^{3}/sec) against a pressure of 1000 psi at an extrusion temperature of 400° F. The physical properties of the polyethylene are as follows:

$$\begin{aligned} \mu &= 0.14 \text{ lb } (f)\text{-sec/in.}^2 \quad (\text{at } 400^\circ \text{ F}) \\ \mu &= 0.50 \text{ lb } (f)\text{-sec/in.}^2 \quad (\text{at } 250^\circ \text{ F}) \\ (1/b) &= 126 \text{ deg. F} \\ C &= 166 \text{ in.-lb } (f)/\text{in.}^3\text{-deg. F} \end{aligned}$$

A 3.0-inch extruder having a length of 45 in. is to be used. Two screws of different design are available for use with the extruder. A decision as to which screw would have the better performance characteristics for the job is required. Table 11-1 summarizes the dimensions and the calculated constants for the two screws. The extruder flow equation for isothermal operation is written

$$Q = \alpha N - \frac{\beta \Delta P}{\mu Z}$$

It can be applied to the pumping sections of both screws to estimate the required speed. In making the calculation, isothermal operation is assumed and the viscosity at 400° F is used. The required speed for screw I is 52 rpm and for screw II it is 34 rpm.

The stability index, defined by equation (11-27), can be written in the form

$$\delta(\Delta P) = \frac{Q \mu Z}{100 \beta}$$

For screw I it has value of 133 psi, and for screw II its value is 58 psi. These are the pressure fluctuations that will cause a 1% change in the extrusion rate.

The adiabatic heating factor, as defined by equation (11-26), has been calculated for both screws. For screw I it is assumed that the polymer is melted primarily by transferred heat and that it enters the pumping section at about 400° F. For screw II it is assumed that the melting process occurs primarily through mechanical heat generation and that the polymer enters the pumping section

Table 11-1
Extruder Screw Data

Quantity	Symbol	Screw I	Screw II
Diameter (in.)	D	3.0	3.0
Length (in.)	L	45	45
Channel depth in pumping section (in.)	H_p	0.15	0.25
Channel depth in feed section (in.)	H_f	0.40	0.40
Length of pumping section (in.)	L_p	15	30
Length of feed section (in.)	E_f	30	15
Helix angle	θ	17.7	17.7
Drag flow constant (in.^{3})	α	1.94	3.23
Pressure flow constant (in.^{4})	β	2.45×10^{-4}	11.3×10^{-4}
Power constant (in.^{2})	ϵ	6080	3650

at about 250° F. Under these assumptions the adiabatic heating factor for screw I is 81 deg. F and for screw II it is 115 deg. F.

The residence times for the two screws are calculated by dividing the total volume of the channel by the extrusion rate. It is 1.4 min for screw I and 1.3 min for screw II. The results of the performance calculations are summarized below:

	Screw I	Screw II
Required screw speed	52 rpm	34 rpm
Stability factor	133 psi	58 psi
Adiabatic heating factor	81 deg. F	115 deg. F
Residence time	1.4 min	1.3 min

For both screws the residence time is about the same and is low enough so that it will not be a significant factor in polyethylene extrusion at 400° F. The adiabatic heating factor for screw II exceeds that of screw I, but both are low enough so that operation with either screw will be essentially heat transfer controlled. In the case of the stability factor, screw I appears to have a decided advantage over screw II, particularly in the case of those extrusion processes where a very uniform delivery from the screw is required. For this reason screw I would be a better choice.

In the design of plasticating extruders, the usual objective is to specify the dimensions of an extruder that will satisfy a given set of performance specifications at minimum cost. Performance specifications include the physical properties of the material to be extruded, the required extrusion rate, the temperature and pressure of the polymer melt at the delivery end of the screw, and any special requirements with regard to mixing, uniformity, or stability. It is clear that completely systematic design procedures cannot be established and that successful design results from blending theory with experience.

One aspect of design is to minimize extruder power requirements, a direct cost of extrusion. Power is classified either as thermal or mechanical. Thermal power is derived from the extruder heating system, mechanical power from the extruder drive motor. For a given performance specification (the temperature and pressure requirements), the thermodynamic state of the polymer (and, therefore, the total power requirement) is fixed. If ΔH represents the change in enthalpy of the polymer between the feed hopper and the end of the screw, then the thermodynamic analysis presented in Section 8-1 for steady flow processes shows that

$$\Delta H = Q - W_s \tag{11-29}$$

Equation (11-29) can be written in the alternate form

$$m \, \Delta H = p_t + p_m \tag{11-30}$$

where m is the mass extrusion rate, p_t is the rate of heat transfer to the polymer (the thermal power), and p_m is the rate of mechanical energy transfer to the polymer (the mechanical power).

The quantity $m \, \Delta H$ represents the theoretical minimum power requirement. Heat losses, frictional losses in the drive mechanism, and other unavoidable losses require that more than this theoretical minimum power be supplied. This gives rise to an extruder energy efficiency E_e, defined by the equation

$$E_e = \frac{m \, \Delta H}{p_t + p_m} \tag{11-31}$$

In most extruders both mechanical and thermal power are derived from electricity. Total power costs are, therefore, relatively independent of the ratio of thermal to mechanical power and are more closely related to the energy efficiency of the process. Since the energy efficiency depends primarily upon the extrusion temperature (because of its effect on heat losses), one concludes that with a given set of performance specifications, the power costs are nearly independent of the design.

One approach to the design of plasticating extruders involves scale-up of models. The rules derived for the scale-up of geometrically similar screw pumps can be applied under certain conditions to plasticating extruders. The critical point in this procedure is to determine whether mechanical or thermal power or both are the controlling factors in the extrusion process. This will require that careful energy measurements be made during the operation of the model so that an accurate energy balance can be established.

If the process is controlled by mechanical power, the scale-up follows the third power of the diameter ratio. The experiments of Carley and McKelvey[8], where the performance of a model having a diameter of 2 in. was compared with that of a geometrically similar machine having a diameter of 20 in., tend to confirm this rule. The process involved the extrusion of polyethylene under conditions where mechanical power was controlling. The results are shown in Table 11-2.

Table 11-2
Results of Scale-Up Experiment[8]

	2-inch Model	20-inch Extruder	
	Observed	Calculated	Observed
Extrusion rate (lb/hr)	2.5	2500	2400
Power (hp)	0.29	290	245
Extrudate temperature (° C)	169	169	156
Pressure developed (psi)	1850	1850	1600
Screw speed (rpm)	15	—	15

In the event that thermal power is controlling, the scale-up should follow the second power of the diameter ratio, and for processes where both thermal and mechanical power are important, the scale factor will have some intermediate value.

Problems

11-1 Estimate the enthalpy of polyethylene at 150° C and 2000 psi relative to polyethylene at 150° C and atmospheric pressure. Is one normally justified in neglecting the effect of pressure on enthalpy in calculating power requirements for plasticating extruders?

11-2 Explain why it is logical to define the efficiency of a plasticating extruder with the equation

$$E = \frac{m\,\Delta H}{p_t + p_m}$$

but to use the formula

$$E = \frac{Q\,\Delta P}{p_m}$$

to calculate the efficiency of a screw pump.

[8] J. F. Carley and J. M. McKelvey, *Ind. Eng. Chem.*, **45**, 985 (1953).

11-3 Design an extruder and die to be used for the production of 2-inch standard pipe from polyethylene. The required production rate is 400 lb/hr. List the major assumptions involved in the design.

11-4 Design an extruder and die to be used for the production of nylon monofilament having a diameter of 0.010 in. The required production rate is 200 lb/hr. List the major assumptions involved in the design.

11-5 Consider the blown film process for the manufacture of thin-wall tubing in which a tube of molten plastic is expanded with internal air pressure. Suppose that the maximum production rate from a single die (the limiting factor probably being the rate of cooling of the molten tube) for the required tube size is 50 lb/hr and that the total required production is 150 lb/hr. Discuss the economic and engineering aspects of the following proposals: (*a*) have three separate production lines each fed by a small extruder or (*b*) have a single larger extruder feeding a manifold containing three dies.

chapter 12

Mixing

Mixing is an important operation of polymer processing. Typical examples of mixing processes include the dispersion of carbon black in polyethylene with a Banbury, the compounding of a rubber formulation on a roll mill, and the blending of two polymers in an extruder.

In all mixing operations one is concerned with the "efficiency" of mixing, which, in turn, raises questions concerning the "goodness" of a mixture. The main points dealt with in this chapter are, therefore, the characterization of mixtures and the analysis of mixing processes. Analyses of mixing processes are inherently difficult to make. They have been made only for a few special cases, and progress in the art of mixing has been slow, particularly in the field of polymer processing.

12-1 Introduction

A mixture consists of two or more identifiable components. For simplicity, two-component mixtures are considered in this chapter. The component having the higher over-all concentration is the *major component*, the other is the *minor component*. Extension to multi-component systems is possible by considering each component, in turn, as a minor component being mixed into the remaining components.

The term *ultimate particle* is used throughout this chapter in a rather restricted sense. The ultimate particles are the smallest particulate pieces of the minor component normally existing during the course of mixing. Ultimate particles may be of molecular, colloidal, microscopic, or larger size. In a true solution the ultimate particles are the molecules themselves. The dry blending of molding powders provides an example in which the ultimate particles are very large.

The term *simple mixing* is used for any operation employed to increase the randomness of the spatial distribution of the ultimate particles without reducing their size. In simple mixing, the ultimate particles of the system pass from a less probable to a more probable arrangement.

If the configurational entropy of the entire system could be measured with an "entropy gage," then the progress of a simple mixing operation could be readily followed. The entropy would increase as mixing progressed and would reach a maximum when the configuration became random. In practice, measurements of a property other than entropy (one that is directly measurable) are made on samples withdrawn from the mixture or on small elements within the mixture. The data generated by such measurements are then treated by standard statistical methods to estimate the progress of the mixing. It should be noted that a sample does not have to be removed physically from the mixture in order to be analyzed. Agents such as light or electric current can penetrate many mixtures and yield information that can be used for characterization.

The term *dispersive mixing* applies to those mixing processes that reduce the size of the ultimate particles as well as randomize their positions. The preparation of a master batch of carbon black in polyethylene in a Banbury is an example of a dispersive mixing process. In this process the ultimate particles before mixing are the agglomerates of the colloidal carbon black particles. After mixing, they are the particles themselves. The analysis of a dispersive mixing process is similar to that of a simple mixing process except that an additional complication is added—the nature and magnitude of forces required to rupture ultimate particles must be considered.

Many simple mixing processes occur spontaneously or naturally. For example, if a partition separating two gases is removed, the gases will pass into each other and form a uniform mixture. Similarly, two miscible liquids will "mix themselves," although a longer period of time will be required. The mechanism of these natural mixing processes is molecular diffusion. Whenever the ultimate particles are of molecular size, diffusion can be an important factor, although, with high polymers, diffusion rates are slow and diffusive mixing is often relatively unimportant.

In nondiffusive mixing, the particles must be put into motion by some outside agency. If the major component is a fluid, and if it can be made turbulent, mixing proceeds rapidly. With gases and low viscosity liquids, *turbulent mixing* is the usual mode of operation. When the fluid viscosity is high, which is the case with polymer melts and solutions, it is usually impossible to achieve turbulence; one must be content with *laminar mixing*, which proceeds at a slower rate.

A complete description of the state of a mixture would involve the specification of the size and position of each ultimate particle. Obviously

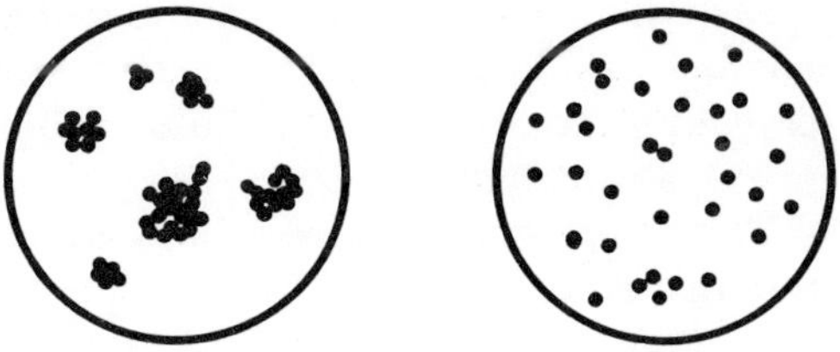

(a) (b)

Figure 12-1. Coarse grained samples of (*a*) poor mixture and (*b*) good mixture.

one must be satisfied with something less than a complete description. Usually a mixture can be adequately characterized by its *gross uniformity*, its *texture*, and its *local structure*.

To illustrate the meaning of the term *gross uniformity*, consider a polyethylene extrusion compound that contains about 2% carbon black and is in the form of pellets having a characteristic dimension of about $\frac{1}{8}$ in. If we have a large batch of this mixture, its gross uniformity could be determined by measuring the concentration of carbon black in a number of pellets selected at random from various parts of the mixture. The distribution of the concentrations would be an indicator of the gross uniformity or lack of gross uniformity of the mixture. If the mixture were grossly uniform, then the concentrations would follow the binomial distribution law. The deviation of the observed distribution from the theoretical can be used as a measure of the degree of uniformity of the mixture, as is discussed in some detail in Section 12-2.

The discussion of the local structure and texture of a mixture is facilitated by the introduction of another term—the *scale of examination*. This refers to the size of samples taken from the mixture for analysis. With a scale of examination that is of the same order as the size of the ultimate particles, a sample will contain a relatively small number of ultimate particles and in appearance will be *coarse grained*. With a scale of examination that is large with respect to the size of the ultimate particles, a sample will contain many ultimate particles and in appearance will be *fine grained*. Figures 12-1 and 12-2 are representations of coarse- and fine-grained samples.

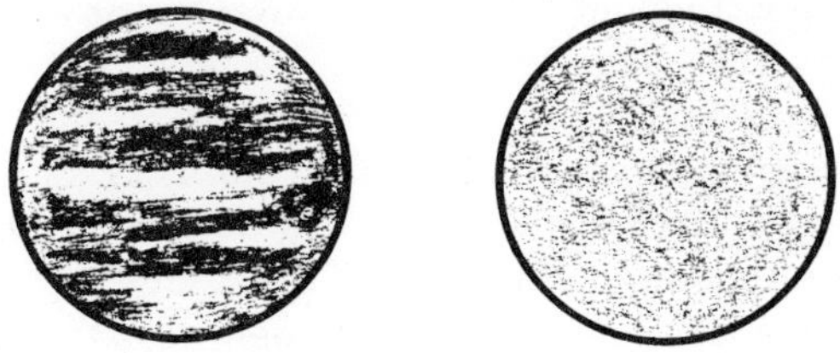

(a) (b)

Figure 12-2. Fine grained samples of (*a*) poor mixture and (*b*) good mixture.

Clearly the graininess of a sample is not an indicator of the nature of a mixture. Samples from any mixture, even a true solution, will be coarse grained if the scale of examination is of the same order as the dimensions of the ultimate particles.

To investigate the texture of a mixture one must use a scale of examination that is large in comparison to the ultimate particles. Hence one will be concerned with fine-grained samples. The characterization of fine-grained samples has been discussed by Danckwerts,[1] who points out that one is dealing not with ultimate particles but with "clumps" of the minor component. As mixing proceeds, the clumps become reduced in size and diluted with the major component, thus changing the texture of the mixture.

In order to characterize completely the texture of the mixture, the size, shape, concentration of the minor component, and the location of each clump must be specified. In practice, it is necessary to resort to statistical methods to do this. By analogy to the statistical theory of turbulence, Danckwerts proposed two statistically defined quantities, the *scale of segregation* and the *intensity of segregation* as a measure, respectively, of the size and concentration of the clumps.

Suppose Figure 12-2 represents fine-grained samples of colored plastic sheets. The sample from mixture *b* is free of spots or streaks and has a uniform texture. The sample from mixture *a* is streaked. Its scale of segregation is a measure of the size of the streaks. Its intensity of segregation is a measure of the relative intensity or brightness of the dark and light areas. Additional mixing will improve the texture of mixture *a*. With sufficient mixing it can be made uniform at the given scale of examination.

Danckwerts has pointed out that the scale of segregation or the intensity of segregation may sometimes be used individually as an indicator of the goodness of a mixture. He believes, however, that a combination of both quantities will frequently be more useful. In the above example, the product of the two quantities might have provided a useful index of mixedness. A high intensity of segregation could be tolerated if the scale were small and, conversely, a larger scale of segregation could be tolerated if the intensity were low. For example, small streaks of pure color would not be objectionable if the streaks were too small to be detected by ordinary sight.

The characterization of the texture of mixtures is discussed in more detail in Section 12-3. The quantities, scale of segregation, and intensity of segregation are defined and various methods of determining them discussed.

To investigate the local structure of a mixture, one must use a scale of examination of the same order as the size of the ultimate particles. Microscopic examination, when possible, is the most direct means of determining

[1] P. V. Danckwerts, *Appl. Sci. Res.*, **3**, 279 (1952).

local structure. Figure 12-1 is a representation of a microscopic examination of two carbon black-polyethylene mixtures. Assuming that the samples shown are representative of the mixtures as a whole, it is clear that the local structure of the one mixture is entirely different from that of the other. However, the gross uniformity and texture of these two mixtures may be indistinguishable. Techniques other than direct microscopic observation are also used to investigate local structure. Optical methods using microdensitometers have, for example, been used. The important point is that, regardless of the method, the scale of examination must be of the same order as the size of the ultimate particles.

In general, one would expect gross uniformity to be important in the characterization of all mixtures. Whether or not the local structure and the texture are both important in the characterization of a mixture depends upon the *function* of the minor component. For example, in some applications of black polyethylene the function of the carbon black is simply to color the polyethylene. Here the local structure is clearly immaterial if the texture of the mixture is adequate. In other applications the function of the carbon black is to protect the polyethylene against attack by ultraviolet light. Here both texture and local structure are important, for it is known that (at the 2% level) adequate protection can be obtained only if the carbon black is not agglomerated. Clearly the local structure must be examined in order to determine whether or not the mixture will be adequate to its purpose.

Best and Tomfohrde[2] have developed a method of determining the goodness of carbon black dispersions in polyethylene that illustrates a practical application of these concepts. They employed a microdensitometer that sends a beam of white light 10 microns in diameter through a 1-mil thick film of the mixture. The variation in light transmission between various points in the mixture was used to characterize the goodness of the dispersion, hence its local structure.

It should be noted that in this application, the scale-of-examination must be very small in order to obtain significant results. The beam of light examines a cylindrical sample about 25 microns thick and 10 microns in diameter. Its volume, therefore, is about $(5\pi/8) \times 10^{-9}$ cm^3. The concentration of carbon black is about 2.6% and, assuming that the carbon black particle is about 1 micron in diameter, each sample examined by the light beam will have, on the average, about 75 particles. We are, therefore, dealing with coarse-grained samples, and we should be able to learn something about the local structure of the mixture (the degree of agglomeration of the carbon black particles) by this technique.

Assuming that the point-to-point uniformity of light transmission will

[2] W. G. Best and H. F. Tomfohrde, *S.P.E. Journal*, p. 139, Feb. 1959.

improve as the carbon black agglomerates are reduced, the variance of the transmission measurements should provide an indicator of the local structure of the mixture, with the variance decreasing and eventually reaching a minimum as the degree of agglomeration decreases. It is assumed that the mixture is grossly uniform; thus the effect is truly one of agglomerate reduction rather than simple mixing on a very small scale.

The results of Best and Tomfohrde are shown in Table 12-1, in terms of the coefficient of variation, which is defined as the standard deviation divided by the mean. Note that their results are in excellent agreement with visual microscopic ratings.

Table 12-1

Rating Upon Visual Microscopic Examination	Coefficient of Variation of Densitometer Measurements
B	2.22 (2.10)
B	2.90
B	2.94
C	3.55
D	4.75
D	4.38 (6.52)

12-2 Gross Uniformity or Degree of Mixedness

This section discusses statistical methods that can be used to test for gross uniformity of a mixture and that can also be used to define an index of mixedness to characterize mixtures not grossly uniform. We consider only the special case where the minor component consists of uniformly sized ultimate particles and the major component is a fluid. The major component, the fluid, is then divided into hypothetical particles of the same size as the ultimate particles of the minor component. This device permits us to speak of the number of particles of the major component in a sample. The analysis of this special case, of course, applies equally well to solid-solid mixtures in which particles are all the same size.

If the ultimate particles of the minor component are randomly distributed throughout the mixture, the probability of finding one is the same at all points in the mixture and is equal to the volume fraction of the minor component in the mixture as a whole.

If a number of samples of equal size (small in comparison to the size of the mixture as a whole) are withdrawn, the fraction of the minor component will vary from sample to sample. However, it has been shown both theoretically and experimentally that the sample concentrations will vary according to the binomial distribution law.

Consider the case where each sample contains a total of n particles and where the fraction of the ultimate particles of the minor component in the mixture as a whole is p. The probability of drawing a sample having a fraction $x = (b/n)$ of the minor component is

$$P(x) = \left[\frac{n!}{b!(n-b)!}\right] p^b(1-p)^{n-b} \tag{12-1}$$

where b is the number of ultimate particles of the minor component in the sample. This is the binomial frequency function. If the concentrations of a number of samples vary according to this law, then their *variance* is given by the formula

$$\sigma^2 = \frac{p(1-p)}{n} \tag{12-2}$$

To determine whether or not the mixture is grossly uniform, one must withdraw samples, analyze them, and then determine if the sample compositions are distributed according to the binomial distribution law. If they are, the mixture is grossly uniform.

An extremely large number of samples would be required to generate enough data to make a direct comparison with the binomial distribution law. However, certain statistical techniques are available with which one can characterize a distribution with a relatively small number of samples. Weidenbaum[3] points out that for most solids mixing problems, the distribution can be characterized by the *mean* and the *variance*.

The mean $\bar{x}$ of N samples is defined by the equation

$$\bar{x} = \frac{1}{N}\sum_{i=1}^{N} x_i \tag{12-3}$$

where x_i represents the concentration of ultimate particles of the minor component in the ith sample. Note that $\bar{x}$ should not differ significantly from p, the fraction of the minor component in the mixture as a whole. If these two quantities should differ, then the sampling technique is probably faulty.

The variance s^2 of the measurements is defined by the equation

$$s^2 = \frac{1}{N-1}\sum_{i=1}^{N} (x_i - \bar{x})^2 \tag{12-4}$$

and it can also be computed with the following formula:

$$s^2 = \frac{1}{N-1}\sum_{i=1}^{N} x_i^2 - \frac{1}{N(N-1)}\left[\sum_{i=1}^{N} x_i\right]^2 \tag{12-5}$$

[3] S. S. Weidenbaum, "Mixing of Solids," in *Advances in Chemical Engineering*, vol. II, edited by T. B. Drew and J. W. Hoopes, Academic Press, 1958.

In order to make an objective comparison of the mean and the variance of the samples with p and σ^2 certain *statistical tests of significance* are used. These tests enable one to decide, when the quantities being compared differ, whether they are inherently different or only different because of chance.

If the mixture has been properly sampled, then $\bar{x}$ should not be significantly different from p. The normal proportion test can be used to determine if any observed difference is significant. The test is applied by calculating the quantity

$$z = \frac{(\bar{x} - p)\sqrt{N}}{s} \tag{12-6}$$

and by reference to the table of the standardized normal distribution, one can determine the probability that the observed difference occurred through chance.

The *chi-square* test is the statistical test of significance used in comparing s^2 with σ^2. If the ratio (s^2/σ^2) is nearly unity, one can conclude that the concentrations of the samples vary according to the binomial distribution and that the mixture is random. In general (s^2/σ^2) will not be unity, and the chi-square test can be used to determine the probability that the difference is due to chance or to an inherent difference between the distributions.

The end product of a simple mixing operation is a random distribution of the ultimate particles. However, in many processes a completely random mixture is not required; it is useful to have a measure of the approach to randomness or degree of mixedness. Such a measure is also useful in analyzing the performance of mixing equipment.

Many methods have been proposed for specifying the degree of mixedness. They are discussed in a comprehensive review by Weidenbaum.[3] It would appear that those which are most generally useful should be based upon quantities having statistical significance, such as the variance or standard deviation.

The variance s^2 approaches a limiting minimum value σ^2 as the mixture approaches a random state. Since σ^2 can be calculated directly with (12-2), a useful index can be defined by the equation

$$M = \frac{\sigma^2}{s^2} \tag{12-7}$$

The standard deviation is the square root of the variance and can also be used to define the index of mixedness

$$M = \left(\frac{\sigma}{s}\right) \tag{12-8}$$

Lacey[4] has defined another index by the equation

$$M = \frac{s_0^2 - s^2}{s_0^2 - \sigma^2} \qquad (12\text{-}9)$$

where s_0^2 represents the variance of the perfectly unmixed state. It can be calculated with the equation

$$s_0^2 = p(1 - p) \qquad (12\text{-}10)$$

Note that the index defined by (12-9) has the advantage of ranging from zero for the perfectly unmixed state to unity for the perfectly mixed state.

Example 12-1.

This example illustrates the application of the statistical concepts and tests discussed above. Hypothetical "two-dimensional" mixtures of black particles and white particles, shown in Figure 12-3, are considered. The over-all concentration of black particles is the same in all of these mixtures and is equal to 12% ($p = 0.12$). Each mixture was "made" by filling in 600 squares of a 5000 square grid.

Mixture *a* represents the perfectly unmixed state. In it the 600 black particles are grouped in a single clump. Mixture *d* represents the perfectly mixed state. It was made by choosing the coordinates of the 600 black particles from a table of random numbers. The distribution of the black particles in Mixtures *b* and *c*, which represent imperfect mixtures, was given a bias to the left.

Samples from each mixture were "taken" in the following manner: A point on the grid was located by choosing its coordinates from a table of random numbers. The 100 particles ($n = 100$) surrounding the point were taken as the sample. The number of black particles *b* in the sample were counted. This was done 20 times for each mixture. The data are shown in Table 12-2.

The mean number of black particles $\bar{b}$ was calculated with equation (12-3) and the mean concentration $\bar{x}$ was found by dividing $\bar{b}$ by 100. The variance s^2 was calculated with equation (12-5). The theoretical variance σ^2 calculated with (12-6) was 0.00106. The results of these calculations are shown in the table.

Reference to normal tables shows that the differences between $\bar{b}$ and p are well within the limits set by chance and are therefore not significant. The chi-square test is used to determine if the differences between s^2 and σ^2 values are significant. For Mixture *d* the difference is obviously not significant (the tables show that the value 1.03 occurs 99 times in 100 by chance), and it is concluded that the black particles in Mixture *d* are randomly distributed. For Mixture *c*, reference to the chi-square tables shows that a value of 2.01 will occur less than 4 times in 100 by chance, from which one concludes that it is very unlikely that Mixture *c* is a random mixture. Similarly, the data show that the probability of either Mixture *a* or *b* being random is practically nil.

The degree of mixedness or randomness increases from Mixture *a* to Mixture *d*.

[4] P. M. C. Lacey, *J. Appl. Chem.* (London), **4**, 257 (1954).

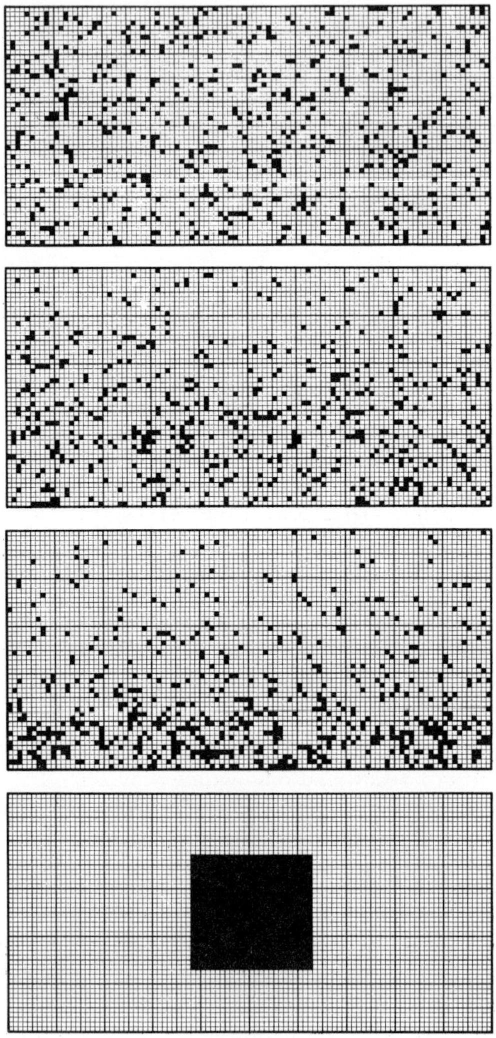

(a) (b) (c) (d)

Figure 12-3. Two-dimensional coarse-grained mixtures. Over-all concentration of black particles is 12%. Particles in (*d*) are randomly distributed.

Table 12-2

Number of Black Particles

Sample No.	Mixture *a*	Mixture *b*	Mixture *c*	Mixture *d*
1	0	5	8	8
2	0	28	8	18
3	0	6	5	15
4	0	22	16	13
5	100	8	11	11
6	0	13	10	12
7	0	18	22	7
8	0	5	17	14
9	0	10	14	6
10	0	4	17	15
11	0	3	11	16
12	0	3	3	16
13	100	3	12	11
14	0	0	10	9
15	20	1	3	10
16	0	1	8	10
17	0	36	2	15
18	0	3	24	12
19	0	5	12	10
20	0	34	17	10
$\bar{b}$	11.0	10.4	11.5	12.1
$s^2 \times 20^4$	950	125	21.3	10.9
t	0.14	0.64	0.50	0.14
(s^2/σ^2)	89.6	11.8	2.01	1.03

Values of the indices defined by equations (12-7), (12-8), and (12-9) have been calculated and are shown below:

Defining Equation	Mixture *a*	Mixture *b*	Mixture *c*	Mixture *d*
(12-7)	0.011	0.084	0.498	0.973
(12-8)	0.106	0.290	0.705	0.986
(12-9)	0.105	0.890	0.990	1.000

The choice of an index of mixedness is arbitrary. It would appear, however, that indices based upon (12-7) or (12-8) would be more useful for mixtures that are nearly random, while the index based on (12-9) would be useful for characterizing mixtures during the initial stages of the mixing process.

12-3 Texture

This section discusses the characterization of the texture of a mixture through its scale and intensity of segregation, as proposed by Danckwerts.[1]

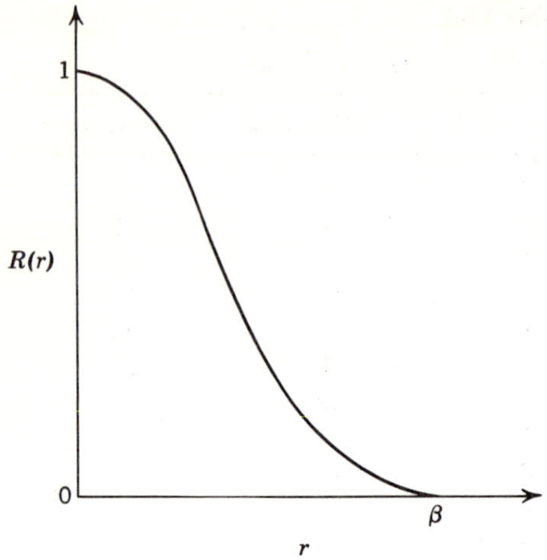

Figure 12-4. Typical plot of the coefficient of correlation $R(r)$ as a function of the distance r between point pairs. When $R(r) = 0$, the relationship between concentrations in point pairs is random. Area under the curve is the linear scale of segregation.

It is assumed that the scale of examination is very large compared to the size of the ultimate particles, but is small in comparison to the mixture as a whole. In this case it is meaningful to speak of the concentration at a point, which is designated with the symbol x.

If the concentration is determined at two points separated by the distance r, then the concentration pair can be represented as (x', x''). The coefficient of correlation $R(r)$ is defined by the equation

$$R(r) = \frac{1}{Ns^2} \sum_{i=1}^{N} (x_i' - p)(x_i'' - p) \tag{12-11}$$

where s^2 represents the variance of the concentration measurements and N is the number of pairs of points. The over-all concentration of the minor component in the mixture is p.

In general, $R(r)$ may have any value between zero and unity. For small values of r, $R(r)$ is usually near unity because points close together are likely to be in the same clump. In the limiting case, $R(0) = 1$. For large values of r, the correlation coefficient reduces to zero, indicating a random relationship between the concentrations in point pairs. If $R(r)$ is plotted against r, curves similar to the one shown in Figure 12-4 result.

This curve is called the correlogram of the mixture. The linear scale of segregation S_L is defined as the area under the curve and is given by the equation

$$S_L = \int_0^\beta R(r)\,dr \tag{12-12}$$

Danckwerts also defines a volumetric scale of segregation S_V, which is defined by the equation

$$S_V = 2\pi \int_0^\beta r^2 R(r)\, dr \tag{12-13}$$

The ease with which each of these quantities can be determined experimentally is likely to determine which definition will be used in any particular case. It can be shown that a random collection of spheres of diameter D gives the following result:

$$S_L = 0.38D \qquad S_V = 1.79D$$

The intensity of segregation I is defined by the equation

$$\mathrm{I} = \frac{s^2}{\bar{x}(1-\bar{x})} = \frac{s^2}{p(1-p)} \tag{12-14}$$

Referring to equation (12-10), it can be seen that I is a ratio of the variance of the mixed state to the variance of the unmixed state. If segregation is complete, then I = 1 because the concentration at any point will be either zero or unity. When the concentration becomes uniform at all points, then $s^2 = 0$ and the intensity of segregation is zero. Note that a given mixture might appear to be perfectly smooth (I = 0) at a given scale of examination and quite grainy if one used a smaller scale of examination. In the limiting case, all mixtures will appear grainy if the scale of examination is of molecular size.

Danckwerts discusses several methods that may be used to derive the values of these quantities from concentration measurements. For I, any set of measurements from which s^2 can be computed is satisfactory. Note that the sample size or probe size (the scale of examination) must be of the same order of magnitude as the graininess that one is trying to characterize.

Regarding the scale of segregation, one could measure concentrations at a large number of point pairs and then apply equation (12-11) directly to the data to calculate the correlation coefficient. An alternate method discussed by Danckwerts would be to determine values of the quantity J, defined by the equation

$$J = \int_0^Y x\, dy \tag{12-15}$$

where y is the distance measured along a line of length Y. If J is determined for a large number of lines in the sample, it can be related, through its variance, $s^2(J)$, to the scale of segregation by the approximate equation

$$S_L = \frac{s^2(J)}{2Ys^2(x)} \tag{12-16}$$

The error in using (12-16) is of the order of (β/Y). For the usual case where β is of the same order as S_L, Y must be considerably larger than S_L if this method is to be used. However, the larger Y is, the more difficult it becomes to determine $s^2(J)$ accurately, and a compromise must be made.

The volumetric scale of segregation S_V can be determined by using a method based on the analysis of samples. If a number of samples of volume v are taken from the mixture and the amount K of the minor component determined, then

$$S_V = \frac{s^2(K)}{2vs^2(x)} \tag{12-17}$$

provided that $v \gg V$.

12-4 Laminar Simple Mixing

In most mixing problems involving a fluid phase, the rate of mixing is markedly increased by making the fluid turbulent. This method of promoting mixing is widely used with gases and low viscosity liquids. However, when the fluid phase is a high viscosity polymer melt or solution, it is not practicable to create turbulence, and mixing proceeds at an inherently slower rate. An important distinction is made, therefore, between laminar and turbulent mixing.

Diffusion is another important factor in fluid mixing. With gases and low-viscosity liquids, diffusion quickly reduces the intensity of segregation, giving the mixture a smooth texture. With polymers, diffusion occurs at such a slow rate that it is of little help in smoothing the texture of a mixture. This section is concerned with laminar, nondiffusive, simple mixing—the most important type in polymer processing.

A laminar, nondiffusive mixing process is illustrated by the fluid particle paths shown in Figure 12-5, where a black fluid is being mixed into a white fluid in the annular space formed between concentric cylinders. In Figure 12-5*a* the black fluid is represented initially by a radial line stretching from the inner stationary cylinder to the outer rotating cylinder.

Since the flow is laminar, each fluid particle moves in a closed circular path. When the outer cylinder completes one revolution, the particle of black fluid immediately adjacent to it also completes one revolution, but the particle of black fluid immediately adjacent to the stationary inner cylinder is stationary. Black particles intermediate to these extremes move intermediate distances. After one complete revolution of the outer cylinder, the line of black fluid is deformed into the shape shown in the second diagram, and its configuration after 2, 3, and 1000 revolutions is shown in the succeeding diagrams.

If L represents the annular spacing and r the distance separating adjacent lines, then

$$r = \frac{L}{n + 1} \approx \frac{L}{n} \tag{12-18}$$

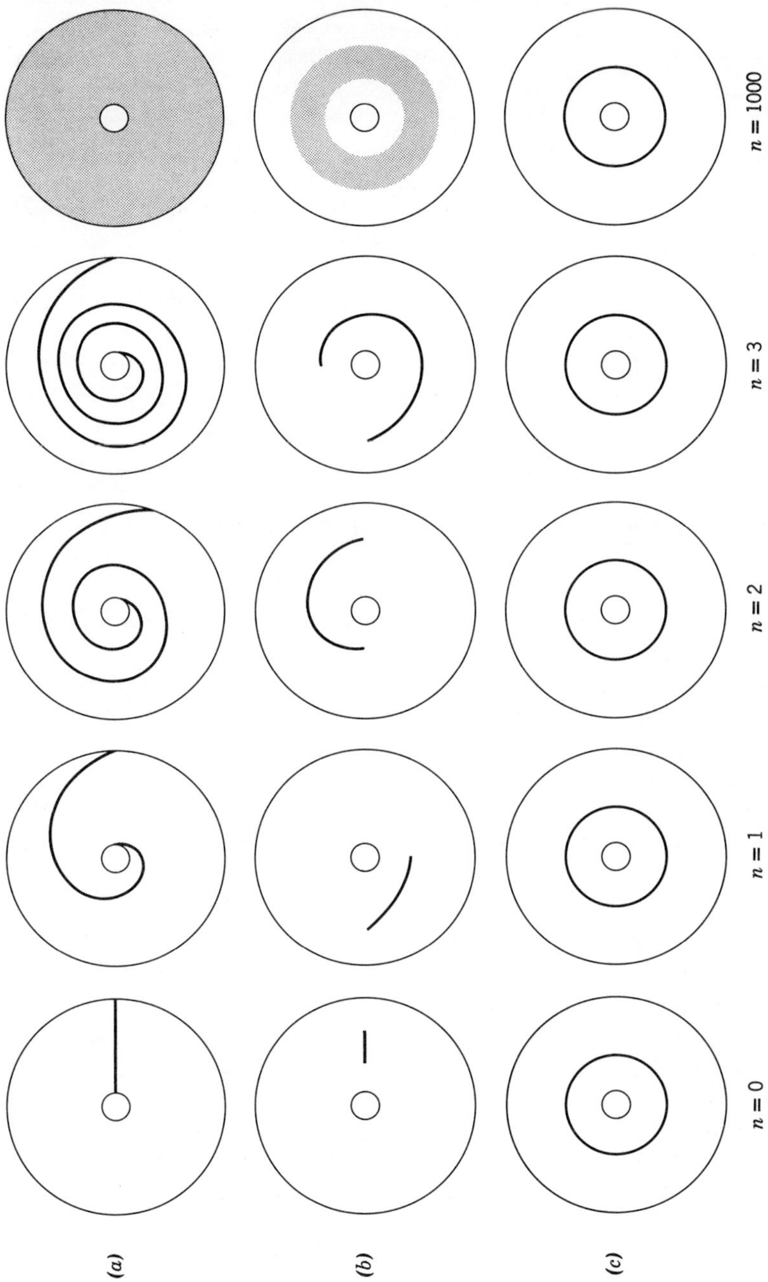

Figure 12-5. Representation of a laminar, nondiffusive mixing process between concentric cylinders. Initial orientation of components and total shear strain determine the extent of mixing.

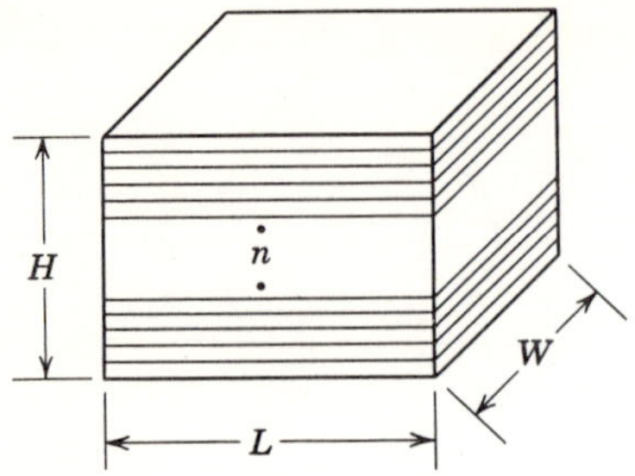

Figure 12-6. Mass of fluid having n equally spaced striations of the minor component. Total interfacial area is $2nLW$.

where n is the number of revolutions. The quantity r has been called the *striation thickness* by Mohr.[5]

If, for example, $L = 1$ cm, then, after 1000 revolutions, $r = 10$ microns and the fluid would appear to the eye to be well mixed and to have a smooth texture. On the other hand, if the scale of examination were of the order of 1 micron, then the mixture would be judged to be nonuniform and to have a coarse or grainy texture.

The effect of the initial orientation of the minor component on the efficiency of the mixing process is illustrated in Figure 12-5*b*. Here the black fluid is a line that straddles the center third of the annular area. After 1000 revolutions, the black fluid is distributed only throughout this region. In Figure 12-5*c* the black fluid is again a line, this time lying on a streamline. Since all fluid particles on a streamline move at the same speed, no mixing occurs, and after 1000 revolutions the state of mixedness of the system is exactly the same as it was initially.

In the above two-dimensional example, the decrease in striation thickness is associated with an increase in interfacial area. A relationship between striation thickness and interfacial area can be derived in the following manner:

Consider a mass of fluid of length L, width W, and height H, as shown in Figure 12-6. It has n equally spaced striations of the minor component, and the total interfacial area A is $2nLW$. The total volume $V = LWH$ is related to r and A by the equation

$$V = (LW)(H) = \frac{A}{2n}(nr) = \frac{rA}{2} \tag{12-19}$$

which shows that the striation thickness is inversely proportional to the interfacial area and directly proportional to the total volume of the mixture. The reduction of the striation thickness is associated with the generation of new interfacial area. If initially the striation thickness is r_i and the

[5] W. D. Mohr, "Mixing and Dispersing," ch. 3 of *Processing of Thermoplastic Materials*, edited by E. C. Bernhardt, Reinhold Publishing Co., 1959.

interfacial area is A_i, then a decrease in the striation thickness to r will require an increase in interfacial area to A, these quantities being related by the equation

$$\left(\frac{r}{r_i}\right) = \left(\frac{A_i}{A}\right) \tag{12-20}$$

Now consider the deformation of a fluid interface under the action of a simple unidirectional shear strain. It is shown below that if γ is the shear strain then

$$\frac{A'}{A} = \sqrt{1 - 2\gamma \cos\alpha \cos\beta + \gamma^2 \cos^2\alpha} \tag{12-21}$$

where the angles α and β define the initial orientation of the interface in space. By combining the above two equations, one obtains

$$r = \frac{r_i}{\sqrt{1 - 2\gamma \cos\alpha \cos\beta + \gamma^2 \cos^2\alpha}} \tag{12-22}$$

which relates the striation thickness after a given amount of unidirectional shear to the initial striation thickness and the initial orientation of the interface in space.

The following analysis, based on that of Mohr,[5] relates the change in area of a plane surface (contained in a region that is subjected to a unidirectional shear strain) to the initial orientation of the surface and to the magnitude of the strain. Consider the triangular surface having area A that is formed by the vectors $\boldsymbol{a}$ and $\boldsymbol{b}$, as shown in Figure 12-7. The vector $\boldsymbol{c}$ is defined by the equation

$$\boldsymbol{c} = [\boldsymbol{a} \times \boldsymbol{b}] \tag{12-23}$$

hence it is perpendicular to the plane containing $\boldsymbol{a}$ and $\boldsymbol{b}$ and its magnitude is exactly twice the triangular area A. Therefore

$$|\boldsymbol{c}| = 2A$$

The following relations among the components of the vectors $\boldsymbol{a}$, $\boldsymbol{b}$, and $\boldsymbol{c}$ are easily established by expanding the right side of equation (12-23):

$$\begin{aligned} c_x &= a_y b_z \\ c_y &= -a_x b_z \\ c_z &= -a_y b_x \end{aligned} \tag{12-24}$$

Suppose the space containing the triangular surface is subjected to a unidirectional shear strain in the x-direction. The vector $\boldsymbol{b}$ would be unaffected but vectors $\boldsymbol{a}$ and $\boldsymbol{c}$ would become $\boldsymbol{a}'$ and $\boldsymbol{c}'$, where $\boldsymbol{c}'$ is the cross product of $\boldsymbol{a}'$ and $\boldsymbol{b}$. Hence

$$|\boldsymbol{c}'| = 2A'$$

where A' represents the area of the deformed surface.

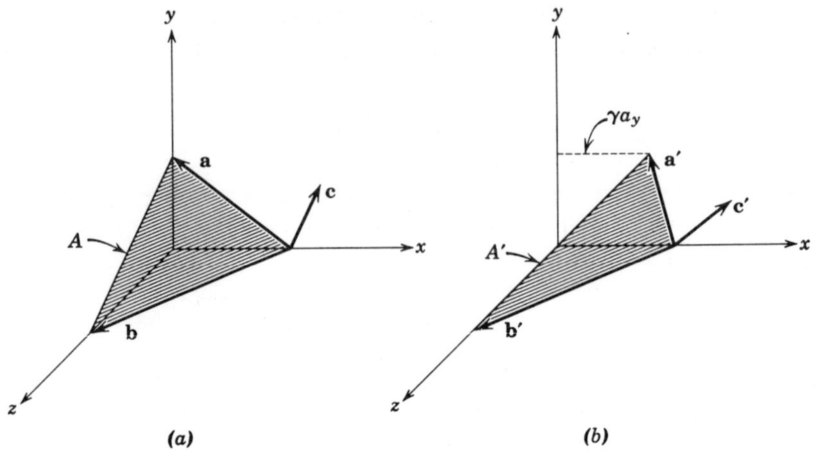

Figure 12-7. Change in interfacial area upon unidirectional shear strain. In unstrained state (*a*) surface has area A and in strained state (*b*) area is A'.

Expressing the ratio of the areas of the two surfaces in terms of the vectors $\boldsymbol{c}$ and $\boldsymbol{c}'$, we have

$$\frac{A'}{A} = \frac{|\mathbf{c}'|}{|\mathbf{c}|} = \sqrt{\frac{c_{x'}{}^2 + c_{y'}{}^2 + c_{z'}{}^2}{|\mathbf{c}|^2}} \tag{12-25}$$

Noting that

$$\boldsymbol{a}' = (a_x + \gamma a_y)\boldsymbol{i} + a_y\boldsymbol{j} \tag{12-26}$$

$$\boldsymbol{b} = b_x\boldsymbol{i} + b_z\boldsymbol{k} \tag{12-27}$$

it is easily verified that

$$\boldsymbol{c}' = [\boldsymbol{a}' \times \boldsymbol{b}] = (a_y b_z)\boldsymbol{i} - (a_x b_z + \gamma a_y b_z)\boldsymbol{j} - (a_y b_x)\boldsymbol{k} \tag{12-28}$$

Upon introducing equations (12-24) into (12-28), one obtains

$$\boldsymbol{c}' = c_x'\boldsymbol{i} + c_y'\boldsymbol{j} + c_z'\boldsymbol{k} = c_x\boldsymbol{i} + (c_y - \gamma c_x)\boldsymbol{j} + c_z\boldsymbol{k} \tag{12-29}$$

from which it can be seen that

$$\begin{aligned} c_x' &= c_x \\ c_y' &= (c_y - \gamma c_x) \\ c_z' &= c_z \end{aligned} \tag{12-30}$$

If equations (12-30) are introduced into (12-25) and the resulting expression is rearranged, one obtains

$$\frac{A'}{A} = \sqrt{1 - 2\gamma\left[\frac{c_x}{|\mathbf{c}|}\right]\left[\frac{c_y}{|\mathbf{c}|}\right] + \gamma^2\left[\frac{c_x}{|\mathbf{c}|}\right]^2} \tag{12-31}$$

Noting that the direction cosines of the vector $\boldsymbol{c}$ are defined by the equations

$$\cos\alpha = \frac{c_x}{|\mathbf{c}|} \qquad \cos\beta = \frac{c_y}{|\mathbf{c}|}$$

we can write equation (12-31) in the equivalent form:

$$\frac{A'}{A} = \sqrt{1 - 2\gamma \cos\alpha \cos\beta + \gamma^2 \cos^2\alpha} \tag{12-32}$$

Equation (12-32) shows that the increase in interfacial area depends upon the initial orientation of the surface, as characterized by the direction cosines of a normal vector, and upon the magnitude of the strain. When $\cos\alpha = 0$, the x-component of the vector $\mathbf{c}$ is zero and $\mathbf{c}$ lies in a plane parallel to the y-z plane. Under shear deformation, a surface element will become distorted but its area will remain constant. When $\cos\beta = 0$, the y-component of $\mathbf{c}$ is zero, and $\mathbf{c}$ lies in a plane parallel to the x-z plane. Shear deformation will cause the surface to increase in area, with the maximum increase occurring when $\cos\alpha = 1$, that is, when the surface lies in the y-z plane.

Example 12-2.

This example considers the reduction in striation thickness of a fluid system located between parallel plates and with interfaces oriented as shown in the lower diagram of Figure 12-8.

The plates are separated by a distance H. If the lower plate is stationary and the upper plate displaced an amount L to the right, the shear strain γ is (L/H) or $\cot(\phi)$. This strain reduces the striation thickness from its initial value of r_i, as shown in the lower diagram, to r, as shown in the upper diagram. It is easily seen from the geometry of the system that

$$\frac{r}{r_i} = \sin(\phi)$$

showing that as the angle ϕ decreases, the striation thickness becomes smaller.

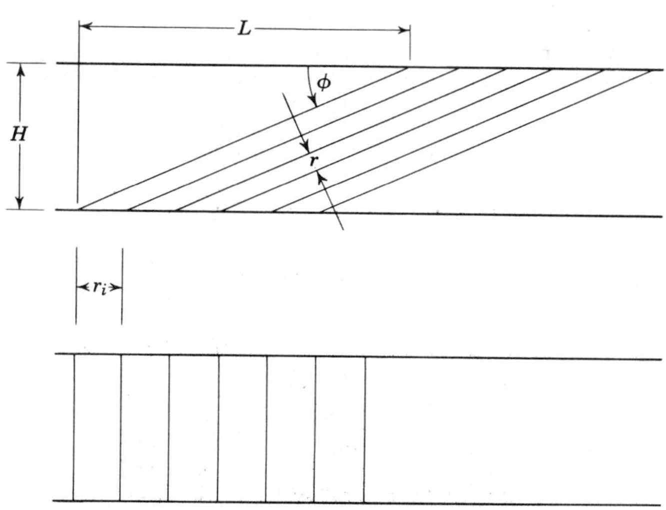

Figure 12-8. Reduction in striation thickness from r_i to r due to shear strain $\gamma = L/H$.

This same result can be derived directly from equation (12-22). Since the interfaces are perpendicular to the x-axis, $\cos \alpha = 1$ and $\cos \beta = 0$ and (12-22) reduces to

$$r = \frac{r_i}{\sqrt{1 - \gamma^2}} = \frac{r_i}{\sqrt{1 - \cot^2 \phi}} = r_i \sin \phi \tag{12-33}$$

For large strains, $\sin(\phi) \approx \tan(\phi)$ and the above equation can be written

$$r = r_i \tan(\phi) = r_i \left(\frac{L}{H}\right) \qquad (L \gg H)$$

In fluid-fluid mixing problems, the ratio of the viscosities of the major and minor components can be an important factor. For example, if the viscosity of the minor component is much greater than the viscosity of the major component, the major component can undergo extensive shear deformation, with the minor component remaining relatively undeformed. The following example is a simple illustration of this point:

Example 12-3.

The major and minor components are immiscible Newtonian liquids having viscosities μ' and μ, respectively. For simplicity, they are located in layers between parallel plates, as shown in Figure 12-9, where the minor component is the center layer. Its thickness is ΔH, the thickness of the layers of the major component is A, and the total plate separation is H.

The upper plate is moved to the right at a speed V and the lower plate is stationary. At steady state, the shear stresses in the three layers must be equal. Therefore

$$\mu' \dot{\gamma}' = \mu \dot{\gamma} = \mu' \dot{\gamma}'' \tag{12-34}$$

where $\dot{\gamma}'$ represents the shear rate in the lower layer and $\dot{\gamma}''$ in the upper layer.

If V' is the speed of the lower interface and V'' the speed of the upper interface, then the shear rates in each layer are given by

$$\dot{\gamma}' = \frac{V'}{A} \qquad \dot{\gamma} = \frac{(V'' - V')}{H} \qquad \dot{\gamma}'' = \frac{(V - V'')}{(H - A)} \tag{12-35}$$

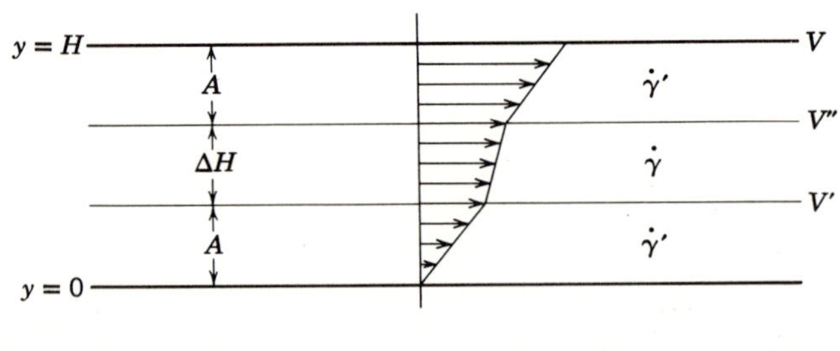

Figure 12-9. Velocity profile for drag flow between parallel plates of layers of immiscible Newtonian liquids.

If the above expressions for shear rate are introduced into (12-34), one can solve for V' and V'' and from these obtain the shear rate in the minor component

$$\dot{\gamma} = \left(\frac{V}{H}\right)\left[\frac{1}{(\Delta H/H) + (2A/H)(\mu/\mu')}\right] \tag{12-36}$$

For the special case where the viscosities of the major and minor components are equal, equation (12-36) reduces to

$$\dot{\gamma} = \left(\frac{V}{H}\right)$$

If the viscosity of the major component should be very small or the viscosity of the minor component very large, the ratio (μ/μ') becomes very large, approaching ∞ in the limit and $\dot{\gamma}$ approaches zero. On the other hand, when the ratio (μ/μ') becomes very small, the shear rate in the minor component becomes

$$\dot{\gamma} = \left(\frac{V}{\Delta H}\right)$$

showing that it is determined solely by the geometrical configuration of the apparatus and the speed at which the boundaries are deformed.

12-5 Mixing in Extruders

The screw extruder is used as a continuous mixer in many polymer processing operations. This section is limited to a discussion of *simple* mixing processes occurring in single screw extruders. Dispersive mixing is considered in the next section.

Since the motion of the fluid in the channel of the screw is laminar, the process is one of laminar simple mixing. As discussed in the previous section, the striation thickness is a measure of the extent of the mixing process. For a constant volume system, the reduction in striation thickness is related to the creation of new interfacial area, which, in turn, is related to the shear strain of the fluid and the initial orientation of the interface. This section is concerned primarily with the calculation of the shear strain of a fluid element as it passes through the channel of the extruder screw.

The following analysis is based upon the isothermal flat plate model of the extruder screw. Newtonian flow is assumed. In principle, the analysis is the same as that of Mohr, Saxton, and Jepson,[6] but differs in the method used to calculate the average shear rates.

A fluid particle passing through the screw channel describes a type of distorted helix, as was shown in Figure 10-8. During the time that the particle is in the lower part of the channel, it moves in the negative x-direction. When it is in the upper part of the channel, it moves in the positive x-direction. The motion of the particle will be made clearer by referring to Figure 10-7, where the various velocity components are shown. Note that the position $a = \frac{2}{3}$ separates the positive and negative

[6] W. D. Mohr, R. L. Saxton, and C. H. Jepson, *Ind. Eng. Chem.*, **49**, 1855 (1957).

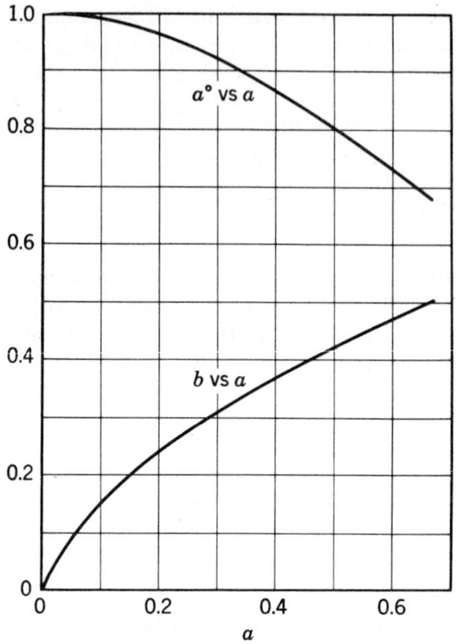

Figure 12-10. Graphs of the functions a° and b versus a.

components, and that this plane of separation is independent of all operating and design variables.

A particle moving through the lower region at position a will move through the upper region at the complementary position a°. If its velocity at any position is v_x, then a and a° are related by the equation

$$\int_0^a v_x\,da = -\int_{a^\circ}^1 v_x\,da \tag{12-37}$$

which, upon integration, gives the result

$$[a^2 - a^3] - [(a^\circ)^2 - (a^\circ)^3] = 0 \tag{12-38}$$
$$(0 \le a \le \tfrac{2}{3})$$
$$(\tfrac{2}{3} \le a^\circ \le 1)$$

showing that a° is a unique function of a. Figure 12-10 is a plot of a° versus a.

Let b represent the fraction of time that a particle is in the upper part of the channel. If v°_x is its velocity in the upper part of the channel, then

$$b = \frac{t^\circ}{t^\circ + t} = \frac{1}{1 + (t/t^\circ)} \tag{12-39}$$
$$= \frac{1}{1 + |v^\circ_x/v_x|} = \frac{1}{1 - (v^\circ_x/v_x)}$$

where the velocity v_x, given by equation (10-53), is

$$v_x = V_x(2a - 3a^2) \tag{12-40}$$

which was derived in Section 10-3. If equation (12-40) is introduced into (12-39), one obtains

$$b = \frac{1}{1 - \dfrac{a^\circ(2 - 3a^\circ)}{a(2 - 3a)}} \tag{12-41}$$

which shows that b is a unique function of a. Figure 12-10 also shows a plot of b versus a.

The time that it takes a particle to pass through the extruder is now calculated. If the axial length of the screw is L and if v_λ is the velocity component in this direction, then

$$t = \frac{L}{\bar{v}_\lambda} \tag{12-42}$$

where $\bar{v}_\lambda$ is the average velocity of a particle in the axial direction.

When the particle is in the lower part of the channel, its position is a and its velocity is v_λ. When it is in the upper part of the channel, its position is a° and its velocity is $v^\circ{}_\lambda$. Since the fraction of time that it is in the upper part of the channel is b, the mean velocity is

$$\bar{v}_\lambda = v_\lambda(1 - b) + v^\circ{}_\lambda b \tag{12-43}$$

In section 10-3 the λ-component of velocity was found to be

$$v_\lambda = 3Va(1 - a)(1 - \phi) \sin\theta \cos\theta \tag{12-44}$$

where ϕ represents the ratio of pressure flow to drag flow. If the above three equations are combined, one obtains

$$t = \frac{L}{3V(1 - \phi) \cos\theta \sin\theta[a(1 - a) + b(a^\circ - a)(1 - a - a^\circ)]} \tag{12-45}$$

Note that the particles which are initially near the center of the channel make the fastest trip through the extruder. The minimum residence time occurs when $a = \frac{2}{3}$. For the special case where $\phi = 0$ (no pressure flow) and $a = \frac{2}{3}$, equation (12-45) reduces to $(3Z/2V_z)$. The relative residence time, defined as (t/t_0), where t_0 is the time for a particle at $a = \frac{2}{3}$ to pass through, is plotted in Figure 12-11 as a function of a. Note that there is a rather broad region in the center of the channel where the residence time is nearly constant. However, as one approaches the top or bottom of the channel, the residence time increases rapidly. At the boundaries it becomes

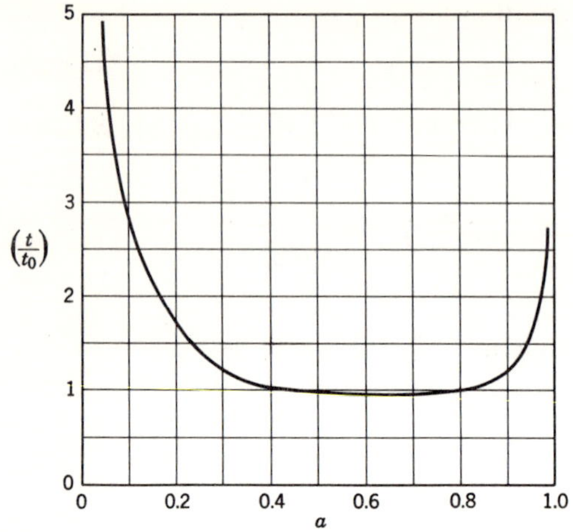

Figure 12-11. Relative residence time as a function of initial position.

infinite. Since particles that are at the boundaries are constantly interchanging their positions with particles in the new fluid layer, all particles will eventually pass through the extruder, although those at the root of the screw may take a considerable period of time to do so.

Now consider the x- and z-components of shear strain in a fluid particle. They are given by the equations

$$\gamma_x = \bar{\dot{\gamma}}_x t \qquad \gamma_z = \bar{\dot{\gamma}}_z t \tag{12-46}$$

where the average shear rates are defined by the following equations:

$$\bar{\dot{\gamma}}_x = |\dot{\gamma}_x|\,(1 - b) + b\dot{\gamma}^{\circ}{}_x \tag{12-47}$$

$$\bar{\dot{\gamma}}_z = (1 - b)\,|\dot{\gamma}_z| + b\,|\dot{\gamma}^{\circ}{}_z| \tag{12-48}$$

Mohr, Saxton, and Jepson[6] defined the average shear rate $\bar{\dot{\gamma}}_x$ as

$$\bar{\dot{\gamma}}_x = \frac{1}{H}\left(\frac{d\bar{u}}{da}\right) \tag{12-49}$$

where the symbol u is used for v_x and $\bar{u}$ is defined by the equation

$$\bar{u} = u(1 - b) - u^{\circ}b \tag{12-50}$$

Introducing (12-50) into (12-49) and differentiating gives

$$\bar{\dot{\gamma}}_x = \left(\frac{1}{H}\right)\left[(1 - b)\left(\frac{du}{da}\right) - b\left(\frac{du^{\circ}}{dz}\right) - (u + u^{\circ})\left(\frac{db}{da}\right)\right] \tag{12-51}$$

If the expression for b, equation (12-39), is introduced into (12-51), it can be simplified to the following:

$$\bar{\gamma}_x = \left(\frac{1}{H}\right)\left[(2b^2 - 2b + 1)\left(\frac{du}{da}\right) - 2b^2\left(\frac{du^\circ}{da}\right)\right] \tag{12-52}$$

This result is to be contrasted with average shear rate defined by equation (12-47), which can be written

$$\bar{\gamma}_x = \left(\frac{1}{H}\right)\left[(1 - b)\left|\frac{du}{da}\right| - b\left(\frac{du}{da}\right)\right] \tag{12-53}$$

It is necessary to specify the absolute value of (du/da) because of the change in sign that occurs at $a = \frac{1}{3}$, as shown in Figure 10-8. It is assumed that the magnitude of the shear strain, rather than its direction, is the significant factor in mixing.

As a specific example, the average strain for an element of fluid located at $a = 0.40$ is calculated. From Figure 12-10, the values of a° and b are found to be 0.86 and 0.38, respectively. The computed shear rates are as follows:

$$\left(\frac{du}{da}\right) = -0.40U \qquad \left(\frac{du^\circ}{da}\right) = -3.16U$$

and the value of $\bar{\gamma}_x$ computed with equation (12-53) is $1.448(U/H)$. Using equation (12-52), its value is $0.70(U/H)$, where U is defined as V_x.

In the x-direction, the average shear rates computed by the two methods differ only by a numerical factor. Since the mixing analysis is applied only qualitatively, the difference is probably of little significance. However, when the shear rates in the z-direction are computed, the one method shows them to be independent of ϕ, while the method used here shows $\bar{\gamma}_z$ to be a function of ϕ. This is believed to be more probable.

The shear rate for a particle in the lower part of the channel is obtained through differentiation of equation (12-40). One obtains

$$\dot{\gamma}_x = \frac{dv_x}{dy} = \frac{1}{H}\left(\frac{dv_x}{da}\right) = \frac{2V_x}{H}(1 - 3a) \tag{12-54}$$

and the average shear rate in the x-direction is

$$\bar{\gamma}_x = \frac{2V_x}{H}[(1 - b)\,|1 - 3a| - b(1 - 3a^\circ)] \tag{12-55}$$

If equations (12-55) and (12-45) are introduced into (12-46), one obtains

$$\gamma_x = \left(\frac{2}{3}\right)\left(\frac{L}{H}\right)\left(\frac{1}{1 - \phi}\right)\left(\frac{1}{\cos\theta}\right) F(a) \tag{12-56}$$

where the function $F(a)$ is defined by the equation

$$F(a) = \frac{(1 - b)\,|1 - 3a| - b(1 - 3a^\circ)}{a(1 - a) + b(a^\circ - a)(1 - a - a^\circ)} \tag{12-57}$$

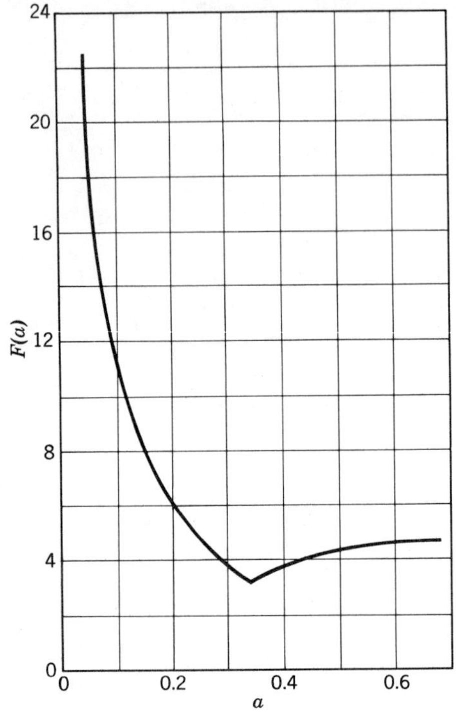

Figure 12-12. Graph of the function $F(a)$ versus a.

Note that $F(a)$ depends only upon the value of a, since a° and b are also functions of a. $F(a)$ is plotted in Figure 12-12 as a function of a.

Now consider the z-component of the shear strain. In Section 10-3 it was shown that

$$v_z = V_z(a - 3a\phi + 3a^2\phi) \tag{12-58}$$

Therefore, the shear rate in the z-direction is

$$\dot{\gamma}_z = \frac{dv_z}{dy} = \frac{1}{H}\left(\frac{dv_z}{da}\right) = \left(\frac{V_z}{H}\right)(1 - 3\phi + 6a\phi) \tag{12-59}$$

and the average shear rate, calculated with (12-48), is

$$\bar{\dot{\gamma}}_z = \left(\frac{V_z}{H}\right)[(1 - b)\,|1 - 3\phi + 6a\phi| + b(1 - 3\phi + 6a^\circ\phi)] \tag{12-60}$$

If the above equations are combined with (12-46), one obtains

$$\gamma_z = \left(\frac{1}{3}\right)\left(\frac{L}{H}\right)\left(\frac{1}{1-\phi}\right)\left(\frac{1}{\sin\theta}\right) G(a, \phi) \tag{12-61}$$

where the function $G(a, \phi)$ is defined by the equation

$$G(a, \phi) = \frac{(1 - b)\,|1 - 3\phi + 6a\phi| + b(1 - 3\phi + 6a^\circ\phi)}{a(1 - a) + b(a^\circ - a)(1 - a - a^\circ)} \tag{12-62}$$

The total strain γ, obtained by adding (12-61) and (12-56), is given by the equation

$$\gamma = \left(\frac{L}{H}\right)\left(\frac{1}{1-\phi}\right)\left[\frac{2F(a)}{3\cos\theta} + \frac{G(a,\phi)}{3\sin\theta}\right] \tag{12-63}$$

Note that the total shear strain becomes very large for long screws with shallow channels. For any screw, it can be made large by operating with a high ratio of pressure to drag flow, in other words, by drastically restricting the output. Note, also, that the strain becomes very large for screws having very small helix angles and for those having very large angles because the pumping efficiency becomes very low at these extremes. Note, in addition, that all fluid elements passing through the extruder are not strained to the same extent, since the strain γ is a function of the position of the element.

Example 12-4.

In this example we consider the performance of a mixing screw that has a 20° helix angle and an (L/H) ratio of 500. The screw is operated first at free discharge ($\phi = 0$) and then with its output restricted to half of its free discharge ($\phi = \frac{1}{2}$). The total shear strain of a fluid element is computed with equation (12-62) as a function of position in the screw channel.

For the special case where $\phi = 0$, equation (12-62) reduces to

$$\gamma = \left(\frac{L}{H}\right)\left[\frac{2F(a)}{3\cos\theta} + \frac{G(a,0)}{3\sin\theta}\right]$$

where

$$G(a,0) = \left[\frac{1}{a(1-a) + b(a^\circ - a)(1 - a - a^\circ)}\right]$$

Values of γ have been computed and are plotted in Figure 12-13.

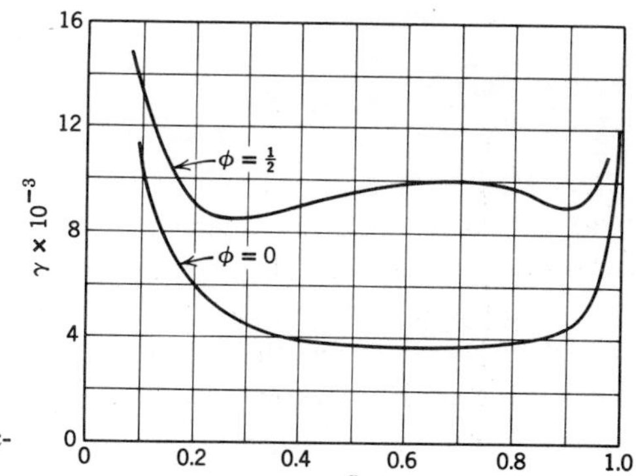

Figure 12-13. Total strain as a function of initial position.

When $\phi = \frac{1}{2}$, equation (12-62) becomes

$$\gamma = 2\left(\frac{L}{H}\right)\left[\frac{2F(a)}{3\cos\theta} + \frac{G(a, 0.5)}{3\sin\theta}\right]$$

where

$$G(a, 0.5) = \frac{(1 - b)\,|3a - 0.5| + b(3a^\circ - 0.5)}{a(1 - a) + b(a^\circ - a)(1 - a - a^\circ)}$$

Values of γ for this case are also plotted in Figure 12-13.

Restricting the output to half of its drag flow value nearly doubles the strain, as can be seen in Figure 12-13. Note, also, that both curves show that the fluid in the center of the channel is strained almost uniformly, but at a very low level compared to that of the material which circulates through the top and bottom sections of the channel, indicating that the fluid in the center of the channel will be least well mixed.

The mixing experiments of Maddock[7] confirm, qualitatively, the conclusions drawn from the above analysis. His work showed the following:

1. Screws having shallow channels are more effective mixers than those with deep channels. Equation (12-63) shows that the strain of all fluid elements is inversely proportional to the channel depth.
2. Circulation of cooling water through the core of the screw improves mixing. This has the effect of reducing the effective channel depth and is therefore explained on the same basis as (1) above.
3. Increasing the pressure at constant screw speed improves the mixing action of the screw. Note that at constant screw speed the value of ϕ (which is the ratio of pressure to drag flow) is directly proportional to the pressure. Equation (12-63) shows that the strain will be increased if the value of ϕ is increased.
4. Increasing the pressure at constant extrusion rate improves the mixing action of the screw but not as much as the same increase at constant screw speed. Note that if the extrusion rate is constant, the screw speed will have to be increased. Since this increases the drag flow rate, the increase in ϕ under these conditions, for the same increase in pressure, is not as great as when the speed is constant.

12-6 Forces in Dispersive Mixing

Dispersive mixing, as defined earlier in this chapter, involves the rupture of clumps or agglomerates of ultimate particles of the minor component. Among the more important applications of dispersive mixing is the incorporation of pigments, fillers, and other ingredients into molten polymers. Dispersive mixing of this type is carried out primarily in Banbury mixers

[7] B. H. Maddock, *S.P.E. Journal*, **15,** 383 (1959).

and roll mills and is a batch operation although some dispersion operations have been put on a continuous basis with specially designed extrusion equipment.

In order to gain some insight into dispersive mixing we consider a system in which the minor component exists as two-particle agglomerates suspended in a viscous liquid. The concentration of the agglomerates is assumed to be low; thus no interaction between agglomerates occurs. The forces operative in such a system include the cohesive force holding the agglomerate together and the surface forces exerted on the particles when relative motion between a particle and the fluid exists.

The following analysis utilizes a moving coordinate system, having its origin at the center of one of the particles of the agglomerate. This is shown in Figure 12-14. The problem is to determine the position (the x-y coordinates) of the center of the second particle during the course of the mixing operation.

Both particles of the agglomerate have radius R; hence the closest the particle centers can approach is $2R$. The center-to-center separation of the particles is designated r, where

$$r = \sqrt{x^2 + y^2}$$

In the region ($2R \leq r \leq r^*$), where r^* is called the critical radius of separation, it is assumed that the attractive force holding the particles together is constant. In the region ($r > r^*$) it is assumed that the attractive force between the particles is negligible. The quantity r^* depends upon the nature of the forces causing agglomeration and is an important parameter in all dispersive mixing operations.

Figure 12-14 also shows that the coordinate system is located so that the liquid surrounding the particles undergoes unidirectional shear flow to the right. The streamline passing through the origin is taken as the reference streamline. The fluid velocity u relative to that on the reference streamline is given by the equation

$$u = \dot{\gamma} y$$

where the shear rate $\dot{\gamma}$ is assumed to be constant over the region of interest.

Whenever there is relative motion between the fluid and the particle, drag forces act on the particle surface. If the fluid velocity is greater than the particle velocity, the drag force tends to accelerate the particle. If the particle velocity is greater than the fluid velocity, the converse is true. If the particles are spheres and if the Reynolds number for the flow about the particle is low, the drag force exerted on the particle, according to Stokes law, is $6\pi R\mu v$, where v is the relative velocity between the fluid and the particle.

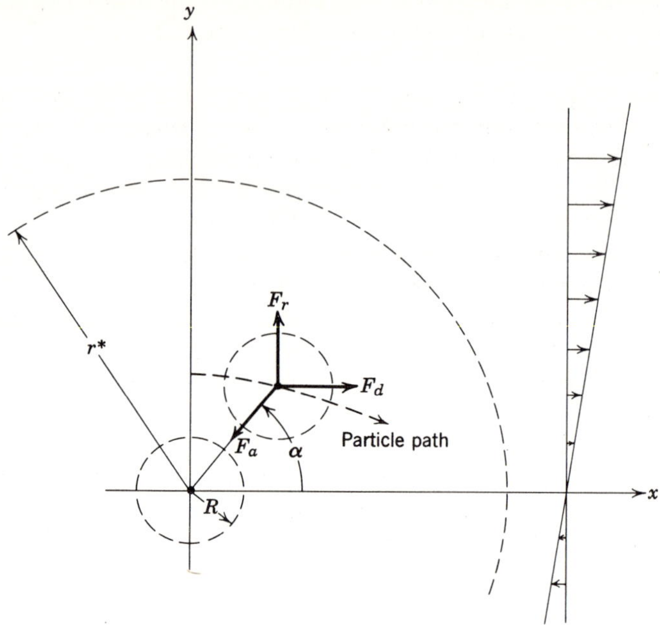

Figure 12-14. Forces operative in the dispersion of a two-particle agglomerate.

Now refer to Figure 12-14. The force resulting from the interaction of the fluid and the surface of the sphere is resolved into the components F_d and F_r, defined by

$$F_d = 6\pi R\mu\left[\dot{\gamma}y - \left(\frac{dx}{dt}\right)\right] \tag{12-64}$$

$$F_r = -6\pi R\mu\left(\frac{dy}{dt}\right) \tag{12-65}$$

where (dx/dt) represents the particle speed in the x-direction and (dy/dt) the particle speed in the y-direction.

The following pair of simultaneous equations is obtained by a force balance that neglects inertia (a reasonable assumption in a system of small particles in very viscous liquids):

$$F_d = F_a \cos \alpha \tag{12-66}$$

$$F_r = F_a \sin \alpha \tag{12-67}$$

where α is the angle that the line connecting the particle centers makes with the abscissa and F_a is the magnitude of the attractive force operative between the particles.

Introducing the expressions for F_d and F_r into the above equations and expressing the trigonometric functions in terms of x and y yields the following equations:

$$6\pi R\mu\left[\dot{\gamma}y - \left(\frac{dx}{dt}\right)\right] = F_a\left(\frac{x}{r}\right) \tag{12-68}$$

$$-6\pi R\mu\left(\frac{dy}{dt}\right) = F_a\left(\frac{y}{r}\right) \tag{12-69}$$

Equations (12-68) and (12-69) constitute a pair of simultaneous ordinary differential equations with time t as the independent variable and x and y as the dependent variables. By eliminating the quantity dt between them, a single equation in x and y having the form

$$\left(\frac{dx}{dy}\right) - \left(\frac{x}{y}\right) = -Kr \tag{12-70}$$

is obtained. The parameter K is defined by the equation

$$K = \frac{6\pi R\dot{\gamma}\mu}{F_a} = \frac{6\pi R\tau}{F_a} \tag{12-71}$$

where τ represents the shear stress in the liquid and is introduced into the expression for K through the defining equation for viscosity, $\mu = (\tau/\dot{\gamma})$.

Because of its right-hand side, (12-70) is nonlinear. However, if the quantity r is replaced with its approximate equivalent $(x + y)$, equation (12-70) becomes

$$\left(\frac{dx}{dy}\right) + x\left[K - \left(\frac{1}{y}\right)\right] = -Ky \tag{12-72}$$

a first order linear differential equation with x as the dependent variable. Its general solution is written

$$\left(\frac{x + y}{y}\right) = Ce^{-Ky} \tag{12-73}$$

where C is the constant of integration.

It is easily shown that the approximation $r = x + y$ is equivalent to assuming that the attractive force F_a is the following function of position in the region $(2R \leq r \leq r^*)$:

$$F_a' = F_a\left(\frac{r}{x + y}\right) \tag{12-74}$$

Note that the ratio $r/(x + y)$ depends only upon the angle α, decreasing from unity when $\alpha = 90°$ to 0.707 when $\alpha = 45°$, and then increasing again to unity when $\alpha = 0$. The approximate nature of the analysis warrants this simplification.

The constant of integration is evaluated by specifying that the particle center passes through the point (x_0, y_0). In this case equation (12-73) becomes

$$\left(\frac{x + y}{x_0 + y_0}\right)\left(\frac{y_0}{y}\right) = \exp\left\{Ky_0\left[1 - \left(\frac{y}{y_0}\right)\right]\right\} \quad (2R \leq r \leq r^*) \quad (12\text{-}75)$$

Equation (12-75) shows that the particle path (and consequently the course of the dispersive mixing operations) depends upon the parameter K, the critical radius r^*, and the initial orientation of the agglomerate. This is illustrated by the following example:

Consider an agglomerated pair of particles oriented as shown in Figure 12-15, where

$$R = 0.5 \qquad x_0 = 0$$
$$r^* = 3 \qquad y_0 = 1$$

The fluid surrounding the particles is subjected to the shear stress that causes unidirectional flow to the right.

Consider first the case where K is very large. This occurs when the attractive force between the particles is small or when the shear stress in the fluid is large. In the limiting case, when $K \to \infty$, the particles follow the streamlines of the fluid, indicating that the dispersive mixing process has degenerated to one of simple mixing. Note, also, that outside of the critical radius r^*, $F_a = 0$ and $K = \infty$, so that all particle paths in this region are streamlines.

The effect of the attractive force operative between the particles becomes noticeable whenever K is finite. For example, when $K = 4$, there is, initially, a slight rotation of the agglomerate and then the upper particle moves downward and outward on a curved path until it reaches the critical radius, at which point the attractive force becomes zero and the particle moves on a streamline to the right.

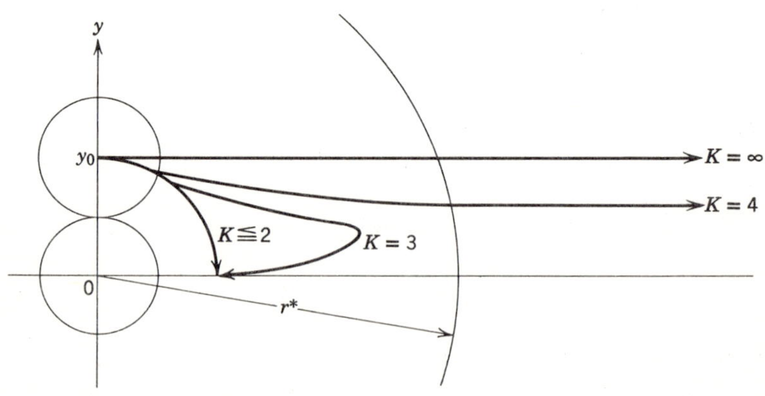

Figure 12-15. Particle paths for dispersion of two-particle agglomerate.

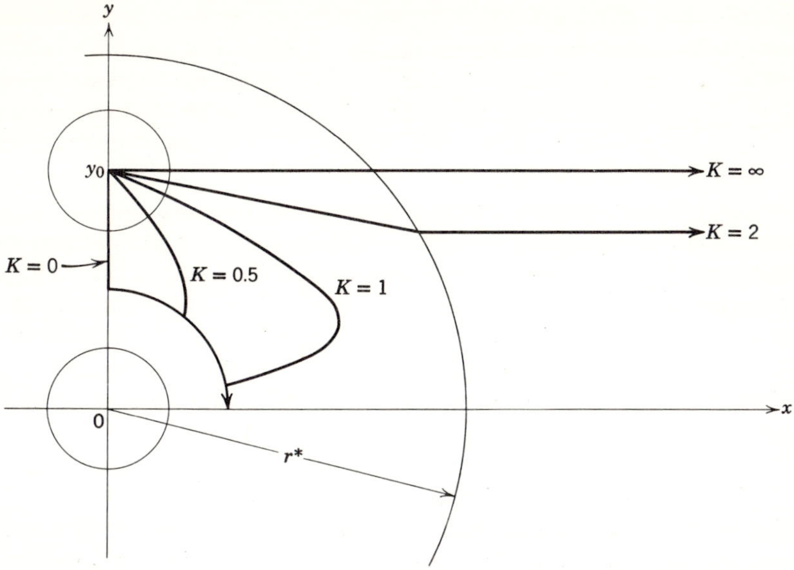

Figure 12-16. Particle paths for dispersion of two-particle agglomerate.

This same process starts when $K = 3$, but in this case the particle eventually reaches a point where the force balance becomes unfavorable, because of reduced fluid drag, and it reverses its direction and reagglomerates itself with the other particle.

For $K = 2$ the attractive force, relative to the viscous force, is so strong that the center-to-center distance does not change. The path of the center of the upper particle is a quarter circle and the net result is a rotation of the agglomerate so that its long axis lies on a streamline.

In this example then, the agglomerate will be ruptured only when the parameter $K \geqq 4$. If the initial orientation were more favorable, a smaller value of K would be sufficient to cause dispersion. For example, if the initial orientation of the particles were as shown in Figure 12-16, dispersion would occur for $K = 2$, indicating that roughly half the shear required before would now suffice.

The first conclusion to be drawn from this analysis of dispersion, one that is perhaps intuitively obvious, is that high shear stresses promote dispersion. Furthermore, with a given system there is some stress below which no dispersion can occur, and when the shear stress is only slightly greater than the critical, only those agglomerates that are initially favorably oriented (with their long axis perpendicular to the streamlines) will be dispersed.

It is also clear that if the flow in the mixer is completely unidirectional, only those agglomerates that have a favorable initial orientation will be ruptured. The others will simply be aligned with the flow, and continued mixing will result in no further dispersion. However, if the mixer makes the flow pattern continually change direction, continued mixing should result in further dispersion. Given sufficient time, each agglomerate will eventually be caught by the shear stresses in a position of favorable orientation and rupture will occur.

Finally, a tentative third conclusion regarding dispersive mixing can be drawn. If one assumes that the attractive force F_a is relatively independent of particle size, then equation (12-71) shows that K is directly proportional to R, indicating that larger particles are more easily deagglomerated than small ones. The assumption that F_a is independent of R is reasonable if the area of contact between the particles is relatively independent of particle size and if the binding forces are of the same type as those operative in adhesion (see Chapter 6).

12-7 Stress Generation in Dispersive Mixing

The problem of generating high shear stresses in non-Newtonian liquids, which may well be the central problem of dispersive mixing in polymer processing, is now considered. The idealized dispersive mixer, shown in Figure 12-17, is used in the following discussion, in which shear stresses are calculated using the method of Bolen and Colwell.[8]

The liquid is contained in the annulus formed by concentric cylinders. The main part of the annulus, called the channel, has a height H and the other part of the annulus, called the tip region, has a height h. The length of the tip region is z and the length of the channel is Z. Unit length of the mixer perpendicular to the plane of the paper is considered. The outer cylinder rotates in a clockwise direction with surface speed V, where $V = \pi DN$ with N representing the frequency of rotation and D the inside diameter of the outer cylinder. The problem is to estimate the average shear stress $\bar{\tau}$ in the channel and $\bar{\tau}'$ in the tip region.

The calculation of the shear stresses is based on the velocity weighted average shear rates defined by equation (10-127). The average shear stresses in the two regions are given by the equations

$$\bar{\tau} = \mu\bar{\bar{\gamma}} \qquad \bar{\tau}' = \mu'\bar{\bar{\gamma}}' \tag{12-76}$$

where $\bar{\bar{\gamma}}$ and $\bar{\bar{\gamma}}'$ represent the velocity weighted average shear rate in the channel and in the tip region, and the μ's are effective viscosities.

If the functional relationship between viscosity, temperature, and shear rate is written in the form

$$\mu = g(\bar{\bar{\gamma}}, T) \qquad \mu' = g(\bar{\bar{\gamma}}', T') \tag{12-77}$$

[8] W. R. Bolen and R. E. Colwell, *S.P.E. Tech. Papers*, **IV**, 1004 (1958).

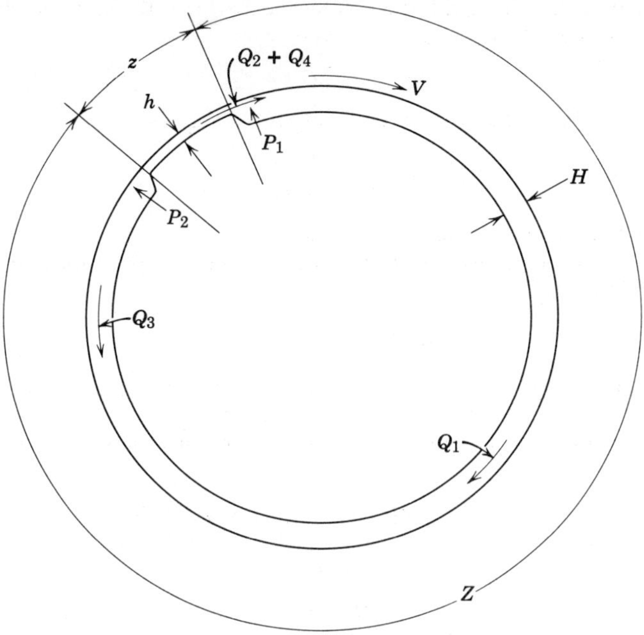

Figure 12-17. Idealized dispersive mixer (Ref. 8).

then the shear stresses are given by the equations

$$\bar{\tau} = \bar{\dot{\gamma}} g(\bar{\dot{\gamma}}, T) \qquad \bar{\tau}' = \bar{\dot{\gamma}}' g(\bar{\dot{\gamma}}', T') \tag{12-78}$$

so that the problem of calculating the shear stresses is resolved into one of calculating the velocity weighted average shear rates and the temperatures in the channel and the tip region. The calculation of the shear rates is outlined below:

The rotation of the outer cylinder causes drag flow of rate Q_d in the channel and Q'_d in the tip region. Because $Q_d > Q'_d$, there is a build-up of pressure in front of the tip and $P_2 > P_1$. This causes a reverse pressure flow of rate Q_p in the channel and a forward pressure flow of rate Q'_p in the tip region.

The calculation of the drag and pressure flow rates is based on the assumption of isothermal Newtonian flow. Furthermore, it is assumed that the dimensions h and H are small in comparison to the diameter of the cylinders so that the annulus can be approximated by flat parallel plates. Under these assumptions, the drag and pressure flow rates are given by the following formulas:

$$Q_d = \frac{VH}{2} \qquad Q'_d = \frac{Vh}{2} \tag{12-79}$$

$$Q_p = \frac{H^3 \Delta P}{12\mu Z} \qquad Q'_p = \frac{h^3 \Delta P}{12\mu z} \tag{12-80}$$

where $\Delta P = P_2 - P_1$.

For an incompressible liquid, the volumetric flow rate through the channel is the same as through the tip region. Therefore

$$Q_d - Q_p = Q'_d + Q'_p \tag{12-81}$$

Combining the above equations and solving for ΔP gives

$$\Delta P = \frac{6V(H - h)}{(h^3/\mu' z) + (H^3/\mu Z)} \tag{12-82}$$

Note that if $h = H$, there is no pressure build-up.

The velocity weighted average shear rates, defined by equation (10-127) are expressed in the form

$$\bar{\dot{\gamma}} = \frac{V}{H} f(\phi) \qquad \bar{\dot{\gamma}}' = \frac{V}{h} f(\phi') \tag{12-83}$$

where $f(\phi)$ is plotted as a function of ϕ in Figure 10-17. The quantity ϕ represents the ratio of pressure to drag flow in the channel and ϕ' the same ratio in the tip region. From equations (12-79), (12-80), and (12-81) it can be seen that

$$\phi = \frac{Q_p}{Q_d} = \frac{H^2 \Delta P}{6\mu Z V} = \frac{1 - (h/H)}{1 + (h/H)^3 (Z/z)(\mu/\mu')} \tag{12-84}$$

Similarly,

$$\phi' = \frac{(H/h) - 1}{1 + (H/h)^3 (z/Z)(\mu'/\mu)} \tag{12-85}$$

Referring to equation (12-77), the viscosity ratio can be written

$$\frac{\mu}{\mu'} = \frac{g(\bar{\dot{\gamma}}, T)}{g(\bar{\dot{\gamma}}', T')} \tag{12-86}$$

and because of (12-83)

$$\frac{\mu}{\mu'} = \frac{g[(V/H)f(\phi), T]}{g[(V/h)f(\phi'), T']} = G \tag{12-87}$$

where G depends upon ϕ, ϕ', H, h, T, T', and V.

Upon introducing (12-87) into (12-84) and (12-85), one obtains

$$\phi = \frac{1 - (h/H)}{1 + G(h/H)^3 (Z/z)} \tag{12-88}$$

and

$$\phi' = \frac{(H/h) - 1}{1 + (H/h)^3 (z/Z) G} \tag{12-89}$$

If it is assumed that the temperature of the liquid in the channel and the tip region is known, then (12-88) and (12-89) constitute a pair of simultaneous equations that can be solved for ϕ and ϕ'. These values can then be substituted into (12-83) and the values of the velocity weighted average shear rates determined.

Bolen and Colwell,[8] using digital computing equipment, calculated shear stresses in molten polyethylene from the flow data Philippoff and Gaskins[9]

[9] W. Philippoff and F. H. Gaskins, *J. Poly. Sci.*, **21**, 205 (1956).

obtained for polyethylene. They assumed that the temperature in the tip region would be the same as in the channel and that it would vary with speed, according to the table below, which also gives the dimensions of the mixer.

N (rps)	T (°K)	
0	400	$h = 0.060$
1	405	$H = 0.437$
2	420	$z = 0.375$
3	440	$Z = 9.045$
4	460	$D = 3.000$

Their results are shown in Figure 12-18. Note that the shear stress for all speeds is greater in the tip region than in the channel, as would be expected. The interesting result is that the average shear stress rises rapidly to a maximum at a speed of less than 1 rps and then decreases.

Jones and Snyder[10] investigated the dispersion of zinc oxide pigments in natural and synthetic rubbers with a Banbury mixer. Some of their results, those showing the effect of temperature and rotor speed on the goodness of dispersion of an untreated pigment in *GR-S* rubber, are given in Table 12-1.

The first three runs show that at constant rotor speed (constant $\dot{\gamma}$ in the tip region) there is an inverse relationship between temperature and goodness of dispersion. The effect of an increase in temperature is to reduce the viscosity. Because of the relation

$$\tau = \mu\dot{\gamma} = a\bar{e}^{bT}\,\dot{\gamma}$$

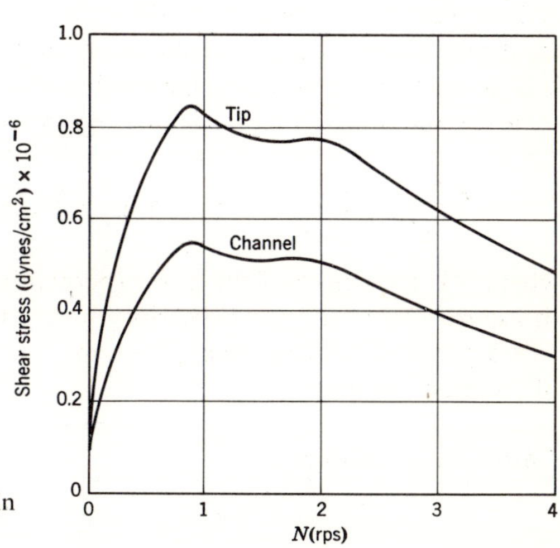

Figure 12-18. Stress generation in idealized dispersive mixer (Ref. 8).

[10] H. C. Jones and E. G. Snyder, *Ind. Eng. Chem.*, **43,** 2602 (1951).

Table 12-1

Data for Dispersion of Zinc Oxide in *GR-S* Rubber*

Jacket Temperature (°C)	Final Temperature (°C)	Time of Mixing (min)	Rotational Speed (rpm)	Energy Consumed (whr)	Rotor Tip Speed (ft/min)	Relative Rating
50	92.5	3	69	398	89	1
65	100.0	3	69	380	89	2
80	107.5	3	69	379	89	3
50	80.0	4	35	286	45	3
50	92.5	3	69	398	89	1
50	105.0	2	137	532	174	2

* Data from Ref. 10.

it can be seen that the stresses generated in the tip region are reduced when the temperature rises. This presumably is the cause of poorer dispersion with increasing temperature.

The last three runs show the effect of rotor speed on the goodness of dispersion. As the rotor speed increases, the temperature and energy consumption rise sharply and the goodness of dispersion passes through a maximum and then decreases. This is the expected result if one assumes that the stress level in the tip region is the controlling factor. These results support the analysis of dispersion presented in Section 12-6.

The scale-up of dispersive mixers can be a valuable design technique and is discussed briefly here. Consider dispersive mixing occurring in a pair of idealized mixers of the type shown in Figure 12-17, where the one mixer is a geometrically similar scale model of the other. Let X represent the scale factor for the mixers, defined by the equation

$$X = \left(\frac{R}{R_0}\right) = \left(\frac{L}{L_0}\right) = \left(\frac{Z}{Z_0}\right) = \left(\frac{z}{z_0}\right) = \left(\frac{H}{H_0}\right) = \left(\frac{h}{h_0}\right) = \left(\frac{B}{B_0}\right) \quad (12\text{-}90)$$

where B is the length of the mixer (perpendicular to the plane of the paper in Figure 12-17) and the subscript 0 refers to the model.

We consider the case where both mixers operate at the same frequency of rotation, so that

$$N = N_0 \quad (12\text{-}91)$$

Furthermore, assume that

$$T = T_0 \quad \text{and} \quad T' = T_0' \quad (12\text{-}92)$$

The validity of this assumption is examined later.

Reference to equations (12-87), (12-88), and (12-89) shows that, under the above assumption,

$$\phi = \phi_0 \qquad \text{and} \qquad \phi' = \phi_0' \tag{12-93}$$

and, because of equations (12-83)

$$\bar{\dot{\gamma}} = \bar{\dot{\gamma}}_0 \qquad \bar{\dot{\gamma}}' = \bar{\dot{\gamma}}_0' \tag{12-94}$$

Because of equations (12-78),

$$\bar{\tau} = \bar{\tau}_0 \qquad \bar{\tau}' = \bar{\tau}_0' \tag{12-95}$$

which shows that the stress level is the same in both mixers. The large mixer should, therefore, do an equal job of dispersion as the model. The power required per unit volume of charge is easily shown to be the same for both mixers.

Let p represent the total power input and p_c and p_t the power consumed in the channel and the tip region, respectively. Using the same method of calculation as in Chapter 10, the power is given by the equation

$$p = p_c + p_t = 2\pi NR[\tau ZB + \tau' zB] \tag{12-96}$$

and on a unit volume basis we have

$$\frac{p}{V} = \left(\frac{2\pi NRB}{V}\right)[\tau Z + \tau' z] \tag{12-97}$$

and because of the above scale relations, we see that

$$\left(\frac{p}{V}\right) = \left(\frac{p_0}{V_0}\right)$$

It can also be shown that a fluid particle in the large mixer makes the same number of passes through the tip region as a fluid particle in the model mixer. Let η represent the number of passes per unit time in the large mixer. Then

$$\eta = \left(\frac{\bar{v}}{2\pi R}\right)$$

and

$$\frac{\eta}{\eta_0} = \left(\frac{\bar{v}}{\bar{v}_0}\right)\left(\frac{R_0}{R}\right) = \left(\frac{1}{X}\right)\left(\frac{\bar{v}}{\bar{v}_0}\right)$$

where $\bar{v}$ represents the average speed of a fluid particle in the channel of the large mixer. Because of equations (12-79) and (12-80), we have

$$\bar{v} = \frac{Q_d - Q_p}{H} = \left(\frac{V}{2}\right) - \left(\frac{H^2\,\Delta P}{12\mu Z}\right)$$

Equation (12-82) shows that $\Delta P = \Delta P_0$ and it is clear that

$$\left(\frac{\bar{v}}{\bar{v}_0}\right) = X$$

and that

$$\eta = \eta_0$$

The assumption represented by equations (12-92) is now examined. It is clear that in order to maintain isothermal conditions in a mixer, the rate of heat transfer from the mixer must be equal to the rate of mechanical power consumption. All other factors being equal, the rate of heat transfer is proportional to the area, which increases according to the second power of X, while the volume increases in proportion to the third power of X. Thus, the maintenance of isothermal conditions is more difficult in large than in small mixers, and unless special precautions are taken, the temperature in a large mixer will rise higher, causing the shear stresses to be reduced and the mixing performance to be less satisfactory than in a small mixer of the same design.

Problems

12-1 A batch of a 2% pigment-plastic mixture is made by tumbling the pure plastic with the pigment, extruding the mixture, and cutting the extrudate into small pellets having an average linear dimension of 0.2 cm. Assuming that the pigment consists of uniformly sized particles having a linear dimension of 0.01 cm and that the density of the pigment is the same as the plastic, estimate the degree of mixedness for this mixture, using equation (12-8) and the following data, which were obtained by withdrawing 20 randomly selected pellets from the mixture and determining the weight percent pigment in each:

Pellet	*% Pigment*	*Pellet*	*% Pigment*
1	1.2	11	1.8
2	1.0	12	2.0
3	0.8	13	0.3
4	4.6	14	0.1
5	5.8	15	5.0
6	0.6	16	8.0
7	0.2	17	0.2
8	2.2	18	0.4
9	3.8	19	0.6
10	0.2	20	1.8

12-2 The imperfect mixture described in Problem 12-1 is again passed through the extruder and again the pigment content of 20 randomly selected pellets is determined. Compare the degree of mixedness of the re-extruded mixture with the original mixture.

Pellet	*% Pigment*	*Pellet*	*% Pigment*
1	1.9	11	2.2
2	2.0	12	2.2
3	2.0	13	1.8
4	2.2	14	1.9
5	2.0	15	2.1
6	1.9	16	1.8
7	1.9	17	2.1
8	2.2	18	1.9
9	1.9	19	2.1
10	2.0	20	2.2

12-3 Write the differential equations for the idealized dispersive mixer with isothermal boundaries, taking into account internal heat generation and using the power law as the rheological equation of the fluid. Include the effect of temperature on viscosity. List all implicit assumptions.

chapter 13

Injection Molding

Injection molding is the most widely used technique for molding thermoplastics. In it a thermoplastic resin is melted and injected (in the liquid state and under high pressure) into a closed mold where it cools and solidifies. The mold is then opened, the piece ejected, and the sequence repeated.

Methods of melting and injecting the plastic differ from one machine to another and are constantly being improved. Conventional machines use a cylinder and piston to do both jobs. This method simplifies machine construction but makes control of injection temperatures and pressures an inherently difficult problem. Other machines use a plasticating extruder to melt the plastic and a piston to inject it while some have been designed to use a screw for both jobs.

Many of the problems connected with injection molding arise because the densities of polymers change so markedly with temperature and pressure. At high temperatures, the density of a polymer is considerably lower than at room temperature, provided the pressure is the same. Therefore, if molds were filled at atmospheric pressure, "shrinkage" would make the moldings deviate from the shape of the mold. Their quality would be poor. To compensate for this effect, molds are filled at high pressures. The pressure compresses the polymer and allows more material to flow into the mold. Shrinkage is reduced and better quality moldings are produced. Shrinkage in injection molding is analyzed below in Section 13-1.

The quality and cost of molding depend upon the *molding cycle* and the *moldability* of the resin. The molding cycle is discussed in Section 13-2, and in Section 13-3 it is related to one of the key quality factors in a molding—frozen strains. This chapter concludes with a discussion of the effect of the thermal and rheological properties of a resin on its moldability.

13-1 Shrinkage

The term *shrinkage*, as used in this book, is defined by the equation

$$S_L = \frac{L^* - L}{L^*} \tag{13-1}$$

where L^* is any linear dimension of the mold and L is the corresponding dimension of the molded object when it is at some standard temperature and pressure. If S is zero for all dimensions, the object is a dimensionally perfect copy of the mold. If S should have the same value for all dimensions, then the object would be geometrically similar to the mold.

Shrinkage can be calculated only for extremely simple geometric shapes because it depends upon the rate at which various sections of the object cool. On the other hand, the volumetric shrinkage, defined as

$$S_V = \frac{V^* - V}{V^*} \tag{13-2}$$

can be calculated if the initial and final (equilibrium) states of the polymer are known. Here V^* is the volume of the mold and V is the volume of the molded object under standard conditions.

If the average linear dimensions $\bar{L}$ and $\bar{L}^*$ are defined by the equations

$$\bar{L} = \sqrt[3]{V} \qquad \bar{L}^* = \sqrt[3]{V^*} \tag{13-3}$$

then the *average shrinkage* $\bar{S}$ can be defined as

$$\bar{S} = \frac{\bar{L}^* - \bar{L}}{\bar{L}^*} \tag{13-4}$$

If equations (13-3) are introduced into (13-4), one obtains

$$\bar{S} = 1 - \left(\frac{V}{V^*}\right)^{1/3} \tag{13-5}$$

If m is the mass of polymer in the mold and v is its specific volume under the *molding* conditions, then

$$V^* = mv \tag{13-6}$$

and

$$V = mv^\circ \tag{13-7}$$

where v° is the specific volume of the polymer at the standard temperature and pressure.

Assuming the equation of state

$$(P + \pi)(v - b) = \frac{RT}{M} \tag{13-8}$$

applies to the liquid state of the polymer, then equations (13-6), (13-7), and (13-8) can be combined with (13-5) to give

$$\bar{S} = 1 - \sqrt[3]{y} \approx \frac{(1 - y)}{3} + \frac{(1 - y)^2}{9} \tag{13-9}$$

where
$$y = \frac{Mv^{\circ}(P + \pi)}{RT + Mb(P + \pi)}$$

which gives the average shrinkage in terms of the initial state of the polymer and the constants of the equation of state.

Example 13-1.

The average shrinkage of four thermoplastics for which the constants of the equation of state are known (see Table 4-3) are calculated in this example. The molding temperature is taken as 500° F and the average shrinkage is calculated for four pressures, ranging from atmospheric to 20,000 psi. The standard state is taken as 20° C and 1 atm. The results are shown in Table 13-1.

Table 13-1

Average Shrinkage of Various Thermoplastics

Pressure (psi × 10^{-3})	Average Shrinkage (mils/in.) Polystyrene	Polymethyl Methacrylate	Cellulose Acetate Butyrate	Polyethylene
0	35	35	52	70
5	23	26	43	59
10	14	18	34	51
20	3	7	21	36
v°	0.946	0.840	0.827	1.085

The calculations indicate that polystyrene and polymethyl methacrylate could be injection molded with small average shrinkage at readily obtainable pressures, while cellulose acetate butyrate and polyethylene would show considerable shrinkage even at the upper range of molding pressures.

If shrinkage were the same for all dimensions, the molded object would be geometrically similar to the mold and one could compensate for shrinkage by enlarging the mold by the proper factor. This would make high injection pressures unnecessary and reduce the size and cost of injection molding machines. However, this could not be a successful solution of the shrinkage problem because almost all objects molded in this manner would show nonuniform shrinkage.

Nonuniform shrinkage is connected with nonuniform cooling. Objects cool nonuniformly because of the inherent nature of the heat conduction process (thick sections cool slower than thin ones), rather than due to defects in the cooling system. Therefore the polymer density increases more rapidly in thin sections than in thick ones. This tends to create a pressure differential that causes some of the polymer in the thick section to flow into the thin. This *internal flow* is the direct cause of nonuniform shrinkage.

Example 13-2.

This example considers unsteady cooling and the decay of pressure in slabs. In order to simplify the calculations, it is assumed that the polymer is initially at a uniform temperature T_i, that the mold is at the constant temperature T_s, and that the thermal diffusivity of the polymer does not change with temperature. Methods of solving unsteady state cooling problems for slabs have been discussed in Section 8-3. In this example we shall be concerned with the mean temperature $\bar{T}$ of the slab, where $\bar{T}$ is defined by the equation

$$\bar{T}(t) = \frac{1}{2W}\int_{-W}^{+W} T(x, t)\, dx$$

Note that the mean temperature $\bar{T}$ is a function of time only, but that the local temperature T depends on both time and position. If the expression for the local temperature, given by equation (8-54), is introduced into the above equation and the integration performed, one obtains a result having the form

$$\frac{T_i - \bar{T}}{T_i - T_s} = f\left(\frac{\alpha t}{W^2}\right) \tag{13-10}$$

where T_i is the initial temperature of the slab and T_s is the surface temperature. The derivation of an expression for the function f in the above equation is left as an exercise (Problem 13-2) at the end of this chapter.

The results of calculations based on equation (13-10) for the unsteady cooling of two slabs (each having thermal diffusivity $\alpha = 1.3 \times 10^{-3}$ cm²/sec) from an initial uniform temperature of 490° F with isothermal surfaces at 100° F are shown in Figure 13-1. The half-thickness of the first slab is 0.216 cm and the half-thickness of the second slab is 0.432 cm.

The average temperature of slab *a* drops at a much faster rate than that of slab *b*. It takes the material in the first slab about 15 seconds to reach the point where its average temperature is below the solidification point; it takes the thick slab about 60 seconds to reach the same point. As a rule of thumb, the required cooling time is proportional to the second power of the thickness of the section.

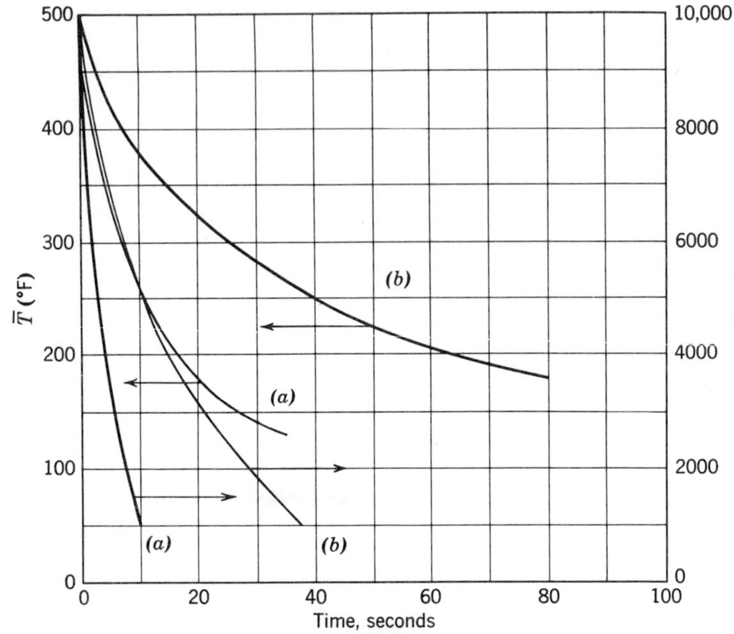

Figure 13-1. Cooling curves and pressure decay curves for polystyrene slabs. The half thickness of slab (*a*) is 0.216 cm and that of slab (*b*) is 0.432 cm.

Thus, if the thickness is doubled, the cooling period would have to be increased by a factor of four.

Now, consider the decay of pressure in the slabs. This can be calculated with the equation of state, (13-8), using the mean temperature of the slab. Note that the specific volume of the polymer must remain constant during the cooling process, as long as pressure is exerted against the sides of the mold. Under these conditions there are no voids; thus the mass and volume remain constant, and equation (13-8) can be written

$$P = C\bar{T} - \pi \tag{13-11}$$

where C and π are constants.

For polystyrene π has a value of 1840 atm (27,000 psi). If the initial pressure in the slabs (at a temperature of 490° F) is 10,000 psi, the value of C in (13-11) must be 38.9. Since $\bar{T}$ is known as a function of time, the pressure can be calculated as a function of time. The results of the calculations are also shown in Figure 13-1.

Note that the pressure in the thin slab decays to zero in less than 15 seconds while the thick slab has positive pressure for more than 50 seconds. If the two slabs were joined together, as in a molding, there would be an internal flow of polymer from the thick to the thin section. This would start as soon as the temperature in the thin slab dropped below that in the thick slab and would continue until the pressure in the thick slab approached zero.

If there is a wide variation in the thickness of various parts of a molding, unequal cooling rates can cause large internal flow and hence sink marks and frozen strains. Sink marks can be prevented by injecting the polymer at a pressure sufficiently high so that positive pressure is always maintained against the mold. However, if the internal pressure at the end of the cycle (the residual pressure) is too high, it would be impossible to remove the object from the mold without damaging it. Therefore there is an optimum range of residual pressures.

13-2 The Molding Cycle

We now consider in some detail the molding cycle itself. Figure 13-2, which is a schematic diagram of an injection mold, is used to illustrate the specialized terminology employed in this discussion. The molten plastic is injected into the mold by the *piston*. It passes through the *nozzle* of the injection cylinder and flows to a central distribution point through the *sprue* and then through *runners* and *gates* to the *cavities* of the mold. Multiple cavity molds, as shown in Figure 13-2, are commonly used.

We consider first the molding cycles resulting when the polymer in the injection cylinder is in the liquid state and when the pressure loss in the cylinder is negligible. The molding cycle is defined when the

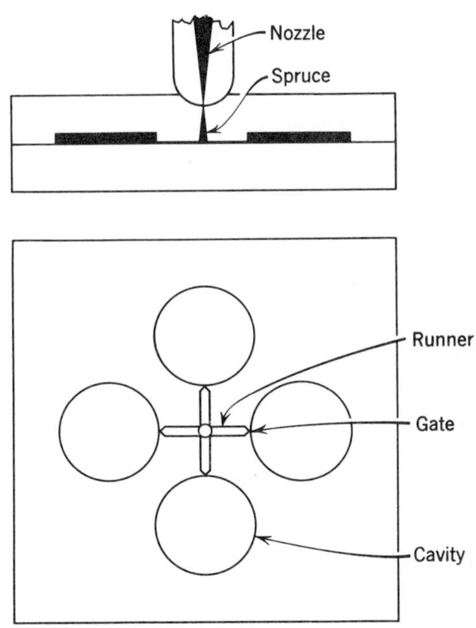

Figure 13-2. Schematic representation of a four-cavity injection mold.

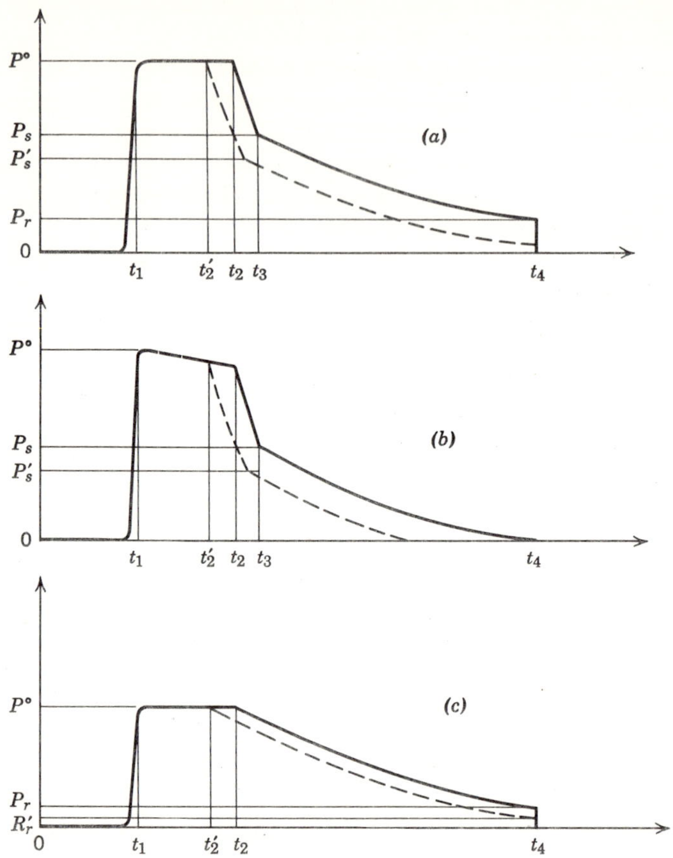

Figure 13-3. Injection molding cycles.

temperature and pressure of the polymer in the mold cavities are specified as a function of time. As we have seen, neither the temperature nor the pressure in the molding is uniform, and it is their average values which are referred to in the following discussion.

The pressure-time diagrams for three modes of operation are shown in Figure 13-3. The first or *injection* stage of the cycle is the same for all three modes. At time zero the piston starts forward and begins to inject polymer into the mold. There is essentially no pressure build up until the cavities are filled, then pressure rises rapidly, reaching a maximum value P° at time t_1. At this point, the polymer in the cavities is at its maximum pressure and nearly at its maximum (injection) temperature.

This is followed by the second stage of the cycle, the *compression* stage.

During this stage, the polymer in the mold cools and its density increases, hence the name compression or packing for this stage of the molding cycle. Note that in diagrams *a* and *c* the pressure of the polymer in the cavity is constant. This is achieved by maintaining a constant hydraulic pressure against the reverse side of the piston. As the polymer in the mold cools, the piston moves forward a slight amount to force the additional polymer into the mold and maintain the constant pressure. In diagram *b* there is some pressure decay during the compression stage. This would occur if the piston remained in the same position that it had at the end of the injection stage. In this case all of the polymer in front of the piston can be considered to form a closed system. As its average temperature drops, its pressure must drop. However, there is still an internal flow of polymer from the cylinder to the mold because of the increasing density of the polymer in the mold.

The third stage of the cycle is the *leakage* or discharge stage. At the beginning of this stage the piston is withdrawn. The high pressure in the mold cavities forces some of the polymer out of the mold, thus causing the mold pressure to drop, as shown in diagrams *a* and *b*. In the third mode of operation, shown in diagram *c*, there is a check valve in the nozzle that permits flow only in the forward direction and prevents leakage. Leakage continues in *a* and *b* until the polymer in the gate freezes. At this point, the cavities are sealed and there can be no further flow. The pressure P_s and the temperature T_s at the time of sealing determine the average shrinkage of the molding. Note that when the check valve is used, the sealing pressure becomes $P°$.

The final stage of the cycle is the *sealed cooling* stage. It starts at the time of sealing, t_3 in diagrams *a* and *b* and t_2 in diagram *c*. During this stage, there is no flow into or out of the cavities although interval flow occurs. This stage ends when the mold opens and the molding is ejected. At this time, any pressure remaining in the object is called the residual pressure and is designated as P_r on the diagrams.

The total time that the piston is in the forward position has a major effect on the temperature and pressure at sealing, and thus on the residual pressure and the average shrinkage. This fact is illustrated by the dashed lines in Figure 13-3. They show the effect of reducing the time that the piston is in its forward position from t_2 to t_2'. In diagrams *a* and *b* the sealing pressure drops from P_s to the lower value of P_s'. Because the sealing temperature is somewhat higher, the density of the polymer at the time of sealing is reduced. Consequently, the average shrinkage is increased to the point where P_r becomes negative.

At the other extreme, if the forward time of the piston were increased beyond the time required to seal the gate, the cycles in diagrams *a* and *b* would become similar to that in diagram *c*. There would be no leakage and the residual pressure would be increased. Under these circumstances, the pressure $P°$ would have to be reduced in order to compensate for the increased density at the time of sealing.

The above discussion shows that the weight of a molding is directly related to the average temperature and pressure in the cavity at the time of sealing. If m^* represents the mass of a molding that is a dimensionally perfect copy of the mold, and m the weight of one that is less than perfect, then from equation (13-6), we can write

$$m^* = \frac{V^*}{v^\circ} \qquad m = \frac{V^*}{v}$$

where v is the specific volume of the polymer at the time the gate closes. Therefore

$$\frac{m}{m^*} = \frac{v^\circ}{v} = y \tag{13-12}$$

where y is defined by (13-9).

Table 13-2

Molding Data of Gilmore and Spencer[1]

Run No.	t_2 (sec)	t_3 (sec)	P_s (psi)	Weight (grams)	$\bar{T}$ (° F)	y	(y/y_5)	(w/w_5)
1	5	6	1000	142	300	0.950	0.937	0.928
2	7	8	1600	144	270	0.958	0.946	0.941
3	9	11	3000	146	240	0.973	0.961	0.955
4	13	17	5000	150	200	0.992	0.980	0.980
5	17	21	9000	153	175	1.011	1	1

Equation (13-12) can be used to analyze the data obtained by Gilmore and Spencer[1] in molding a polystyrene box. The wall thickness of the box was 0.17 in. (0.216 cm), the injection temperature was 490° F, and the mold temperature, while not stated specifically in their article, was probably about 100° F. Five runs were made in which the time that the piston was held in the forward position was varied. They measured the pressure in the mold cavity, using a strain gage pressure transducer. Their data are given in Table 13-2.

Assuming that t_1 is small, the average temperature in the cavity can be estimated with curve $\bar{T}_1$ of Figure 13-1, since this curve was constructed

[1] G. D. Gilmore and R. S. Spencer, *Modern Plastics*, **27**, 143 (1950).

for the cooling of a slab having the same thickness as the box and cooling under the same conditions. Knowing the average temperature and the average pressure in the molding at the time the gate seals, and the physical properties of polystyrene, the value of y, defined by equation (13-9), can be calculated. The results are also shown in Table 13-2.

Note that for Run No. 5 y is greater than unity, indicating that the weight of the molded box is slightly larger than that of a dimensionally perfect copy of the mold. If the value of y for each run is divided by the value of y for Run No. 5, then this ratio represents the ratio of the weight of each box to that of box No. 5, i.e., (y/y_5) represents (w/w_5). These ratios are shown in the last two columns of the table. Note that the agreement is excellent.

13-3 Frozen Strains

Other factors, in addition to shrinkage, determine the quality of a molding. One of the most important is the absence of frozen (residual) strains.

A molding contains frozen strains when the polymer molecules in it are in a nonequilibrium configuration. Frozen strains are a phenomenon associated with the viscoelasticity of polymer melts. The following discussion is based on a simplified version of the mechanical model of Figure 2-13, formed by removing the spring that represents the instantaneous elastic response because it has no role in the processes leading to frozen strains.

Consider, first, the application and removal of stress under *isothermal* conditions. As shown in Figure 13-4*a*, the retarded elastic deformation is completely recovered when the stress is removed, so that the model returns to its original equilibrium position. A different result is obtained if the model is cooled as it is deformed. Upon removal of the stress, the recovery of the elastic deformation is slow because the viscosity associated

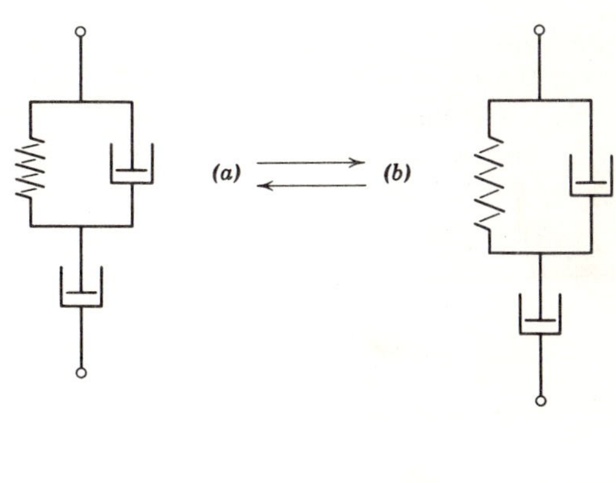

Figure 13-4. Model simulating viscoelastic behavior. Under isothermal conditions model returns to equilibrium state (*a*) upon removal of stress, but if cooling occurs in strained state (*b*), then strain is not recovered upon removal of stress and the strain is said to be frozen.

with this process has increased. If cooling is carried below a certain temperature before the stress is removed, the model is essentially *frozen* in its deformed state. It is said to contain residual or frozen strains.

As discussed briefly in Section 2-7, a retarded elastic deformation of an amorphous linear polymer corresponds, on the molecular level, to a transformation to a less probable state or configuration in which there is a partial orientation or alignment of the polymer chains in some preferred direction. Thus the freezing of the mechanical model in a strained position corresponds to the freezing of polymer chains in a position of partial orientation. Molecular orientation accompanied by cooling, therefore, is the direct cause of frozen strains.

Cooling occurs continuously during the molding cycle. Therefore strains will be frozen in the molding at all points where molecular orientation occurs. First consider orientation arising from flow during the *filling* stage of the cycle. During filling, the polymer melt flows rapidly into the mold and, although the bulk of the polymer cools very little during this step, there is a thin skin of rapidly cooling polymer next to the mold surfaces. This results in the formation of a thin, highly oriented layer on the surface of the molding, a layer that is in a state of frozen strain.

During the *compression* portion of the cycle, the mass of the molding is cooling and, at the same time, there is an inflow of polymer. This results in some molecular orientation throughout the mass of the molding. During the *leakage* stage of the molding cycle, the polymer is at a still lower temperature, and additional molecular orientation occurs during the outflow of the polymer.

Once the gate of the mold cavity closes there is no further inflow or outflow of polymer. However, in moldings containing thick and thin sections there can be internal flow, and this can cause additional molecular orientation and frozen strains.

There are still other causes of frozen strains. Consider the case of a fairly thick molding. As it cools, a stiff outer skin will form while the interior is still fluid, and this skin will prevent the molding from pulling away from the walls of the mold. As a result, stresses will be set up in the direction of heat flow, and these stresses will tend to orient the molecules in the same direction. Thus additional frozen strains will be introduced into the molding.

A frozen strain of another type can result even when the molding is cooled uniformly. As the temperature passes through the second order transition point, the rate of cooling is likely to be much faster than the molecular processes that cause the polymer to contract. Hence the molding will be in a nonequilibrium state at the end of the cycle.

Using monochromatic light, Spencer and Gilmore[2] have studied molecular orientation in polystyrene moldings by examining them between crossed Polaroids. The number of dark bands or rings observed is an indication of the extent of the molecular orientation.

Some of their results are shown in the graphs of Figure 13-5. The first is a plot of the weight of the molding versus the time of compression, which is the period between t_1 and t_2 in the diagrams of Figure 13-3. Two curves are shown. The upper curve applies when a check valve is used in the nozzle and the lower curve for conventional operation in which the plunger is simply held in position and then withdrawn at the end of the compression period. Note that a much heavier and, therefore, better molding is obtained with the check valve because all leakage is prevented.

The graph of Figure 13-5*b* shows the number of dark bands as a function of time of compression. At short times of compression, the inflow and outflow take place before sufficient cooling can occur to cause orientation. As the compression time is increased, the weight of the molding increases as does the amount of orientation. The beneficial effects of the check valve are illustrated by considering the conditions required to produce a molding having a weight of 150 grams. With the check valve, a

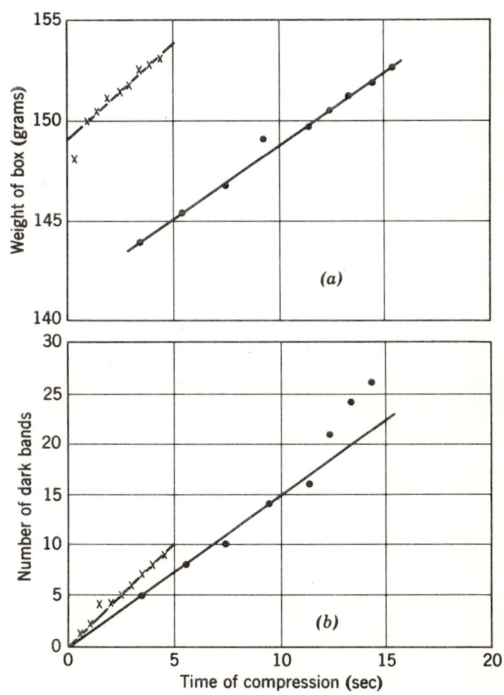

Figure 13-5. Effect of time of compression on (*a*) the weight of the molding and (*b*) the extent of molecular orientation as indicated by the number of dark bands observed when the molding is examined with crossed Polaroids. Lower curve applies to conventional operation, upper to operation with check valve in nozzle (Ref. 2).

compression period of about 2 seconds would be required and orientation would be low, resulting in about 5 dark bands. Without the check valve, a compression period of about 12 seconds would be required and much greater orientation (20 dark bands) would be produced.

We now consider some of the effects of frozen strains upon the quality of the molding. The first effect to be considered is crazing—the development of small cracks in the surface of the plastic. In moldings, craze lines are oriented along the direction of flow and their formation is undoubtedly connected with molecular orientation. Craze lines can be produced by rapidly chilling a sample. The uneven cooling (and thus shrinkage) causes tensile stresses to develop in the surface of the molding so that the surface fails in its weakest direction. In an object having an oriented surface skin, failure will often occur along lines parallel to the direction of molecular orientation. The results of Spencer and Gilmore[2] show that reducing molecular orientation in polystyrene definitely reduces crazing and improves the mechanical properties of the moldings.

Frozen strains also reduce the temperature at which distortion of the molding occurs. If frozen strains were absent, a molding could be heated without distortion to the point where gravity would cause flow. When frozen strains are present, the moldings will begin to distort as soon as the temperature is raised to the point where the rate of recovery of the elastic deformations becomes noticeable.

Strength properties of oriented moldings are likely to be directional. This fact applies particularly to those properties, such as flexural strength, that depend upon the surface rather than on the volume. For example, Spencer and Gilmore measured the flexural strength of a molded polystyrene sample and obtained values of 16,000 psi perpendicular to the direction of flow in the mold and about 6000 psi parallel to the direction of flow. Measurements made on moldings produced with a check valve were compared with those produced in a conventional machine and showed little difference. This is explained on the assumption that a check valve prevents orientation primarily in the bulk of the molding and that flexural strength depends primarily upon surface orientation. Hence it cannot be reduced with a check valve.

13-4 Moldability

The performance of any injection molding process depends upon the design of the equipment (mold and injection machine), the molding cycle, and the physical properties of the thermoplastic molding compound. The term *performance* is used loosely to indicate the quality of the moldings and the rate at which they can be produced.

[2] R. S. Spencer and G. D. Gilmore, *Modern Plastics*, **28,** 97 (1950).

The machine controls, such as the cycle timer and the temperature controllers, are the direct operating variables. These variables determine the nature of the molding cycle (the temperature and pressure in the cavities as a function of time). The physical properties that are important in injection molding can be classified broadly as either rheological or thermal properties. The previous discussion has shown that the *P-V-T* behavior of the polymer, its viscoelastic behavior, and its steady flow behavior all have a profound influence on the performance of the molding process. These all depend upon the rheological properties of the polymer melts. The thermal property of primary importance in injection molding is thermal diffusivity. It governs the rate of unsteady state heating and cooling and thus has a direct bearing on the length of cooling period required. Another thermal property of importance is the temperature at which the polymer begins to undergo thermal degradation. This sets an upper limit to the temperatures that can be used in melting the resin.

The "moldability" of a polymer depends basically upon these rheological and thermal properties. An ideal material for injection molding would melt sharply at a low temperature to produce an inelastic liquid of low viscosity and high thermal diffusivity. Furthermore, the density would be little affected by changes in temperature or pressure, and the material could be heated to high temperatures without thermal degradation. Some types of metal alloys used in special casting applications almost meet the above requirements. Most thermoplastics, however, have properties that are distantly removed from the ideal.

One aspect of moldability is the "ease" with which a given polymer melt will flow into a mold. Consider, for example, the injection of a molten polymer into a long, straight cylindrical cavity (maintained at the constant temperature T_m) with an injection pressure P_i and an injection temperature T_i. Since $T_m < T_i$, the polymer cools as it flows into the mold, with flow continuing until the polymer temperature drops below its freezing point.

For a given mold and with the operating variables (T_i, P_i, and T_m) fixed, the quantity of polymer that can flow into the mold before freezing is a measure of its "ease" of flow under conditions simulating injection molding. The quantity is proportional to the limiting distance of penetration X as shown in Figure 13-6.

The value of X depends primarily upon the viscosity-temperature relationship for the polymer and upon its thermal diffusivity. Large values of X are obtained with polymers having low viscosities that are relatively insensitive to temperature or with polymers having low thermal diffusivities. It should be noted, however, that a large X value does not necessarily mean good over-all moldability. For example, if a large X is obtained because of low thermal diffusivity, then, although molds may be

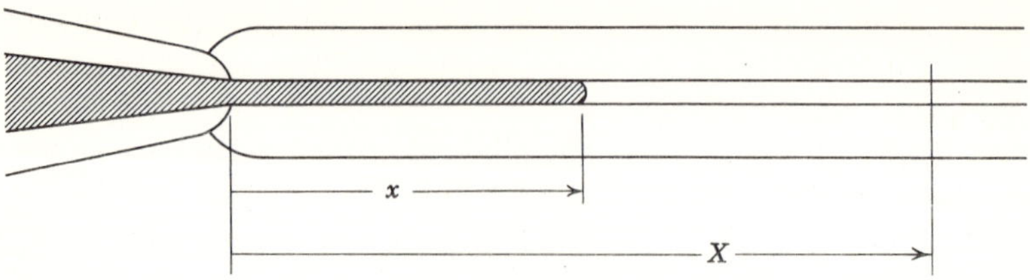

Figure 13-6. Injection of molten polymer into a straight cylindrical cavity maintained at constant temperature. Distance of penetration x is a function of injection time. Limiting penetration distance X is reached when polymer freezes.

filled rapidly and easily, the over-all cycle would have to be excessively long to permit solidification of the polymer before ejection.

Ballman, Shusman, and Toor[3] carried out an extensive investigation of flow into the cavity of an injection mold. With a fixed set of operating variables (T_i, P_i, and T_m), they determined the distance of penetration X of polystyrene and polyethylene as a function of time. From these data they then derived the mean flow speed u as a function of time t and found that log (u) was proportional to t. Figure 13-7 is a typical straight line semilogarithmic plot of their data, implying that the functional relationship between u and t is of the form

$$u = \frac{dx}{dt} = u_0 e^{-t/\beta} \tag{13-13}$$

where u_0 and β depend upon mold design, the physical properties of the polymer, and the operating variables.

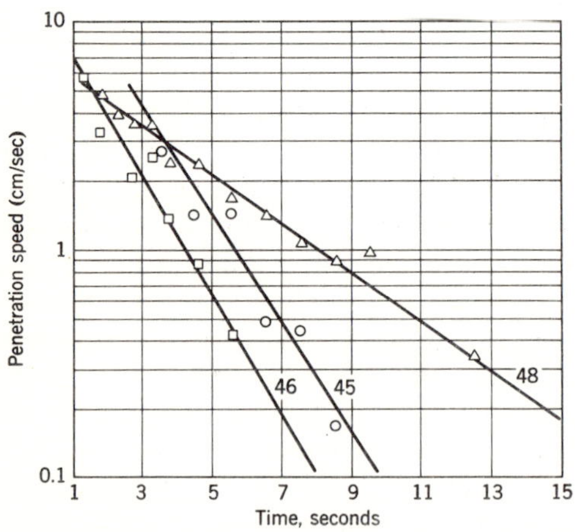

Figure 13-7. Semilogarithmic plot of penetration speed as a function of time (Ref. 3).

[3] R. L. Ballman, T. Shusman, and H. L. Toor, *Modern Plastics*, p. 115, October, 1959.

Integrating equation (13-13) and applying the initial condition $x(0) = 0$ gives the following relationship between x and t:

$$x = u_0\beta[1 - e^{-t/\beta}] \tag{13-14}$$

Equation (13-14) reduces to

$$X = u_0\beta$$

as $t \to \infty$, indicating that the limiting penetration distance depends upon both u_0 and β.

Ballman et al.[3] determined u-t data in a series of 48 runs with polystyrene and polyethylene, using different mold geometries and two different molding machines that covered a rather wide range of the operating variables. They concluded that the simple exponential relationship (13-13) may have general validity for injection molding. It should be noted, however, that the parameters u_0 and β are not necessarily indicative of the moldability of a given polymer because they are dependent upon the operating variables and the mold design. It is clear that a theoretical analysis of nonisothermal, unsteady flow in which the energy equation is coupled with the equation of motion (see Chapter 2) is needed to elucidate the fundamentals of the injection process.

Problems

13-1 Calculate the average shrinkage $\bar{S}$ for a polyethylene object molded at 300° F and 10,000 psi pressure. What pressure would be required to reduce the average shrinkage to 0.2%?

13-2 Derive an analytical expression for the function $f(\alpha t/W^2)$ of equation (13-10). Verify your result by referring to one of the standard heat transfer texts.

13-3 Referring to Table 13-2, estimate the weight of a box molded with an injection temperature of 400° F and with $t_3 = 10$ sec and $P_s = 5000$ psi.

13-4 Define the compressive weight increment Δw of a molding as the difference between the weight of the molding and the weight that the molding would have if the compression time of the molding cycle were zero. Referring to Figure 13-5, calculate Δw for a compression time of 2.5 sec and 5 sec, for each mode of operation. Then plot the number of dark bands against these values of Δw. What conclusions concerning molecular orientation in molded objects can you draw from the plot? Compare your plot and your conclusions with those of Spencer and Gilmore.[2]

chapter 14

Bonding Operations

This chapter discusses two operations in which adhesion (bonding) is a key factor—heat sealing and extrusion coating. The essential difference between these operations is that in heat sealing, the material on both sides of the interface becomes molten or softens during the operation while in extrusion coating, the material (the substrate) on one side of the interface remains in the solid state during the entire operation. Usually, heat sealing is a batch operation whereas extrusion coating is a continuous operation. Plastic bags and packages are made by heat sealing; flexible laminated structures, such as aluminum foil-polyethylene laminates, are usually made through extrusion coating operations.

14-1 Heat Sealing

Thermoplastic films are used extensively for shower curtains, packaging materials, raincoats, automobile seat covers, and many other products. An important factor contributing to their success in these applications is the development of fast, simple, and inexpensive joining techniques, of which heat sealing is by far the most important.

In heat sealing, two films are pressed together and heated for a short period of time. As the plastic at the interface softens, the films stick together and become sealed. The strength of the seal depends upon the contact time, the film temperature at the interface, the contact pressure, and upon the nature of the films themselves.

Heat sealing operations can be classified on the basis of the method used to heat the film. In one type of operation, heat is conducted through the films from heated platens. Another type utilizes dielectric heating to generate heat within the plastic. In operations of the first type, the primary operating variables are contact time, contact pressure, and platen temperature. In operations utilizing dielectric heating, the primary operating variables (in addition to contact time and pressure) are the applied voltage and frequency.

The first step in the analysis of a heat sealing operation is to relate the temperature and pressure at the interface to the operating variables. The pressure is obtained by dividing the force applied to the platens by the area of the seal. It does not change with time. The interface temperature is calculated by using the methods of unsteady state conduction outlined in Chapter 8. It generally increases with time.

The final step in the analysis of a heat sealing operation is to relate the seal strength to the interface temperature, the contact pressure, the contact time, and to the physical properties of the film. This requires an understanding of the basic molecular processes operative in heat sealing.

14-2 Fusion and Surface Seals

The investigations of McKelvey and Strome[1] give some insight into the molecular processes responsible for heat sealing. They concluded, on the basis of experiments with polyethylene film, that two distinct types of heat seals can be formed. The first type, called a *surface seal*, forms almost instantaneously and is due primarily to intermolecular attractive forces acting across the interface. The strength of this type of seal is strongly influenced by the chemical nature of the film surface.

The second type, called a *fusion* seal, forms over a finite period of time and is due primarily to the diffusion of polymer chains across the interface. If the process of diffusion proceeds long enough, the interface completely loses its identity and the films cannot be separated at the seal. The diffusion process is strongly affected by temperature and by anything that tends to reduce the mobility of polymer chains, such as cross-linking.

The mechanism of formation of fusion seals was described by McKelvey and Strome[1] in the following manner:

> Consider two films brought into intimate contact at a temperature above their softening point. Due to molecular motion, parts of polymer chains from the one film will cross the interface and enter the other film, and vice versa. If this diffusional process continues for a long enough period of time, the interface loses its identity and the films are said to be fused. The time required for this condition to be reached is obviously a strong function of temperature. At high temperatures, where the viscosity is low, a state of near fusion can be reached in a matter of seconds. On the other hand, at temperatures in the neighborhood of the softening point, fusion might require days or weeks.

[1] J. M. McKelvey and T. H. Strome, *Modern Plastics*, p. 107, June, 1959.

The fusion temperature τ_f is defined as the temperature at which seals of such high strength are formed (at a given contact time and pressure) that the films cannot be separated even with very careful and slow pulling and peeling. The fusion temperature, therefore, is that temperature at which the molecules of the film become sufficiently mobile that molecular diffusion makes the interface lose its identity during the period of the sealing cycle.

14-3 Sealability

For polyethylene, fusion temperatures (measured at 1 second contact time and 20 psi pressure) are in the range of 300 to 600° F. The value of τ_f is a sensitive indicator of degradative changes occurring in a film, as shown by the data in Table 14-1.

The increase in τ_f with aging is attributed to cross-linking reactions that reduce the mobility of the polymer chains and thus retard the diffusional process leading to fusion seals.

Peel tests are commonly used to evaluate surface seals. In a peel test, the free ends of the films forming the seal are clamped in the jaws of a testing machine and the seal "peeled" apart at a given rate. Figure 14-1 is a

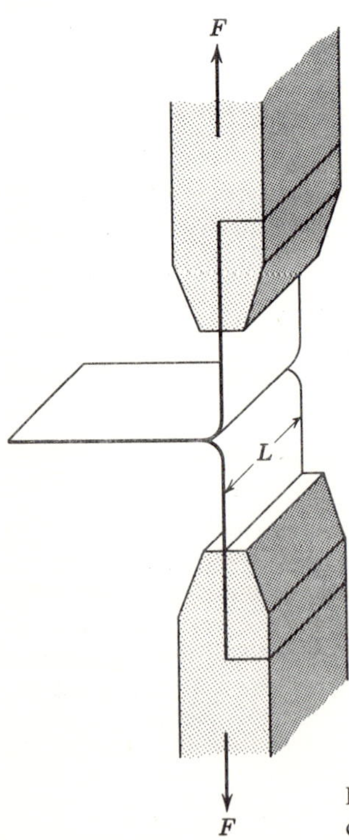

Figure 14-1. Schematic representation of peel test commonly used to evaluate seal strengths.

Table 14-1

Effect of Aging on Fusion Temperature of Polyethylene*

Storage Conditions	Fusion Temperature (°F)
13 days at room temperature	322
34 days at room temperature	380
13 days at room temperature plus 21 days at 150° F	520

* Data of McKelvey and Strome.[1]

schematic representation of the test. The force F required to peel the seal apart, divided by the width of the sealed area L, is taken as the peel strength of the seal.

The peel strength depends upon the rate of peeling, the temperature, the geometry of the seal, and upon other factors that must be standardized in order to make valid comparisons among seals. Because the peel strength ordinarily varies from point to point in a given seal, the average, the maximum, or some integrated average value of strength must be recorded for each test. Considerable variation from test to test can be expected, and sufficient samples must be tested in order to establish a reliable result.

Low strength seals (those with peel strengths below about 1000 grams/in.) are primarily surface seals. A surface seal is essentially an adhesive bond. It forms almost instantaneously whenever two surfaces are brought into intimate contact. The strength of the seal depends upon the area of effective contact and upon the magnitude of the intermolecular attractive forces. The effective contact area is increased by heating the film to soften it and by applying pressure to make the two surfaces conform to a common interface. The intermolecular attractive forces depend strongly upon the chemical nature of the surface and can be increased by introducing polar groups into the surface.

The effect of interfacial temperature on the seal strength of a polyethylene film is shown in Figure 14-2, where the interfacial temperature covers a range of about 40 deg. F. The interfacial temperature was controlled by varying the contact time from $\frac{1}{4}$ to 5 seconds, a 20-fold variation. All the points fall on a single curve, which extends about 30 deg. F below and about 10 deg F above the melting point of the film. At lower temperatures the films are too rigid to be pressed into close contact, and at higher temperatures the molecular mobility is so high that fusion seals are formed almost instantaneously.

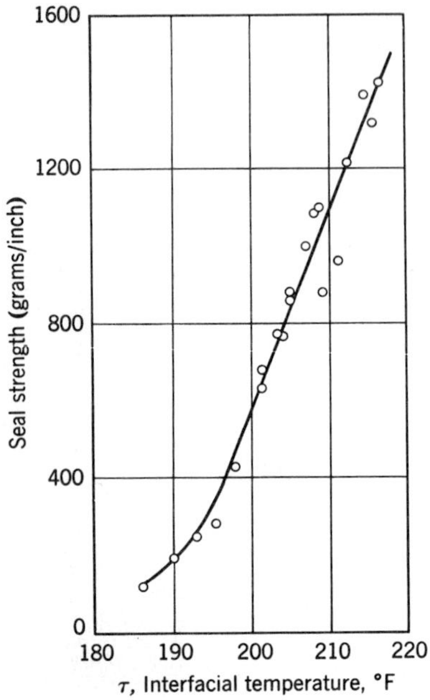

Figure 14-2. Seal strengths obtained at various contact times and platen temperatures show common aependence on the interface temperature (Ref. 1).

Figure 14-3 shows that an increase in contact pressure causes the seal strength to increase. It appears reasonable that as the contact pressure increases, the conformation of the surfaces improves, thereby increasing the effective contact area. The result is increased adhesional strength.

The interfacial temperature at which a surface seal of 500 grams/in. strength is formed (at a given contact time and pressure) is designated τ_{500}. For the film of Figure 14-2, the value of τ_{500} is about 198° F. This quantity should be a rather sensitive indicator of the surface characteristics of a given film, as is indicated by the data in Table 14-2.

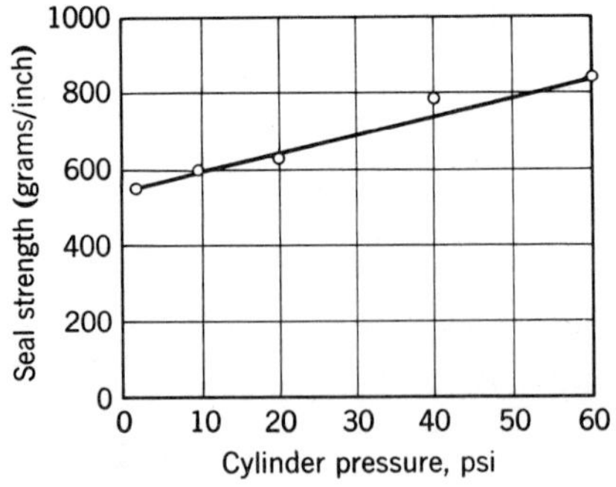

Figure 14-3. Effect of pressure on seal strength at a fixed temperature and contact time (Ref. 1).

Table 14-2

Effect of Surface Activation on the Formation of Heat Seals*

Film	τ_{500} (° F)
No. 1, unactivated	223
No. 1, activated	205
No. 2, unactivated	225
No. 2, activated	190

* Data of McKelvey and Strome.[1]

Note that for each film, surface activation reduces the value of τ_{500}. It is probable that seals formed with the activated films have lower effective contact areas (because of reduced sealing temperatures) but stronger intermolecular attractive forces.

14-4 Conduction Heat Sealing

The methods of unsteady state conduction, outlined in Chapter 8, are used to calculate the temperature distributions in sheets or films during heat sealing. In the following discussion it is assumed that the thermal diffusivity of the material is constant, that there is negligible resistance to heat transfer between the press platens and the film, and that the film is initially at a uniform temperature T_i.

Three special cases, shown schematically in Figure 14-4, are considered. In cases (*a*) and (*b*) the upper platen is maintained at the constant temperature T_p. In case (*a*) the lower platen is also at temperature T_p, but in case (*b*) the lower platen is maintained at temperature T_i, the initial temperature of the material being sealed. This maintains the outer surface of the lower film in the unmelted state throughout the sealing operation.

In case (*c*) the lower platen is replaced with a rubber slab that is neither heated nor cooled. In solving the heat equation for this case, it is assumed that the thermal diffusivity of the rubber base is the same as the material being sealed, so that together they form a single homogeneous mass into which the heat flows.

For all three cases the applicable differential equation is

$$\left(\frac{\partial T}{\partial t}\right) = \alpha\left(\frac{\partial^2 T}{\partial x^2}\right) \tag{14-1}$$

The initial condition for all three cases is written

$$T(x, 0) = T_i$$

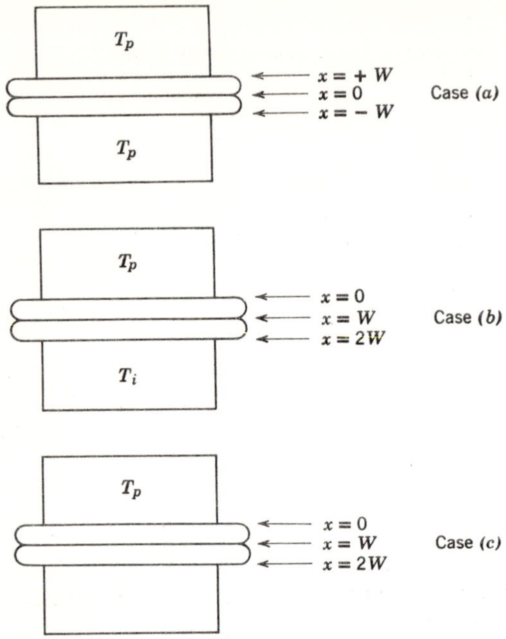

Figure 14-4. Conduction heat sealing: (*a*) lower platen heated, (*b*) lower platen cooled, and (*c*) lower platen neutral.

and the boundary conditions are written

$$\begin{array}{ll} \text{case } (a) & T(\pm W, t) = T_p \\ \text{case } (b) & T(0, t) = T_p \\ & T(2W, t) = T_i \\ \text{case } (c) & T(0, t) = T_p \end{array}$$

Note that in cases (*b*) and (*c*), the origin of the coordinate system has been shifted from the interface to the surface of the upper platen.

The temperature distribution for case (*a*) is given by equation (8-54) and for case (*c*) by (8-49). The solution of the heat equation meeting the boundary conditions of case (*b*), as given by Carslaw and Jaeger,[2] is written

$$\left(\frac{T_p - T}{T_p - T_i}\right) = \left(\frac{x}{2W}\right) + \frac{2}{\pi} \sum_{n=1,2,\cdots} \frac{1}{n} \exp\left(-\frac{n^2\pi^2\alpha t}{4W^2}\right) \sin\frac{n\pi x}{2W} \tag{14-2}$$

Equation (14-2) can be specialized to give the temperature τ at the interface by replacing x with W. One obtains

$$\left(\frac{T_p + T_i - 2\tau}{T_p - T_i}\right) = \frac{4}{\pi} \sum_{n=1,3,\cdots} \frac{(-1)^{(n-1)/2}}{n} \exp\left(-\frac{n^2\pi^2\alpha t}{4W^2}\right) \tag{14-3}$$

[2] H. S. Carslaw and J. C. Jaeger, *Conduction of Heat in Solids*, 2nd ed., Oxford University Press, 1959, p. 313.

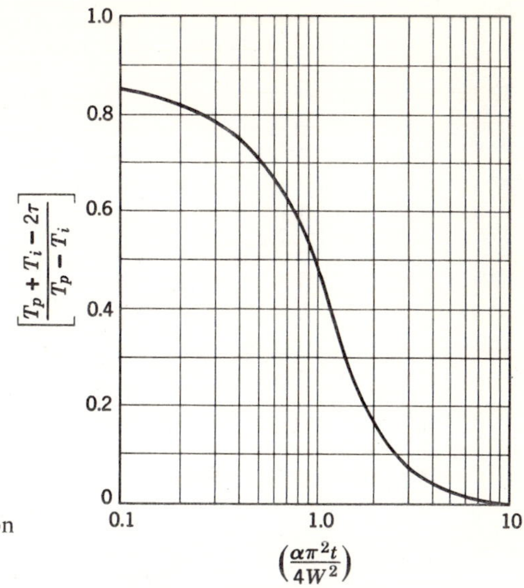

Figure 14-5. Solution of equation (14-3).

Figure 14-5 is a graphical solution of equation (14-3) in which the dimensionless temperature ratio is plotted against the quantity ($\alpha\pi^2 t/4W^2$).

As a specific example, consider a heat sealing operation in which

$$T_i = 80°\text{ F} \qquad W = 2 \text{ mils } (0.005 \text{ cm})$$
$$T_p = 400°\text{ F} \qquad \alpha = 1.6 \times 10^{-3} \text{ cm}^2/\text{sec}$$

and in which the required sealing temperature at the interface is 220° F. The calculation of the required time for sealing for each of the three cases is outlined below:

case (a):

$$\frac{T_p - \tau}{T_p - T_i} = 0.56$$
$$\left(\frac{\alpha t}{W^2}\right) = 0.34 \quad \text{(from Figure 8-6)}$$
$$t = 5.3 \times 10^{-3} \text{ sec}$$

case (b):

$$\left(\frac{T_p + T_i - 2\tau}{T_p - T_i}\right) = 0.125$$
$$\left(\frac{\alpha\pi^2 t}{4W^2}\right) = 2.4 \quad \text{(from Figure 14-5)}$$
$$t = 15.2 \times 10^{-3} \text{ sec}$$

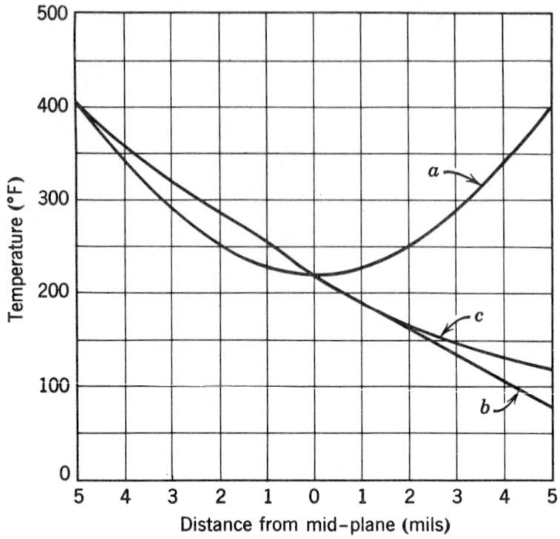

Figure 14-6. Temperature profiles at end of heating period for (*a*) platen heated, (*b*) lower platen cooled, and (*c*) lower platen neutral.

case (*c*):

$$\left(\frac{T_p - \tau}{T_p - T_i}\right) = 0.56$$

$$\frac{W}{2\sqrt{\alpha t}} = 0.55 \quad \text{(from Figure 8-4)}$$

$$t = 12.9 \times 10^{-3} \text{ sec}$$

The temperature distribution at the end of the cycle, as well as the cycle time, is of interest in heat sealing. For case (*a*) it can be calculated with (8-54), for case (*b*) with (14-2), and for case (*c*) with (8-50). The results are shown in Figure 14-6.

14-5 Dielectric Heat Sealing

In dielectric heat sealing, the sheets or films to be sealed are pressed between two electrodes to which a high frequency alternating potential is applied. Because of dielectric losses within the material, the temperature of the material rises. The time required to reach the sealing temperature depends upon the physical (thermal and dielectric) properties of the material, the thickness of the material, the applied voltage, the frequency, and the rate at which heat is lost from the material.

Heat losses occur primarily by conduction to the electrodes. The rate of heat loss is high when the electrodes are cold and when the thermal conductivity of the material is high. It is clear that if the heating cycle continues for a relatively long period of time, heat generation will just balance heat loss by conduction to the electrodes, and a steady temperature distribution will be established in the material.

Under steady state conditions, the heat equation reduces to

$$k\frac{d^2T}{dx^2} = -G \tag{14-4}$$

If the electrodes are maintained at the constant temperature T_p and if the origin of the coordinate system is located in the mid-plane between the electrodes, the boundary conditions become

$$T(\pm W) = T_p$$

where W is the thickness of one of the sheets or films. The solution of equation (14-4) meeting the above boundary condition is written

$$(T - T_p) = \frac{W^2G}{2k}\left[1 - \left(\frac{x}{W}\right)^2\right] \tag{14-5}$$

where G is the heat generation rate (assumed constant) and k is the thermal conductivity of the material.

Equation (14-5) shows that the temperature profile, a parabola, is symmetrical about the mid-plane. The magnitude of the equilibrium temperature rise depends upon the parameter $(W^2G/2k)$. Figure 14-7 shows equilibrium temperature profiles as a function of this parameter.

The theory of dielectric heating was outlined in Chapter 7, where the following equation for the heat generation rate was derived:

$$G = 13.3 \times 10^{-14}\left(\frac{V^2f\epsilon''}{4W^2}\right) \tag{14-6}$$

If the loss factor ϵ'' is constant, then the heat generation rate is readily calculated with the above equation.

As an example, consider the case where

$W = 2$ mils (0.005 cm) $\qquad \epsilon'' = 0.24$
$V = 500$ volts $\qquad k = 10^{-4}$ cal/cm-sec-deg C
$f = 10$ megacycles

The value of G, calculated with equation (14-7), is 0.8×10^3 cal/cm³-sec and the parameter $(W^2G/2k)$ has the value of 100 deg C. The equilibrium temperature profile is shown in Figure 14-7. It should be noted that the loss factor ϵ'' is a function of temperature, and thus the actual temperature profile would be distorted from the parabola shown in the figure.

The maximum attainable temperature rise is obtained when the material is thermally insulated from the electrodes. For perfect insulation, the rate of temperature rise is $(G/c_p\rho)$ deg C/sec. If, in the above example, a temperature rise of 80 deg C is required and if the $(c_p\rho)$ product has a value

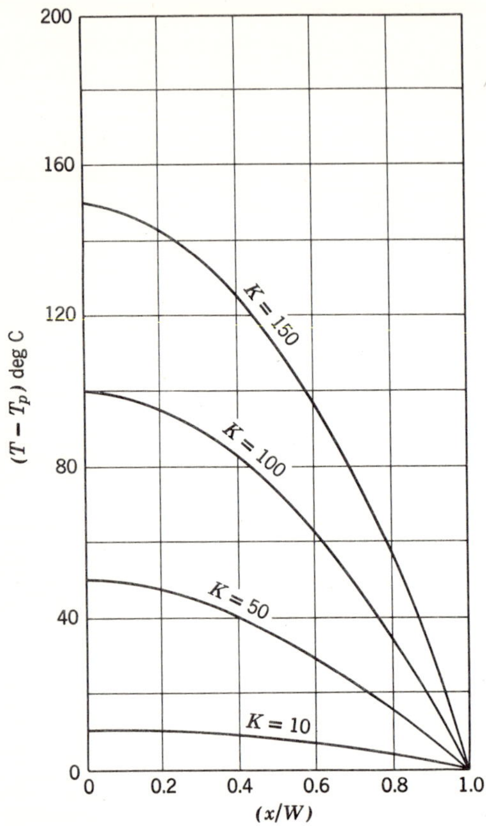

Figure 14-7. Steady state temperature profiles for various heat generation rates ($K = W^2G/2k$ deg C).

of 0.4 cal/cm^3-deg C, then the rate of temperature rise is 2000 deg C/sec and the required heat cycle time is 0.04 sec.

Of more practical interest is the case where conduction losses are taken into account. The special case is considered in which the electrodes are maintained at room temperature, which is also taken as the initial temperature of the material. This is a problem in unsteady state heating with internal heat generation and conduction losses from the surfaces. The applicable differential equation, (8-38), is written

$$\left(\frac{\partial T}{\partial t}\right) = \alpha\left(\frac{\partial^2 T}{\partial x^2}\right) + \left(\frac{G}{c_p \rho}\right) \tag{14-7}$$

and the boundary conditions for the problem are as follows:

$$T(x, 0) = T_p$$
$$T(\pm W, t) = T_p$$

The solution of equation (14-7) meeting the above boundary conditions is given by Carslaw and Jaeger.[3] The mid-plane temperature τ, obtained by placing x equal to zero, is given by the equation

$$(\tau - T_p) = \frac{GW^2}{2k}\left[1 - \frac{32}{\pi^3}\sum_{n=1,2,\ldots}^{\infty}\frac{(-1)^n}{(2n+1)^3}\exp\left(-\frac{\alpha(2n+1)^2\pi^2 t}{4W^2}\right)\right] \tag{14-8}$$

which can be written in the form

$$(\tau - T_p) = \frac{GW^2}{2k} F\left(\frac{\alpha t}{W^2}\right) \tag{14-9}$$

Introducing equation (14-6) into (14-9) and rearranging gives the following result:

$$(\tau - T_p) = KF\left(\frac{\alpha t}{W^2}\right) \tag{14-10}$$

where K is a parameter defined by the equation

$$K = 1.66 \times 10^{-14}\left(\frac{V^2 f\epsilon''}{k}\right) \tag{14-11}$$

Figure 14-8 is a plot of $(\tau - T_p)$ versus the dimensionless group $(\alpha t/W^2)$ for various values of the parameter K.

Returning to the previous example, the value of the parameter K calculated with equation (14-11) is 100, and for a temperature rise of 80 deg C, the value of the quantity $(\alpha t/W^2)$ is about 0.7, as found from Figure 14-8. Therefore the required heating time is 0.07 sec, almost double that required if the plastic is thermally insulated from the electrodes.

If the thickness of the films should be doubled and the sealing carried out under the same conditions of frequency and voltage as before, then, when the electrodes are thermally insulated, the required heating time—being independent of thickness—is the same as before. When conduction losses occur, the parameter K has the same value and, therefore, the quantity $(\alpha t/W^2)$; thus the required time increases by a factor of 4, and becomes 0.28 sec.

The power required for the sealing operation can be computed by multiplying the heat generation rate G by the volume of the material and converting the result to electrical units. If, in the above example, the sealed area were 1 cm wide and had a length of 10 cm, the volume of the material would be 0.10 cm^3 and the total heat generation rate would be 80 cal/sec, which, when expressed in electrical units, is 334 watts.

[3] Carslaw and Jaeger, *op. cit.*, p. 130.

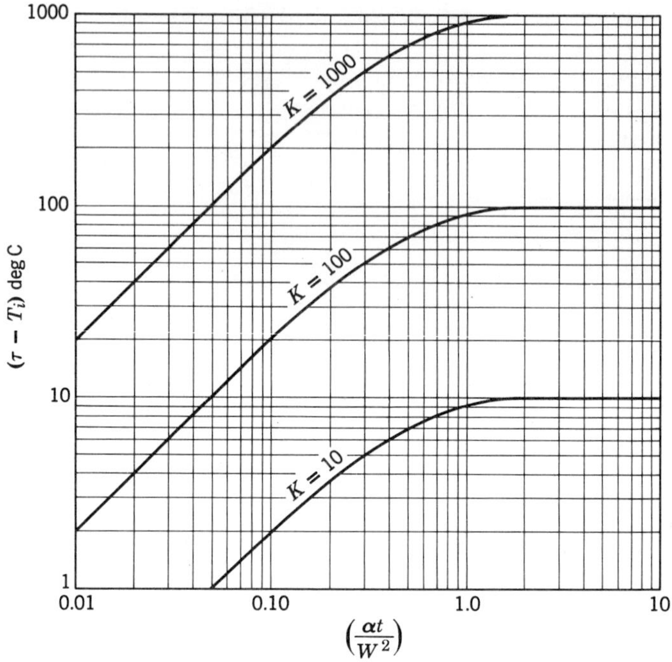

Figure 14-8. Solution of equation (14-10).

14-6 Extrusion Coating

Extrusion coating is a high speed continuous operation in which a film of molten polymer is spread upon a substrate and then chilled to form a solid coating. Usually, an extruder is used to melt the polymer and to pump it through the die. Contacting usually occurs at the nip of a pair of rolls. Figure 14-9 is a schematic diagram illustrating an operation in which an aluminum foil substrate is being coated with polyethylene. Other substrates commonly used in extrusion coating include paper, cellophane, "Mylar," and other plastic films.

A key factor in determining the quality of an extrusion coating is the strength of a bond between the substrate and the coating. With polyethylene this is a particularly acute problem because polyethylene normally has little affinity for other materials. It has been determined, however, that a polyethylene coating will adhere to aluminum and other substrates if it is extruded at temperatures in the range of 600° F. At extrusion temperatures below 600° F, adhesion is often poor and satisfactory coatings cannot be made. This effect is attributed to surface oxidation occurring in the air gap between the die and nip of the rolls. In a sense, the surface of the molten film becomes partially *activated* (see Section 6-5) and the adhesive forces are substantially increased. The extent of this surface activation appears to be affected by the time-temperature history of the surface of the molten polymer in the air gap.

This section is concerned with the calculation of the surface temperature of the molten film as a function of position in the air gap. In the formulation of the problem, the necking-down of the film as it leaves the die is neglected; thus the film becomes a slab bounded by parallel plane surfaces separated by distance $2W$. Furthermore, it is assumed that the polymer leaves the die at a uniform temperature, designated T_i, and that the temperature of the ambient air T_f is constant.

Heat losses from the film occur by convection and radiation. The heat transfer coefficient h depends upon the convection coefficient h_c and the radiation coefficient h_r, and because convection and radiation losses occur in parallel,

$$h = h_c + h_r \tag{14-12}$$

In the following analysis it is assumed that h is independent of temperature.

The applicable differential equation, the heat equation, reduces to

$$\left(\frac{\partial T}{\partial t}\right) = \alpha\left(\frac{\partial^2 T}{\partial x^2}\right) \tag{14-13}$$

if conduction in the direction of motion (the y-direction) is neglected.

In the coating operation, the film moves at a constant speed V, and the position of a point on the surface relative to the die would normally be used

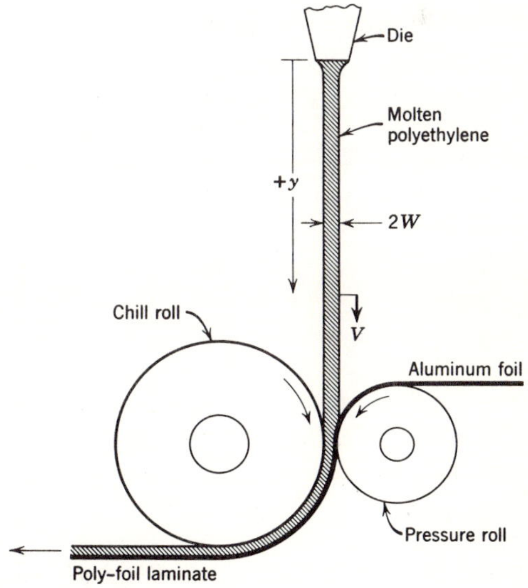

Figure 14-9. Extrusion coating process.

as an independent variable instead of time. These quantities are related by the equation

$$y = Vt \tag{14-14}$$

where y represents the distance from the die lips. Upon transformation from t to y as an independent variable, equation (14-13) becomes

$$V\left(\frac{\partial T}{\partial y}\right) = \alpha\left(\frac{\partial^2 T}{\partial x^2}\right) \tag{14-15}$$

The boundary conditions for the problem are written

$$\begin{aligned} T(x, 0) &= T_i \\ -kT'(+W, y) &= h[T(+W, y) - T_f] \\ kT'(-W, y) &= h[T(-W, y) - T_f] \end{aligned} \tag{14-16}$$

Equation (8-52) is the solution to this boundary value problem provided one substitutes (y/V) for t. The resulting equation can then be specialized for the surface temperature T_s at the nip by replacing x with W and y with Y. One obtains

$$\frac{T_f - T_s}{T_f - T_i} = \sum_{n=1,2,\cdots}^{\infty} f(M) \exp\left(-\frac{M^2 \alpha Y}{W^2 V}\right) \cos M \tag{14-17}$$

where Y represents the distance from the die lips to the nip of the rolls. The parameter M and the function $f(M)$ are defined by equations (8-53) and (8-52), respectively.

The convective heat transfer coefficient h_c can be estimated with equation (8-69), which we write in the form

$$h_c = H\sqrt{V/Y}$$

where

$$H = 0.66k\sqrt[3]{N_{\mathrm{Pr}}}\sqrt{\rho/\mu}$$

and the heat transfer coefficient h becomes

$$h = h_r + H\sqrt{V/Y}$$

As will be shown later, the quantity (hW/k) is very small, and the first term only of the series of equation (14-17) need be considered. Furthermore, the following approximations can be used:

$$\begin{aligned} \sin M_1 &= \tan M_1 = M_1 \\ \sin 2M_1 &= 2M_1 \\ \cos M_1 &= 1 \\ f(M_1) &= 1 \\ M_1^2 &= hW/k \end{aligned}$$

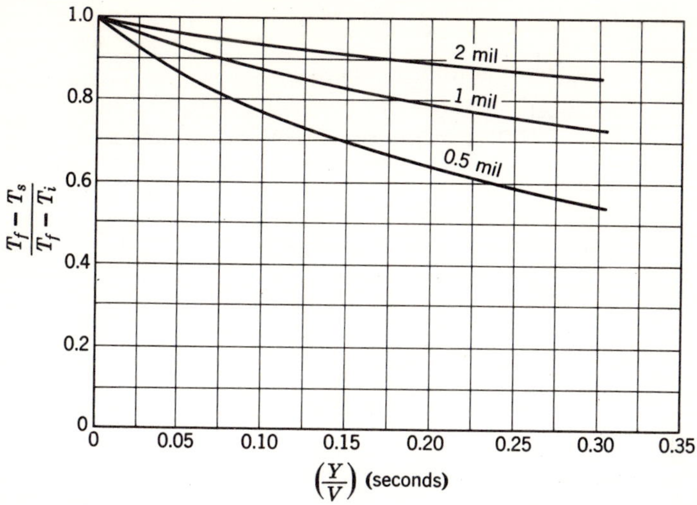

Figure 14-10. Solution of equation (14-18) for extrusion coating of polyethylene with $T_i = 600°$ F, $T_f = 80°$ F, and $h_r = 49.5 \times 10^{-5}$ cal/cm²-sec-deg C.

and equation (14-17) reduces to

$$\frac{T_f - T_s}{T_f - T_i} = \exp\left[-\frac{h_r}{Wc_p\rho}\left(\frac{Y}{V}\right) - \frac{H}{Wc_p\rho}\sqrt{\frac{Y}{V}}\right] \tag{14-18}$$

Equation (14-18) shows that the surface temperature of the film at the nip, for a given film thickness, extrusion temperature, and ambient air temperature, depends only upon the quantity (Y/V), which is the residence time of an element of the film in the air gap.

As a specific example, consider a coating operation in which the polyethylene extrusion temperature is 600° F and the ambient air is 80° F. The value of h_r, as calculated in Example 8-2, is 49.5×10^{-5} cal/cm²-sec-deg C. The physical properties of the air are evaluated at the mean temperature of 340° F and those of the polyethylene at 600° F. Figure 14-10 shows a plot of the temperature ratio $(T_f - T_s)/(T_f - T_i)$ versus the quantity (Y/V) for 2 mil, 1 mil, and 0.5 mil film.

Table 14-3 shows the surface temperature of the molten film at the nip of the rolls and the residence time in the air gap when the gap distance is 0.3 ft and the extrusion speed is 60 ft/min and 300 ft/min.

At low extrusion speeds, the surface of the 0.5 mil film cools rapidly. At high speeds it maintains its temperature, but its exposure time is reduced, indicating that an optimum speed and gap distance probably exist for each film thickness.

Table 14-3

Film Thickness (mils)	Extrusion Speed (ft/min)	Residence Time (sec)	Surface Temperature at the Nip (° F)
2	60	0.3	527
1	60	0.3	460
0.5	60	0.3	361
2	300	0.06	580
1	300	0.06	564
0.5	300	0.06	522

Problems

14-1 Consider a conduction heat sealing operation in which the upper platen is heated and the lower platen cooled. Sealing 1 mil thick films, satisfactory seals are formed with a heating cycle of 0.01 sec when the upper platen is at 400° F and the lower at 100° F. (a) Estimate the required heating cycle time if the upper platen temperature were reduced to 300° F. (b) If, in the original operation, the 1 mil thick films were replaced with 2 mil thick films, what change would have to be made in the cycle time? (c) State the key assumptions that are involved in these calculations.

14-2 Verify equation (14-5). Note that the units of the quantity ($W^2G/2k$) are deg C. What is the physical significance of this quantity?

14-3 Consider a dielectric heat sealing operation in which both platens are maintained at the constant temperature of 100° F. Sealing 1 mil thick films, satisfactory seals are formed with a heating cycle of 0.01 seconds. If the 1 mil thick films were replaced with 2 mil thick films and the voltage applied to the electrodes were doubled, what adjustment would have to be made in the cycle time? State the key assumptions involved in arriving at your answer.

14-4 Consider an extrusion coating operation in which 1/2 mil thick polyethylene is being applied to a substrate. If the extrusion temperature is 500° F and the air gap is 3 in., estimate the surface temperature of the polyethylene as it contacts the substrate, when the linear speed of the operation is 50 ft/min.

appendix **A**

Derivation of Fluid Mechanical Equations

This appendix provides a derivation of the fluid mechanical equations discussed in Chapter 2. The first part summarizes the vector-tensor notation and operations utilized in the derivations. Next the stress tensor is discussed and the equations of continuity, momentum, and energy derived. A general rate of strain tensor is then developed and separated into one part that characterizes volumetric strain effects and into another characterizing shear strain effects. By postulating a linear relationship between the strain rates and the appropriate stresses, the rheological equation of the Newtonian fluid is obtained.

A-1 Vector-Tensor Notation and Operations[1]

The following treatment is restricted to rectangular Cartesian coordinates in which the position of a point in space is specified by the three numbers

[1] The symbols and operational notation used here and throughout most of this book are borrowed from *Transport Phenomena* (by R. B. Bird, W. E. Stewart, and E. N. Lightfoot and published by John Wiley and Sons, in 1960), where a more detailed discussion of vector-tensor operations can be found. Other useful references include *An Introduction to Matrix Tensor Methods in Theoretical and Applied Mechanics* by Sidney F. Borg, J. W. Edwards, Ann Arbor, Mich., 1956, and *Mechanics of Solids and Fluids*, by R. R. Long, Prentice-Hall, 1961.

Following the notation of *Transport Phenomena*, tensors are printed as boldface Greek letters ($\boldsymbol{\sigma}$, $\boldsymbol{\tau}$, $\boldsymbol{\epsilon}$, etc.), all other boldface italic letters represent vectors ($\boldsymbol{A}$, $\boldsymbol{B}$, $\boldsymbol{v}$, etc.), and the remaining symbols represent scalars (P, ρ, T, etc.). The type of quantity produced by the various multiplicative operations is indicated by the form of the parentheses. For example, $(\boldsymbol{A} \cdot \boldsymbol{B})$ represents a scalar, $[\boldsymbol{A} \times \boldsymbol{B}]$ a vector, and $\{\boldsymbol{\sigma} \cdot \boldsymbol{\tau}\}$ a tensor.

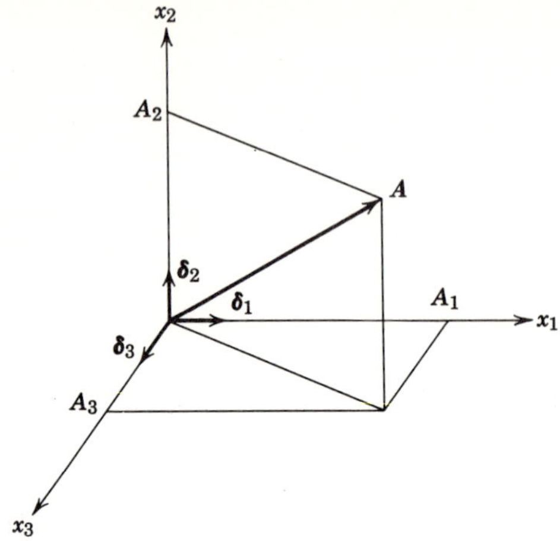

Figure A-1. Rectangular Cartesian coordinate system showing the projections of the vector $\boldsymbol{A}$ on the coordinate axes. $\boldsymbol{\delta}_1$, $\boldsymbol{\delta}_2$, and $\boldsymbol{\delta}_3$ are unit vectors.

x_1, x_2, and x_3. A line connecting the origin with any point in space represents a vector $\boldsymbol{A}$ whose direction is that of the line and whose magnitude is proportional to the length of the line. The projections of $\boldsymbol{A}$ on the coordinate axes are designated A_1, A_2, and A_3 and are shown in Figure A-1.

Line segments starting at the origin and directed along the coordinate axes and having unit length are called unit vectors. They are designated $\boldsymbol{\delta}_1$, $\boldsymbol{\delta}_2$, and $\boldsymbol{\delta}_3$. By utilizing these unit vectors, any vector can be expanded as a sum of its components. For example, the expansion of the vector $\boldsymbol{A}$ is written

$$\boldsymbol{A} = \boldsymbol{\delta}_1 A_1 + \boldsymbol{\delta}_2 A_2 + \boldsymbol{\delta}_3 A_3 = \sum_i \boldsymbol{\delta}_i A_i \tag{A-1}$$

The scalar product of any two vectors $\boldsymbol{A}$ and $\boldsymbol{B}$ is designated $(\boldsymbol{A} \cdot \boldsymbol{B})$ and is often called the dot product. It is defined as the product of the magnitudes of the two vectors multiplied by the cosine of their angle θ of intersection. Therefore

$$(\boldsymbol{A} \cdot \boldsymbol{B}) = AB \cos \theta \tag{A-2}$$

The angle of intersection of any two different unit vectors is always 90°. Hence their dot product is always zero. The dot product of a unit vector with itself is unity because the angle of intersection is zero. The nine dot products that can be formed from the three unit vectors can be written in the compact form

$$(\boldsymbol{\delta}_i \cdot \boldsymbol{\delta}_j) = \delta_{ij} \qquad \begin{array}{lll} \delta_{ij} = 0 & \text{if} & i \neq j \\ \delta_{ij} = 1 & \text{if} & i = j \end{array} \tag{A-3}$$

and the dot product of any two vectors $\boldsymbol{A}$ and $\boldsymbol{B}$ can be written

$$(\boldsymbol{A} \cdot \boldsymbol{B}) = \sum_i A_i B_i \tag{A-4}$$

Example A-1.

The heat flux vector $\boldsymbol{q}$ represents the rate (per unit area) of heat flow through a surface that is perpendicular to the vector $\boldsymbol{q}$. Metric units of $\boldsymbol{q}$ are cal/cm^2-sec. In this example, an expression for the rate (per unit area) is derived for heat flow through a surface element that is oriented arbitrarily with respect to the vector $\boldsymbol{q}$.

The orientation of a surface can be characterized by a unit vector $\boldsymbol{n}$ that is normal to the surface under consideration. In Figure A-2 the slant face of the tetrahedron is perpendicular to $\boldsymbol{n}$ and intersects $\boldsymbol{n}$ at a distance Δn from the origin. Conservation of energy requires that the rate of accumulation of heat within the tetrahedral control element shown in the figure be equal to the net rate at which heat flows into the element through its four sides. This is stated mathematically by the equation

$$\rho C\,\Delta V\left(\frac{dT}{dt}\right) = q_n\,\Delta S + \sum_i q_i\,\Delta S_i \tag{A-5}$$

where ΔV is the volume of the element and C, ρ, and T are average values of specific heat, density, and temperature within the element. On the right side of the equation, q_i represents the x_i component of the vector $\boldsymbol{q}$ and ΔS_i is the area of the face that is perpendicular to the x_i axis. The flux q_n through the slant face is the quantity of interest.

If the area of the slant face is ΔS, then we can define a new vector $\Delta \boldsymbol{S}$ by the equation

$$\Delta \boldsymbol{S} = \boldsymbol{n}\,\Delta S$$

where the magnitude of $\Delta \boldsymbol{S}$ is the area of the slant face and the direction of $\Delta \boldsymbol{S}$ is the same as $\boldsymbol{n}$. The magnitude of the x_i component of $\Delta \boldsymbol{S}$ is given by the dot product of $\boldsymbol{\delta}_i$ and $\Delta \boldsymbol{S}$. Hence we have

$$\Delta S_i = (\boldsymbol{\delta}_i \cdot \Delta \boldsymbol{S}) = (\boldsymbol{\delta}_i \cdot \boldsymbol{n}\,\Delta S) = n_i\,\Delta S \tag{A-6}$$

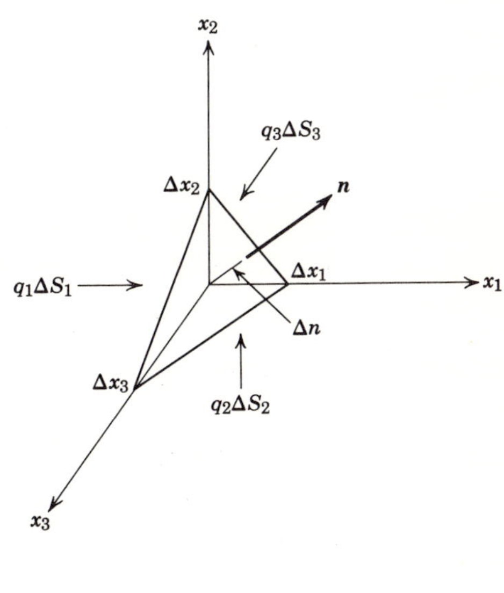

Figure A-2. Tetrahedral element used in the analysis of heat flow through a surface normal to the vector $\boldsymbol{n}$.

where n_i is the x_i component of the normal vector $\boldsymbol{n}$. Therefore, equation (A-5) can be written

$$\rho C \,\Delta V \left(\frac{dT}{dt}\right) = q_n \,\Delta S + \Delta S \sum_i n_i q_i$$

which, upon rearranging, becomes

$$q_n = -\sum_i n_i q_i + \rho C \left(\frac{\Delta V}{\Delta S}\right)\left(\frac{dT}{dt}\right)$$

It is clear that, in the limit as $\Delta n \rightarrow 0$ and the tetrahedron shrinks to a point, the second term on the right becomes zero and the first term on the right represents the dot product of the vectors $\boldsymbol{n}$ and $\boldsymbol{q}$. Hence

$$q_n = -(\boldsymbol{n} \cdot \boldsymbol{q}) \tag{A-7}$$

and we see that $(\boldsymbol{n} \cdot \boldsymbol{q})$ represents the outer directed flux of heat at a point where the flux vector is $\boldsymbol{q}$ and the surface unit normal is $\boldsymbol{n}$.

A similar result is obtained when one considers mass flow rather than heat flow. If the vector $\boldsymbol{v}$ represents the velocity of the fluid at any point in space and the scalar ρ is its density at that point, then $(\rho\boldsymbol{v})$ represents the mass flux vector at that point. The units of the vector $(\rho\boldsymbol{v})$ are grams/cm^2-sec. The rate (per unit area) of mass flow in through a surface whose orientation is characterized by the outer directed unit vector $\boldsymbol{n}$ is therefore $-(\boldsymbol{n} \cdot \rho\boldsymbol{v})$.

The vector product of any two vectors $\boldsymbol{A}$ and $\boldsymbol{B}$ is designated $[\boldsymbol{A} \times \boldsymbol{B}]$ and is also called the cross product. It is a vector perpendicular to the plane of $\boldsymbol{A}$ and $\boldsymbol{B}$ and directed so that $\boldsymbol{A}$-$\boldsymbol{B}$-$\boldsymbol{C}$ form a right-handed system. Its magnitude is equal to the area of a parallelogram having sides $\boldsymbol{A}$ and $\boldsymbol{B}$. The cross product of any two unit vectors is always the third unit vector. The nine cross products involving unit vectors can be written in the compact form

$$[\boldsymbol{\delta}_i \times \boldsymbol{\delta}_j] = \sum_k \epsilon_{ijk}\,\boldsymbol{\delta}_k \tag{A-8}$$

where ϵ_{ijk} has the following meaning:

$$\begin{aligned} \epsilon_{ijk} &= 1 \quad &&\text{if} \quad ijk = 123,\ 231,\ 312 \\ \epsilon_{ijk} &= 0 \quad &&\text{if any two indices are alike} \\ \epsilon_{ijk} &= -1 \quad &&\text{if} \quad ijk = 321,\ 132,\ 213 \end{aligned}$$

The cross product of any two vectors $\boldsymbol{A}$ and $\boldsymbol{B}$ is given by the equation

$$[\boldsymbol{A} \times \boldsymbol{B}] = \sum_i \sum_j \sum_k A_i B_j \epsilon_{ijk}\,\boldsymbol{\delta}_k \tag{A-9}$$

and also by the determinant

$$[\boldsymbol{A} \times \boldsymbol{B}] = \begin{vmatrix} \boldsymbol{\delta}_1 & \boldsymbol{\delta}_2 & \boldsymbol{\delta}_3 \\ A_1 & A_2 & A_3 \\ B_1 & B_2 & B_3 \end{vmatrix} \tag{A-10}$$

A tensor of the second order has nine components and is often represented as a matrix. For example, if $\boldsymbol{\epsilon}$ is a second order tensor, then its matrix representation is

$$\boldsymbol{\epsilon} = \begin{pmatrix} \epsilon_{11} & \epsilon_{12} & \epsilon_{13} \\ \epsilon_{21} & \epsilon_{22} & \epsilon_{23} \\ \epsilon_{31} & \epsilon_{32} & \epsilon_{33} \end{pmatrix} \tag{A-11}$$

If the components ϵ_{ij} of the tensor $\boldsymbol{\epsilon}$ obey the relation

$$\epsilon_{ij} = \epsilon_{ji} \tag{A-12}$$

then $\boldsymbol{\epsilon}$ is said to be a *symmetrical* tensor. On the other hand, if

$$\epsilon_{ij} = -\epsilon_{ji} \tag{A-13}$$

then the tensor is *antisymmetrical.*

Any second order tensor $\boldsymbol{\epsilon}$ can be expressed as the sum of a symmetrical and an antisymmetrical tensor by the device of adding and subtracting half of the transpose tensor. For example, if $\boldsymbol{\epsilon}$ is a second order tensor, then

$$\boldsymbol{\epsilon} = \frac{\boldsymbol{\epsilon} + \boldsymbol{\epsilon}^*}{2} + \frac{\boldsymbol{\epsilon} - \boldsymbol{\epsilon}^*}{2} \tag{A-14}$$

where $\boldsymbol{\epsilon}^*$, the transpose of $\boldsymbol{\epsilon}$, is obtained by interchanging rows and columns, and it is easily established that $\frac{1}{2}(\boldsymbol{\epsilon} + \boldsymbol{\epsilon}^*)$ is a symmetrical tensor and $\frac{1}{2}(\boldsymbol{\epsilon} - \boldsymbol{\epsilon}^*)$ an antisymmetrical tensor.

It can be shown that for any symmetrical tensor $\boldsymbol{\sigma}$ whose matrix representation with respect to the axes x_1, x_2, x_3 is

$$\boldsymbol{\sigma} = \begin{pmatrix} \sigma_{11} & \sigma_{12} & \sigma_{13} \\ \sigma_{21} & \sigma_{22} & \sigma_{23} \\ \sigma_{31} & \sigma_{32} & \sigma_{33} \end{pmatrix} \tag{A-15}$$

there exists another set of reference axes x_1', x_2', x_3' for which

$$\boldsymbol{\sigma} = \begin{pmatrix} \sigma_{11}' & 0 & 0 \\ 0 & \sigma_{22}' & 0 \\ 0 & 0 & \sigma_{33}' \end{pmatrix} \tag{A-16}$$

These axes are called the *principle axes* and the matrix (A-16) is the diagonal form of the matrix (A-15).

It can also be shown that there are three quantities, called the invariants, associated with any tensor that do not change upon a rotation of the coordinate axes. The invariants of the tensor $\boldsymbol{\epsilon}$, designated I_1, I_2, and I_3, are defined by the equations

$$I_1 = \epsilon_{11} + \epsilon_{22} + \epsilon_{33} \tag{A-17}$$

$$I_2 = \begin{vmatrix} \epsilon_{22} & \epsilon_{23} \\ \epsilon_{32} & \epsilon_{33} \end{vmatrix} + \begin{vmatrix} \epsilon_{11} & \epsilon_{13} \\ \epsilon_{31} & \epsilon_{33} \end{vmatrix} + \begin{vmatrix} \epsilon_{11} & \epsilon_{12} \\ \epsilon_{21} & \epsilon_{22} \end{vmatrix} = \tfrac{1}{2}I_1^{\,2} - \tfrac{1}{2}\sum_i\sum_j \epsilon_{ij}\,\epsilon_{ji} \tag{A-18}$$

$$I_3 = \begin{vmatrix} \epsilon_{11} & \epsilon_{12} & \epsilon_{13} \\ \epsilon_{21} & \epsilon_{22} & \epsilon_{23} \\ \epsilon_{31} & \epsilon_{32} & \epsilon_{33} \end{vmatrix} \tag{A-19}$$

Two tensors are added by adding corresponding components. For example, if $\boldsymbol{\sigma}$ and $\boldsymbol{\tau}$ are tensors, then their sum, which is designated $\boldsymbol{\lambda}$, has components of the form

$$\lambda_{ij} = (\sigma_{ij} + \tau_{ij}) \tag{A-20}$$

The *dyadic product* of any two vectors $\boldsymbol{A}$ and $\boldsymbol{B}$ is a tensor of the second order having components of the form A_iB_j, of which there are nine. The dyadic product of $\boldsymbol{A}$ and $\boldsymbol{B}$ is designated $\boldsymbol{AB}$ and has the matrix representation

$$\boldsymbol{AB} = \begin{pmatrix} A_1B_1 & A_1B_2 & A_1B_3 \\ A_2B_1 & A_2B_2 & A_2B_3 \\ A_3B_1 & A_3B_2 & A_3B_3 \end{pmatrix} \tag{A-21}$$

There are nine dyadic products that can be formed from the unit vectors. These *unit dyads*, designated $\boldsymbol{\delta}_i\boldsymbol{\delta}_j$, are second order tensors. All of the components of a unit dyad are zero, except one. For example, the matrix form of the unit dyad $\boldsymbol{\delta}_2\boldsymbol{\delta}_3$ is

$$\boldsymbol{\delta}_2\boldsymbol{\delta}_3 = \begin{pmatrix} 0 & 0 & 0 \\ 0 & 0 & 1 \\ 0 & 0 & 0 \end{pmatrix} \tag{A-22}$$

The unit dyads enable a tensor to be represented as a summation of its components. For example, the summation representation of the tensor $\boldsymbol{\tau}$ is

$$\boldsymbol{\tau} = \sum_i\sum_j \tau_{ij}\,\boldsymbol{\delta}_i\boldsymbol{\delta}_j \tag{A-23}$$

and a similar expression

$$\boldsymbol{AB} = \sum_i\sum_j A_iB_j\,\boldsymbol{\delta}_i\boldsymbol{\delta}_j \tag{A-24}$$

exists for the dyad $\boldsymbol{AB}$.

The following equations[2] define four operations involving unit vectors and unit dyads that are used to establish formulas for various multiplicative operations involving vectors and tensors:

$$(\boldsymbol{\delta}_i \boldsymbol{\delta}_j : \boldsymbol{\delta}_k \boldsymbol{\delta}_l) = \delta_{il}\,\delta_{jk} \tag{A-25}$$

$$[\boldsymbol{\delta}_i \boldsymbol{\delta}_j \cdot \boldsymbol{\delta}_k] = \boldsymbol{\delta}_i\,\delta_{jk} \tag{A-26}$$

$$[\boldsymbol{\delta}_i \cdot \boldsymbol{\delta}_j \boldsymbol{\delta}_k] = \delta_{ij}\,\boldsymbol{\delta}_k \tag{A-27}$$

$$\{\boldsymbol{\delta}_i \boldsymbol{\delta}_j \cdot \boldsymbol{\delta}_k \boldsymbol{\delta}_l\} = \delta_{jk}\,\boldsymbol{\delta}_i \boldsymbol{\delta}_l \tag{A-28}$$

From these formulas the following equations for multiplicative operations with the tensors $\boldsymbol{\tau}$ and $\boldsymbol{\sigma}$ and the vector A are established:

$$(\boldsymbol{\sigma} : \boldsymbol{\tau}) = \sum_i \sum_j \sigma_{ij}\tau_{ji} \tag{A-29}$$

$$\{\boldsymbol{\sigma} \cdot \boldsymbol{\tau}\} = \sum_i \sum_l \boldsymbol{\delta}_i \boldsymbol{\delta}_l \sum_j \sigma_{ij}\tau_{jl} \tag{A-30}$$

$$[\boldsymbol{\tau} \cdot A] = \sum_i \boldsymbol{\delta}_i \sum_j \tau_{ij} A_j \tag{A-31}$$

For rectangular Cartesian coordinates, the differential operator $\boldsymbol{\nabla}$, the Laplacian $\boldsymbol{\nabla}^2$, and the substantial derivative D/Dt are defined by the following equations:

$$\boldsymbol{\nabla} = \sum_i \boldsymbol{\delta}_i \frac{\partial}{\partial x_i} \tag{A-32}$$

$$\boldsymbol{\nabla}^2 = \sum_i \frac{\partial^2}{\partial x_i^{\,2}} \tag{A-33}$$

$$\frac{D}{Dt} = \frac{\partial}{\partial t} + \sum_i v_i \frac{\partial}{\partial x_i} \tag{A-34}$$

In equation (A-34) v_i represents the x_i-component of the fluid velocity vector $\boldsymbol{v}$.

Differential operations on the scalar P are defined by the equations

$$\boldsymbol{\nabla} P = \sum_i \boldsymbol{\delta}_i \left(\frac{\partial P}{\partial x_i}\right) \tag{A-35}$$

$$\boldsymbol{\nabla}^2 P = \sum_i \left(\frac{\partial^2 P}{\partial x_i^{\,2}}\right) \tag{A-36}$$

$$\frac{DP}{Dt} = \frac{\partial P}{\partial t} + \sum_i v_i \left(\frac{\partial P}{\partial x_i}\right) \tag{A-37}$$

[2] R. B. Bird, W. E. Stewart, and E. N. Lightfoot, *Transport Phenomena*, p. 728, John Wiley and Sons, 1960.

Differential operations on the vector A are defined by the following equations:

$$(\nabla \cdot A) = \sum_i \left(\frac{\partial A_i}{\partial x_i}\right) \quad \text{(A-38)}$$

$$[\nabla \times A] = \sum_j \sum_k [\boldsymbol{\delta}_j \times \boldsymbol{\delta}_k]\left(\frac{\partial A_k}{\partial x_j}\right) \quad \text{(A-39)}$$

$$\nabla A = \sum_i \sum_j \boldsymbol{\delta}_i \boldsymbol{\delta}_j \left(\frac{\partial A_j}{\partial x_i}\right) \quad \text{(A-40)}$$

$$\nabla^2 A = \sum_i \boldsymbol{\delta}_i \nabla^2 A_i \quad \text{(A-41)}$$

$$\frac{DA}{Dt} = \sum_i \boldsymbol{\delta}_i \left[\frac{\partial A_i}{\partial t} + (\boldsymbol{v} \cdot \nabla) v_i\right] \quad \text{(A-42)}$$

Differential operations involving the tensor $\boldsymbol{\tau}$ are defined by the equations

$$[\nabla \cdot \boldsymbol{\tau}] = \sum_k \boldsymbol{\delta}_k \left[\sum_i \left(\frac{\partial \tau_{ik}}{\partial x_i}\right)\right] \quad \text{(A-43)}$$

$$[A \cdot \nabla B] = \sum_i \sum_k \boldsymbol{\delta}_k A_i \left(\frac{\partial B_k}{\partial x_i}\right) \quad \text{(A-44)}$$

$$(\boldsymbol{\tau} : \nabla B) = \sum_i \sum_j \tau_{ij} \left(\frac{\partial B_i}{\partial x_j}\right) \quad \text{(A-45)}$$

In the following derivation of the equations of continuity, momentum, and energy, it is necessary to transform surface integrals to volume integrals. The connection between these two types of integrals is given by the divergence theorem, which can be represented by the equation

$$\oint_S (\boldsymbol{n} \cdot A)\, dS = \oint_V (\nabla \cdot A)\, dV \quad \text{(A-46)}$$

where S represents a closed surface that contains volume V of the space. A is any vector function of position.

Two variations[3] of the divergence theorem will also be used. If P is any scalar function of position and $\boldsymbol{\sigma}$ any tensor function of position, then

$$\oint_S (\boldsymbol{n} \cdot \boldsymbol{\sigma})\, dS = \oint_V (\nabla \cdot \boldsymbol{\sigma})\, dV \quad \text{(A-47)}$$

$$\oint_S \boldsymbol{n} P\, dS = \oint_V \nabla P\, dV \quad \text{(A-48)}$$

[3] See p. 732 of Ref. 2.

Also, the following derivations of the equations of continuity, momentum and energy involve finding the time derivative of a volume integral. If the volume element under consideration is fixed in space, then V is not a function of time and the differentiation follows the Leibnitz rule for the differentiation of the definite integral, which can be represented by the equation

$$\frac{d}{dt}\oint_V \rho\, dV = \oint_V \left(\frac{\partial \rho}{\partial t}\right) dV \tag{A-49}$$

Another useful differentiation formula,[3] a variation of the Leibnitz rule, is obtained when the element under consideration *follows the motion of the fluid.* In this case

$$\frac{d}{dt}\oint_V \rho X\, dV = \oint_V \rho\left(\frac{DX}{Dt}\right) dV \tag{A-50}$$

where ρ is the fluid density and X is any scalar function of position. Equation (A-50) applies also to vector functions, in which case it is written

$$\frac{d}{dt}\oint_V \rho \boldsymbol{A}\, dV = \oint_V \rho\left(\frac{D\boldsymbol{A}}{Dt}\right) dV \tag{A-51}$$

A-2 The Stress Tensor

We now derive an expression for the force (per unit area) that a flowing viscous fluid exerts on an imaginary surface oriented arbitrarily in space. A rectangular coordinate system whose origin is in the surface under consideration is established. The orientation of the surface with respect to the coordinate axes is characterized by the unit normal vector $\boldsymbol{n}$ and the fluid motion is characterized by the velocity vector $\boldsymbol{v}$.

Figure A-3 shows a small tetrahedron having a face that is perpendicular to $\boldsymbol{n}$ and at distance Δn from the origin. To determine the force acting on this surface, the sum of the forces acting on the other three sides of the tetrahedron is calculated. Then, in the limit as the slant face approaches the origin, Newton's third law tells us that the sum of the forces acting on the straight sides will be equal but of opposite sign to the force exerted on the slant face.

Consider first the component of the force that acts in the x_1-direction. The tangential force per unit area (stress) acting on the straight side perpendicular to the x_3-axis is designated σ_{31}, where the first subscript indicates the orientation of the surface on which the force acts and the second the direction in which the force acts. The definition of σ_{31} is completed by specifying on which side of the surface the force acts. This is determined by considering a vector acting in the $+x_3$-direction that pierces the surface under consideration. Then σ_{31} acts on that side of the surface in which the

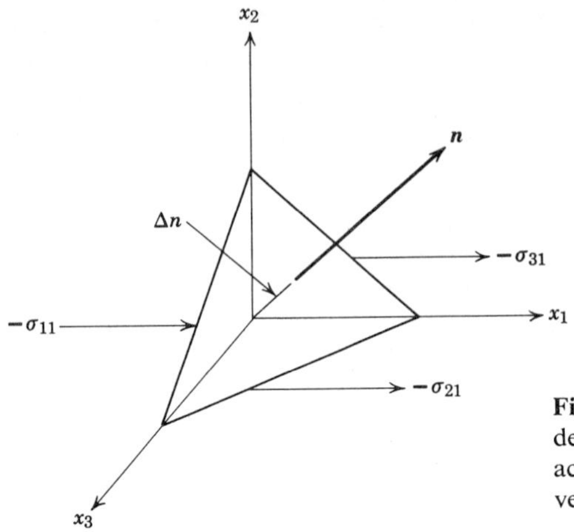

Figure A-3. Tetrahedron used to derive the x_1 component of forces acting on a surface normal to the vector $\boldsymbol{n}$.

head of the vector is located. The stress $-\sigma_{31}$ acts on the opposite side.

Figure A-3 shows the three stresses that cause forces to act in the x_1-direction. The sum of these forces, designated F_1, is given by the equation

$$F_1 = -\Delta S_1\sigma_{11} - \Delta S_2\sigma_{21} - \Delta S_3\sigma_{31} \tag{A-52}$$

which, because of equation (A-6), can be written

$$F_1 = -\Delta S[n_1\sigma_{11} + n_2\sigma_{21} + n_3\sigma_{31}] = -\Delta S \sum n_i\sigma_{i1} \tag{A-53}$$

Similar expressions exist for F_2 and F_3.

The force vector $\boldsymbol{F}$, expressed as the sum of its components, is

$$\boldsymbol{F} = \sum_i \boldsymbol{\delta}_i F_i \tag{A-54}$$

Upon introducing the expressions for the components, equation (A-54) becomes

$$\boldsymbol{F} = -\Delta S \sum_i \boldsymbol{\delta}_i \sum_j n_j\sigma_{ji} \tag{A-55}$$

Referring to equation (A-31), form the dot product of the vector $\boldsymbol{n}$ and a tensor quantity $\boldsymbol{\sigma}$ to obtain, for $\boldsymbol{\sigma}$ symmetrical

$$[\boldsymbol{n} \cdot \boldsymbol{\sigma}] = \sum_i \boldsymbol{\delta}_i \sum_j n_j\sigma_{ji} \tag{A-56}$$

Comparing the above equations, it is clear that the force per unit area acting on the slant surface at the point under consideration can be expressed by the equation

$$\frac{\boldsymbol{F}}{\Delta S} = -[\boldsymbol{n} \cdot \boldsymbol{\sigma}] \tag{A-57}$$

The quantity $\boldsymbol{\sigma}$ is called the *stress tensor*. In the limit, as the tetrahedron shrinks to a point, the quantity $[\boldsymbol{n} \cdot \boldsymbol{\sigma}]$ represents the force per unit area exerted on an imaginary surface whose orientation is characterized by the unit normal vector $\boldsymbol{n}$. In matrix form, the tensor $\boldsymbol{\sigma}$ is written

$$\boldsymbol{\sigma} = \begin{pmatrix} \sigma_{11} & \sigma_{12} & \sigma_{13} \\ \sigma_{21} & \sigma_{22} & \sigma_{23} \\ \sigma_{31} & \sigma_{32} & \sigma_{33} \end{pmatrix} \tag{A-58}$$

It is shown below that the nondiagonal components of the tensor $\boldsymbol{\sigma}$ obey the relation

$$\sigma_{ij} = \sigma_{ji} \tag{A-59}$$

and, therefore, that $\boldsymbol{\sigma}$ is a symmetrical tensor.

Consider a small element of fluid having the shape of a parallelepiped with sides of length Δx_1, Δx_2, and Δx_3. Figure A-4 is a view of the element in a plane perpendicular to the x_3-axis. We now consider the possibility that stresses acting on the various faces of the element could produce torques that would cause the element to rotate about an axis passing through its center and parallel to the x_3-axis.

All of the forces that might cause this type of rotation are shown in the figure. The torque T_1, arising from force F_1, is given by the equation

$$T_1 = \frac{\Delta x_2}{2}\left[-\left\{\sigma_{21} - \left(\frac{\partial \sigma_{21}}{\partial x_2}\right)\left(\frac{\Delta x_2}{2}\right)\right\}\Delta x_1\, \Delta x_3\right] \tag{A-60}$$

and the torque T_2 by the equation

$$T_2 = \frac{\Delta x_2}{2}\left[+\left\{\sigma_{21} + \left(\frac{\partial \sigma_{21}}{\partial x_2}\right)\left(\frac{\Delta x_2}{2}\right)\right\}\Delta x_1\, \Delta x_3\right] \tag{A-61}$$

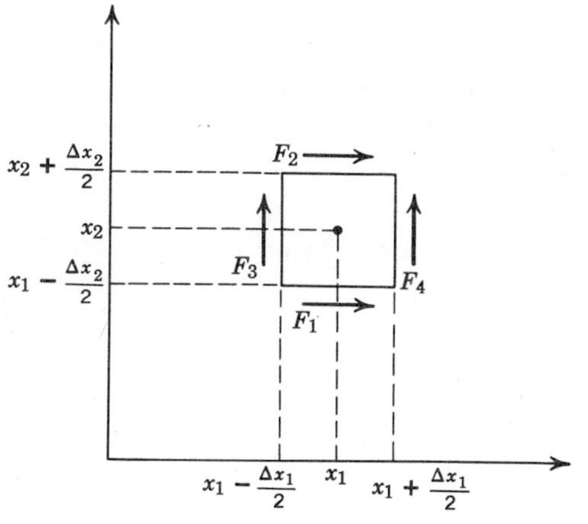

Figure A-4. Forces that tend to produce rotation of fluid element about axis through (x_1,x_2) and parallel to the x_3 reference axis.

The net effect due to these two torques, obtained by subtracting T_2 from T_1, is

$$T_1 - T_2 = -\sigma_{21}(\Delta x_1 \, \Delta x_2 \, \Delta x_3) \tag{A-62}$$

In a similar manner, it can be shown that

$$T_3 - T_4 = +\sigma_{12}(\Delta x_1 \, \Delta x_2 \, \Delta x_3) \tag{A-63}$$

so that the net torque tending to produce rotation about the given axis is

$$T_{\text{net}} = (\sigma_{12} - \sigma_{21}) \, \Delta x_1 \, \Delta x_2 \, \Delta x_3 \tag{A-64}$$

A net torque produces angular acceleration according to the law

$$T = \alpha I \tag{A-65}$$

where I is the moment of inertia and α the angular acceleration. For the element under consideration, the moment of inertia is given by the formula

$$I = \frac{1}{12} \rho \Delta V [(\Delta x_1)^2 + (\Delta x_2)^2] \tag{A-66}$$

Combining the above three equations gives

$$(\sigma_{12} - \sigma_{21}) \Delta V = \frac{\alpha}{12} \rho \Delta V [(\Delta x_1)^2 + (\Delta x_2)^2] \tag{A-67}$$

and we see that

$$\lim_{\substack{\Delta x_1 \to 0 \\ \Delta x_2 \to 0}} (\sigma_{12} - \sigma_{21}) = 0 \tag{A-68}$$

Hence

$$\sigma_{12} = \sigma_{21} \tag{A-69}$$

If we consider the possibility of rotation about the other two orthogonal axes passing through the center of the element, we are led to the conclusion that

$$\sigma_{31} = \sigma_{13} \qquad \sigma_{23} = \sigma_{32} \tag{A-70}$$

and hence that the tensor $\boldsymbol{\sigma}$ is symmetrical.

The effect of hydrostatic pressure P, which can exert only normal forces, can be eliminated from the stress tensor $\boldsymbol{\sigma}$ by subtracting $-P$ from each of the diagonal terms of $\boldsymbol{\sigma}$. (Remember that a positive $\boldsymbol{\sigma}$ corresponds to a negative pressure.) This new stress tensor is designated $\boldsymbol{\tau}$ and is defined by the equation

$$\boldsymbol{\tau} = \boldsymbol{\sigma} + P\boldsymbol{\delta} \tag{A-71}$$

where $\boldsymbol{\delta}$, the unit tensor, has the following matrix representation:

$$\boldsymbol{\delta} = \begin{pmatrix} 1 & 0 & 0 \\ 0 & 1 & 0 \\ 0 & 0 & 1 \end{pmatrix} \tag{A-72}$$

Using subscript notation, equation (A-71) is written

$$\tau_{ij} = \sigma_{ij} + P\,\delta_{ij} \tag{A-73}$$

A-3 The Continuity Equation

The following concise derivation of the equation of continuity considers the accumulation of mass within a closed control surface S of arbitrary size and shape that is *fixed in space* and encloses a fluid element of volume V. At any point on the control surface, the mass flux (see Example A-1) out of the volume element is $(\boldsymbol{n} \cdot \rho \boldsymbol{v})$, where $\boldsymbol{n}$ is an outer-directed unit vector that is normal to the surface S at the point, and ρ and $\boldsymbol{v}$ are the fluid density and velocity at the point.

The net *inward* transport of matter through the control surface is given by the integral

$$-\oint_S (\boldsymbol{n} \cdot \rho \boldsymbol{v})\, dS \tag{A-74}$$

and because the rate of accumulation of mass within the control surface is given by the integral

$$\frac{d}{dt} \oint_V \rho\, dV \tag{A-75}$$

we have

$$\frac{d}{dt} \oint_V \rho\, dV = -\oint_S (\boldsymbol{n} \cdot \rho \boldsymbol{v})\, dS \tag{A-76}$$

The application of the divergence theorem to the right side of equation (A-76) and the Leibnitz rule for differentiation to the left side converts (A-76) to

$$\oint_V \left(\frac{\partial \rho}{\partial t}\right) dV = -\oint_V (\nabla \cdot \rho \boldsymbol{v})\, dV \tag{A-77}$$

which can be written in the equivalent form

$$\oint_V \left[\left(\frac{\partial \rho}{\partial t}\right) + (\nabla \cdot \rho \boldsymbol{v})\right] dV = 0 \tag{A-78}$$

The quantity in the brackets, the integrand, is a point function of position, so that if the volume element should be of size dV, equation (A-78) becomes

$$\left[\left(\frac{\partial \rho}{\partial t}\right) + (\nabla \cdot \rho \boldsymbol{v})\right] dV = 0 \tag{A-79}$$

But dV is not zero; we therefore conclude that

$$\frac{\partial \rho}{\partial t} + (\nabla \cdot \rho \boldsymbol{v}) = 0 \tag{A-80}$$

which is the equation of continuity. By expanding the term containing the divergence and noting the definition of the substantial derivative, the equation of continuity can be written in the equivalent form

$$\boxed{\frac{D\rho}{Dt} + \rho(\boldsymbol{\nabla} \cdot \boldsymbol{v}) = 0} \tag{A-81}$$

A-4 The Momentum Equation

The momentum equation is derived by applying Newton's second law of motion to the fluid particles contained within an arbitrary closed surface that *moves with the fluid.* Because all particles contained within the surface do not move at the same speed or in the same direction (remember that $\boldsymbol{v}$ is a function of time and position), the fluid element may expand, contract, or be distorted from its original shape. However, the boundary of the element—the control surface—always contains the same mass of fluid.

Newton's second law of motion, applied to the fluid element, states that the rate of increase of the momentum of the element must be equal to the sum of the forces acting upon it. The momentum (per unit volume) of any particle is the vector $(\rho\boldsymbol{v})$, so that the momentum of all the particles contained within the control surface is

$$\oint_V (\rho\boldsymbol{v})\, dV \tag{A-82}$$

and the rate of increase of momentum of the fluid element is

$$\frac{d}{dt}\oint_V (\rho\boldsymbol{v})\, dV \tag{A-83}$$

Two kinds of forces can act upon the fluid element: surface forces and body forces. Body forces are those that act at a distance. Examples include gravitational, electrostatic, and magnetic forces. Surface forces arise from the direct action of matter across the boundary. For example, hydrostatic pressure exerts a normal force on any surface, and a viscous fluid flowing parallel to a surface exerts a drag upon it. Designating the total body force acting on the element as the vector $\boldsymbol{G}$ and the total surface force as the vector $\boldsymbol{F}$, the momentum equation is formulated as

$$\frac{d}{dt}\oint_V \rho\boldsymbol{v}\, dV = \boldsymbol{F} + \boldsymbol{G} \tag{A-84}$$

If the vector $\boldsymbol{g}$, a point function of position, represents the body force per unit mass, then the total body force exerted on the fluid element is given by the integral

$$G = \oint_V \rho \boldsymbol{g}\, dV \tag{A-85}$$

Furthermore, it was shown that the surface force (per unit area) at any point on a surface is given by the quantity $[\boldsymbol{n} \cdot \boldsymbol{\sigma}]$, where $\boldsymbol{n}$ is the outer-directed unit normal vector and $\boldsymbol{\sigma}$ the stress tensor at the point under consideration. The total surface force exerted on the fluid element is, therefore, given by the integral

$$F = \oint_S [\boldsymbol{n} \cdot \boldsymbol{\sigma}]\, dS \tag{A-86}$$

In view of equations (A-84), (A-85), and (A-86), the momentum equation becomes

$$\frac{d}{dt} \oint_V \rho \boldsymbol{v}\, dV = \oint_S [\boldsymbol{n} \cdot \boldsymbol{\sigma}]\, dS + \oint_V \rho \boldsymbol{g}\, dV \tag{A-87}$$

Upon applying the modified Leibnitz rule to the left side of equation (A-87) and the applicable form of the divergence theorem to the first term on the right side, and then rearranging, equation (A-87) reduces to

$$\oint_V \left[\rho \frac{D\boldsymbol{v}}{Dt} - [\nabla \cdot \boldsymbol{\sigma}] - \rho \boldsymbol{g} \right] dV = 0 \tag{A-88}$$

and we conclude that

$$\rho \left(\frac{D\boldsymbol{v}}{Dt} \right) = [\nabla \cdot \boldsymbol{\sigma}] + \rho \boldsymbol{g} \tag{A-89}$$

which is one form of the momentum equation.

The introduction of the stress tensor $\boldsymbol{\tau}$, which is defined by the equation

$$\boldsymbol{\tau} = \boldsymbol{\sigma} + P\boldsymbol{\delta} \tag{A-90}$$

transforms equation (A-89) to

$$\boxed{\rho \left(\frac{D\boldsymbol{v}}{Dt} \right) = [\nabla \cdot \boldsymbol{\tau}] - \nabla P + \rho \boldsymbol{g}} \tag{A-91}$$

which is another useful form of the momentum equation.

A-5 The Energy Equation

The following derivation of the energy equation utilizes a fluid element contained within an arbitrarily closed surface that *moves with the fluid.* In a thermodynamic sense, the fluid element is a closed system, and the first law of thermodynamics requires that its rate of increase of energy must be equal to the rate at which heat is transferred to it and work is done on it.

The total energy (per unit mass) of the fluid is E and if Q and W represent heat and work respectively, the mathematical formulation of the first law is

$$\frac{d}{dt}\oint_V \rho E\, dV = \frac{dQ}{dt} - \frac{dW}{dt} \tag{A-92}$$

where the sign convention makes Q positive if heat is transferred to the system and W positive if work is done by the system.

The total energy E is the sum of the internal energy U and the kinetic energy KE. In vector form, the kinetic energy (per unit mass) is given by the equation

$$KE = \tfrac{1}{2}(\boldsymbol{v} \cdot \boldsymbol{v}) \tag{A-93}$$

so that equation (A-92) becomes

$$\frac{d}{dt}\oint_V \rho U\, dV + \frac{d}{dt}\oint_V \tfrac{1}{2}\rho(\boldsymbol{v} \cdot \boldsymbol{v})\, dV = \frac{dQ}{dt} - \frac{dW}{dt} \tag{A-94}$$

If $\boldsymbol{q}$ represents the heat flux vector, then the rate (per unit area) at which heat flows out through the control surface at any point is $(\boldsymbol{n} \cdot \boldsymbol{q})$, and the total rate of heat inflow through the entire surface is given by the integral

$$\frac{dQ}{dt} = -\oint_S (\boldsymbol{n} \cdot \boldsymbol{q})\, dS \tag{A-95}$$

We now consider the work effects associated with both the body and surface forces. The body force (per unit volume) exerted on any particle of fluid contained within the control surface is $\rho \boldsymbol{g}$. If the velocity of this particle is $\boldsymbol{v}$, then the rate at which work is done against the body force is $(\boldsymbol{v} \cdot \rho \boldsymbol{g})$, and the total work for all of the particles contained within the control surface is given by the integral

$$\oint_V (\boldsymbol{v} \cdot \rho \boldsymbol{g})\, dV \tag{A-96}$$

The force (per unit area) exerted at any point on the surface is $[\boldsymbol{n} \cdot \boldsymbol{\sigma}]$, and if the velocity of the fluid at this point is $\boldsymbol{v}$, then the rate at which work is done at this point is $\boldsymbol{v} \cdot [\boldsymbol{n} \cdot \boldsymbol{\sigma}]$. Therefore the rate at which work is done against the total surface S is given by the integral

$$\oint_S (\boldsymbol{v} \cdot [\boldsymbol{n} \cdot \boldsymbol{\sigma}])\, dS \tag{A-97}$$

It is easily shown that

$$\boldsymbol{v} \cdot [\boldsymbol{n} \cdot \boldsymbol{\sigma}] = \boldsymbol{n} \cdot [\boldsymbol{\sigma} \cdot \boldsymbol{v}] \tag{A-98}$$

Hence, the above integral can be written

$$\oint_S \boldsymbol{n} \cdot [\boldsymbol{\sigma} \cdot \boldsymbol{v}]\, dS \tag{A-99}$$

and the total rate at which work is done on the system, obtained by adding equations (A-96) and (A-99), is

$$-\frac{dW}{dt} = \oint_V (\boldsymbol{v} \cdot \rho \boldsymbol{g})\, dV + \oint_S (\boldsymbol{n} \cdot [\boldsymbol{\sigma} \cdot \boldsymbol{v}])\, dS \qquad \text{(A-100)}$$

Introducing equations (A-100) and (A-95) into (A-94) gives

$$\frac{d}{dt}\oint_V \rho U\, dV + \frac{d}{dt}\oint_V \tfrac{1}{2}\rho(\boldsymbol{v} \cdot \boldsymbol{v})\, dV = -\oint_S (\boldsymbol{n} \cdot \boldsymbol{q})\, dS + \oint_V (\boldsymbol{v} \cdot \rho \boldsymbol{g})\, dV + \oint_S (\boldsymbol{n} \cdot [\boldsymbol{\sigma} \cdot \boldsymbol{v}])\, dS \qquad \text{(A-101)}$$

Applying the modified Leibnitz rule to the left side of equation (A-101), the appropriate form of the divergence theorem to the surface integrals on the right side, and rearranging, gives

$$\oint_V \left[\rho\left(\frac{DU}{Dt}\right) + \rho \boldsymbol{v} \cdot \left(\frac{D\boldsymbol{v}}{Dt}\right) + (\nabla \cdot \boldsymbol{q}) - (\boldsymbol{v} \cdot \rho \boldsymbol{g}) - (\nabla \cdot [\boldsymbol{\sigma} \cdot \boldsymbol{v}])\right] dV = 0 \qquad \text{(A-102)}$$

and we conclude that

$$\rho\left(\frac{DU}{Dt}\right) + \rho \boldsymbol{v} \cdot \left(\frac{D\boldsymbol{v}}{Dt}\right) = -(\nabla \cdot \boldsymbol{q}) + (\boldsymbol{v} \cdot \rho \boldsymbol{g}) + (\nabla \cdot [\boldsymbol{\sigma} \cdot \boldsymbol{v}]) \qquad \text{(A-103)}$$

If one takes the dot product with $\boldsymbol{v}$ of each term of the momentum equation, it becomes

$$\rho \boldsymbol{v} \cdot \left(\frac{D\boldsymbol{v}}{Dt}\right) = (\boldsymbol{v} \cdot [\nabla \cdot \boldsymbol{\sigma}]) + (\boldsymbol{v} \cdot \rho \boldsymbol{g}) \qquad \text{(A-104)}$$

Equation (A-104) reduces equation (A-103) to

$$\rho \frac{DU}{Dt} = -(\nabla \cdot \boldsymbol{q}) + (\nabla \cdot [\boldsymbol{\sigma} \cdot \boldsymbol{v}]) - (\boldsymbol{v} \cdot [\nabla \cdot \boldsymbol{\sigma}]) \qquad \text{(A-105)}$$

By means of a tensor identity,[4] equation (A-105) is transformed to

$$\rho \frac{DU}{Dt} = -(\nabla \cdot \boldsymbol{q}) + (\boldsymbol{\sigma} : \nabla \boldsymbol{v}) \qquad \text{(A-106)}$$

and upon introducing the stress tensor $\boldsymbol{\tau}$, which is defined by (A-71), equation (A-106) can be written

$$\rho\left(\frac{DU}{Dt}\right) = -(\nabla \cdot \boldsymbol{q}) + (\boldsymbol{\tau} : \nabla \boldsymbol{v}) - P(\nabla \cdot \boldsymbol{v}) \qquad \text{(A-107)}$$

which is the energy equation in terms of internal energy U.

[4] For proof, see Example A.4-1 (p. 731) of *Transport Phenomena*.

If U is considered a function of the state variables T and ρ, then the total differential of U is

$$dU = \left(\frac{\partial U}{\partial T}\right)_\rho dT + \left(\frac{\partial U}{\partial \rho}\right)_T d\rho \tag{A-108}$$

By introducing standard thermodynamic identities, (A-108) can be put in the equivalent form

$$dU = c_V\, dT - \frac{1}{\rho^2}\left[T\left(\frac{\partial P}{\partial T}\right)_\rho - P\right] d\rho \tag{A-109}$$

If each term of (A-109) is divided by the differential dt, we then obtain the following equation containing total or substantial derivatives of the state variables U, T, and ρ:

$$\frac{DU}{Dt} = c_V\left(\frac{DT}{Dt}\right) - \frac{1}{\rho^2}\left[T\left(\frac{\partial P}{\partial T}\right)_\rho - P\right]\frac{D\rho}{Dt} \tag{A-110}$$

But because of the equation of continuity

$$\frac{D\rho}{Dt} = -\rho(\nabla \cdot v) \tag{A-111}$$

equation (A-110) can be reduced to

$$\rho\left(\frac{DU}{Dt}\right) = \rho c_V\left(\frac{DT}{Dt}\right) + \left[T\left(\frac{\partial P}{\partial T}\right)_\rho - P\right](\nabla \cdot v) \tag{A-112}$$

Upon introducing equation (A-112) into (A-107), we obtain

$$\boxed{\rho c_V\left(\frac{DT}{Dt}\right) = -(\nabla \cdot q) - T\left(\frac{\partial P}{\partial T}\right)_\rho(\nabla \cdot v) + (\tau : \nabla v)} \tag{A-113}$$

which is one of the most useful forms of the energy equation.

A-6 The Rheological Equation of the Newtonian Fluid

We now establish a relationship between the stress tensor and some other tensor quantity that will characterize the rate of strain of the fluid at any given point. To start, let us consider the simple shear flow between parallel plates shown in Figure A-5, where the lower plate is stationary and the upper plate is moving in the x_1-direction.

Consider two fluid particles, labeled a and b. Their initial position is (x_1, x_2) and $(x_1 + \Delta x_1, x_2 + \Delta x_2)$, respectively. After a period of time Δt, both particles are displaced to the right to the new positions shown in

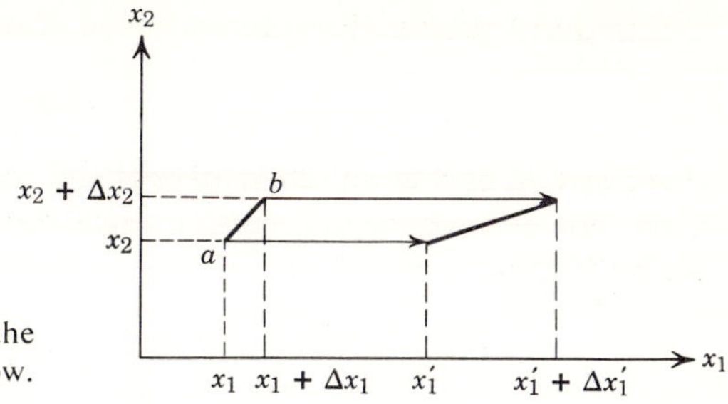

Figure A-5. Characterization of the rate of strain in simple shear flow.

the figure. Particle a has moved a distance $v_1 \Delta t$ and particle b a distance that is given by the expression

$$\left[v_1 + \left(\frac{\partial v_1}{\partial x_2}\right) \Delta x_2\right] \Delta t \tag{A-114}$$

The x_1 separation of the particles after this translation is designated $\Delta x_1'$ and is given by the equation

$$\Delta x_1' = \Delta x_1 + \left[\left(\frac{\partial v_1}{\partial x_2}\right) \Delta x_2\right] \Delta t \tag{A-115}$$

Therefore, the average rate at which the particles are separating in the x_1-direction is

$$\left[\frac{\Delta x_1' - \Delta x_1}{\Delta t}\right] = \left(\frac{\partial v_1}{\partial x_2}\right) \Delta x_2 \tag{A-116}$$

and if this is expressed on the basis of unit initial separation in the x_2-direction, we then have a measure of the average rate of strain. In the limit as $\Delta t \to 0$ and $\Delta x_2 \to 0$, we have the expression

$$\lim_{\substack{\Delta t \to 0 \\ \Delta x_2 \to 0}} \left[\frac{\Delta x_1' - \Delta x_1}{\Delta t\, \Delta x_2}\right] = \left(\frac{\partial v_1}{\partial x_2}\right) \tag{A-117}$$

which represents the rate of strain at a particular point and time. Newton's law of viscosity is established by postulating the linear relationship

$$\tau_{21} = \mu \left(\frac{\partial v_1}{\partial x_2}\right) \tag{A-118}$$

where the proportionality constant is called the coefficient of viscosity.

The above procedure can be generalized by allowing the particles to move in any direction. We then find that their average rate of separation in the x_1-direction is

$$\left[\frac{\Delta x_1' - \Delta x_1}{\Delta t}\right] = \left(\frac{\partial v_1}{\partial x_1}\right) \Delta x_1 + \left(\frac{\partial v_1}{\partial x_2}\right) \Delta x_2 + \left(\frac{\partial v_1}{\partial x_3}\right) \Delta x_3 \tag{A-119}$$

Furthermore, if Δs is the initial distance of separation of the particles, then

$$\Delta x_1 = n_i\,\Delta s \tag{A-120}$$

where n_i is the x_i component of the unit vector $\boldsymbol{n}$ that has the direction of the line connecting the particles in the initial state. Therefore (A-119) can be written

$$\lim_{\substack{\Delta t\to 0\\ \Delta s\to 0}}\left[\frac{\Delta x_1' - \Delta x_1}{\Delta s\,\Delta t}\right] = \left(\frac{\partial v_1}{\partial x_1}\right)n_1 + \left(\frac{\partial v_1}{\partial x_2}\right)n_2 + \left(\frac{\partial v_1}{\partial x_3}\right)n_3 \tag{A-121}$$

The right side of equation (A-121) is the magnitude of the x_1 component of the vector formed by the dot product of the vector $\boldsymbol{n}$ and the dyad $\nabla\boldsymbol{v}$.

The dyad $\nabla\boldsymbol{v}$ has components of the form $\left(\dfrac{\partial v_i}{\partial x_j}\right)$, and hence is clearly not symmetrical. Because the stress tensor is symmetrical, $\nabla\boldsymbol{v}$ cannot be the general rate of strain tensor needed in the formulation of the rheological equation. However, through the device of adding and subtracting half of its transpose tensor, $\nabla\boldsymbol{v}$ can be written as the sum of a symmetrical and an antisymmetrical tensor, as shown by the equation

$$\left(\frac{\partial v_i}{\partial x_j}\right) = \frac{1}{2}\left(\frac{\partial v_i}{\partial x_j} + \frac{\partial v_j}{\partial x_i}\right) + \frac{1}{2}\left(\frac{\partial v_i}{\partial x_j} - \frac{\partial v_j}{\partial x_i}\right) \tag{A-122}$$

Examination[5] of the antisymmetrical part of $\nabla\boldsymbol{v}$ shows that it describes rotation but not strain and is therefore not involved in the formulation of the rheological equation. The symmetrical tensor $\boldsymbol{\Delta}$, defined by

$$\Delta_{ij} = \left(\frac{\partial v_i}{\partial x_j} + \frac{\partial v_j}{\partial x_i}\right) \tag{A-123}$$

is the rate of strain tensor of interest in the formulation of the rheological equation.

A stress applied to a compressible fluid will, in general, result in both volumetric and distortional strains. These are distinctly different processes. Accordingly, the general rate of strain tensor $\boldsymbol{\Delta}$ is separated into two parts, one to characterize volumetric strain rates and the other distortional strain rates. This is done by subtracting the mean normal strain rate Δ_m from each of the diagonal elements of $\boldsymbol{\Delta}$, as shown by the equation

$$\boldsymbol{\Delta} = \boldsymbol{\Delta}' + \Delta_m\boldsymbol{\delta} \tag{A-124}$$

where $\boldsymbol{\Delta}'$ is called the deviatoric strain rate tensor.

[5] See, for example, R. R. Long, *Mechanics of Solids and Fluids*, Prentice-Hall, 1961, p. 52.

The mean normal strain rate is defined by the equation

$$\Delta_m = \tfrac{1}{3}(\Delta_{11} + \Delta_{22} + \Delta_{33}) = \tfrac{2}{3}(\nabla \cdot v) \tag{A-125}$$

and it can be seen that the mean normal strain rate represents $\frac{2}{3}$ of the volumetric expansion rate. It is also clear that the sum of the diagonal elements of the deviatoric tensor $\mathbf{\Delta}'$ is zero and hence that this tensor characterizes distortional strain rates only.

The rheological equation for the Newtonian fluid is obtained by postulating the following linear relationship between stress and the strain rates:

$$\boldsymbol{\sigma} - (-P)\boldsymbol{\delta} = A\mathbf{\Delta}' + B\Delta_m\boldsymbol{\delta} \tag{A-126}$$

where the coefficients A and B are not functions of stress or strain rate and where P represents the hydrostatic pressure. The negative hydrostatic pressure is subtracted from each of the diagonal elements of the stress tensor because the compressive stresses resulting from it oppose the outer directed normal stresses (tensile stresses) that provide the driving force for volumetric expansion.

By introducing equations (A-124) and (A-125) into the above equation, it is possible to eliminate $\mathbf{\Delta}'$ and Δ_m to obtain

$$\boldsymbol{\sigma} = -P\boldsymbol{\delta} + \mu\mathbf{\Delta} + (\kappa - \tfrac{2}{3}\mu)(\nabla \cdot v)\boldsymbol{\delta} \tag{A-127}$$

which is the required rheological equation. The coefficients μ and κ are defined by the equations

$$\mu = A \qquad \kappa = \tfrac{2}{3}B$$

Adding the three diagonal elements of equation (A-127) gives

$$3\sigma_m + 3P = 3\mu\Delta_m + 3(\kappa - \tfrac{2}{3}\mu)(\nabla \cdot v) \tag{A-128}$$

where

$$\sigma_m = \tfrac{1}{3}(\sigma_{11} + \sigma_{22} + \sigma_{33})$$

Equation (A-128) reduces to

$$(\sigma_m + P) = \kappa(\nabla \cdot v)$$

which is the rate equation for the volumetric expansion process and which shows that the driving force for the expansion is the mean normal stress, if the effect of hydrostatic pressure is removed. The proportionality constant κ is called the coefficient of volume viscosity.

If the auxiliary tensor $\boldsymbol{\tau}$, defined by equation (A-71), is introduced into equation (A-127), one obtains

$$\boldsymbol{\tau} = \mu\mathbf{\Delta} + (\kappa - \tfrac{2}{3}\mu)(\nabla \cdot v) \tag{A-129}$$

which, for the special case of an incompressible fluid, reduces to

$$\boldsymbol{\tau} = \mu\mathbf{\Delta} \tag{A-130}$$

appendix B

Notation

The following list summarizes the notation which is used throughout the text (excluding the specialized notation used in Chapter 10, which is listed separately at the end of that chapter) and which employs the following general features:

1. As shown in Figure 2-1, the position variables in the rectangular coordinate system are designated x, y, and z; in the cylindrical system r, θ, and z; and in the spherical system r, θ, and ϕ.
2. Vector and tensor quantities are designated with boldface symbols. For example, $\boldsymbol{q}$ represents the conductive heat flux vector and $\boldsymbol{\tau}$ the viscous stress tensor.
3. Complex quantities are marked with an asterisk. For example, G^* is the complex modulus of rigidity. It has real and imaginary components G' and G'', respectively. Hence $G^* = G' + iG''$, where $i = \sqrt{-1}$.
4. A dot placed over a symbol designates its time derivative. For example, γ represents strain in simple shear and $\dot{\gamma}$ represents the rate of strain in simple shear.
5. A standard (reference) state is designated with a degree sign. For example, the non-Newtonian viscosity of a substance in a standard state characterized by a shear stress τ° is designated η°.
6. The average value of a quantity is designated with an overbar. For example, the mean temperature is designated $\bar{T}$.

Symbol	*Definition*	*Typical Metric Units*
E	energy of activation for viscous flow, defined by equation (2-86)	cal/mole
E	internal energy	cal/gram

E_e	extruder energy efficiency	dimensionless
E_b	intensity of radiant energy	cal/sec-cm^2
F	force	dynes
$\mathbf{F}$	force vector	dynes
F	emissivity factor	dimensionless
G	shear modulus	dynes/cm^2
G	heat generation rate	cal/cm^3-sec
H	enthalpy	cal/gram
H	distance of separation between parallel plates	cm
I	rms current	amps
J	shear compliance $J = 1/G$	cm^2/dyne
L	axial length of screw channel	cm
M	parameter in modified van der Waals' equation of state	grams/mole
N	frequency of rotation	cycles/sec
P	hydrostatic pressure	dynes/cm^2
Q	electric charge	coulombs
Q	transferred heat	cal/gram
Q	volumetric flow rate	cm^3/sec
R	radius	cm
S	applied shear stress	dynes/cm^2
S	peeling force	grams/cm
S_0	yield stress of Bingham plastic	dynes/cm^2
T	temperature	°C
V	volume	cm^3
V	rms potential	volts
V_x	speed in x-direction	cm/sec
W	work	cal/gram
W	width of channel	cm
W	mass flow rate	grams/sec
W_a	work of adhesion	ergs/cm^2
X	displacement	cm
X	scale factor	dimensionless
Z	helical length of screw channel	cm
b	temperature coefficient of viscosity	1/deg C
c_p	specific heat at constant pressure	cal/gram-deg C
f	frequency	cycles/sec
$\mathbf{g}$	vector resultant of body forces	dynes/gram
h_c	convective heat transfer coefficient	cal/cm^2-sec-deg C
k	thermal conductivity	cal/cm-sec-deg C

n	flow index for power law fluids	dimensionless
p	power	ergs/sec
$\boldsymbol{q}$	heat flux vector	cal/cm^2-sec
r	rate of temperature change	deg C/sec
r_*	critical radius in plastic flow	cm
t	time	sec
$\boldsymbol{v}$	velocity vector	cm
v	specific volume	cm^3/gram
v_e	equilibrium specific volume	cm^3/gram
y	degree of surface activation	dimensionless
α	thermal diffusivity	cm^2/sec
β	compressibility	atm^{-1}
β	ratio of radii $\beta = R_0/R_i$	dimensionless
γ	strain in simple shear	dimensionless
γ	surface tension	dynes/cm
$\boldsymbol{\Delta}$	rate of strain tensor	sec^{-1}
$\boldsymbol{\Delta}'$	deviatoric rate of strain tensor	sec^{-1}
Δ_m	mean normal rate of strain	sec^{-1}
$\boldsymbol{\delta}$	unit tensor	dimensionless
δ	phase angle	dimensionless
ϵ	dielectric constant	dimensionless
ϵ	emissivity	dimensionless
ζ	viscosity of Bingham plastic	dyne-sec/cm^2
η	viscosity of a non-Newtonian fluid	dyne-sec/cm^2
θ	contact angle	dimensionless
θ	helix angle of extruder screw	dimensionless
κ	coefficient of volume expansion	deg C^{-1}
λ	time constant for viscoelastic material	sec
μ	viscosity of Newtonian fluid	dyne-sec/cm^2
ν	kinematic viscosity	cm^2/sec
ρ	density	grams/cm^3
$\boldsymbol{\sigma}$	total stress tensor	dynes/cm^2
σ_m	mean normal stress	dynes/cm^2
$\boldsymbol{\tau}$	viscous stress tensor	dynes/cm^2
τ_f	fusion temperature	$^\circ$C
ω	circular frequency	radians/sec

appendix C

Physical Constants and Conversion Factors

Avogadro's number	6.02×10^{23} molecules/mole
Boltzmann's constant	1.38×10^{-16} ergs/molecule-deg C
Gas law constant	1.987 cal/mole(g)-deg C
	82.05 cm^3-atm/mole(g)-deg C
	10.73 ft^3-psi/mole(lb)-deg F
Gravitational acceleration	980 cm/sec^2
Stefan-Boltzmann constant	1.355×10^{-12} cal/cm^2-sec-deg C^4
	0.171×10^{-8} Btu/hr-ft^2-deg F^4

multiply	*by*	*to convert to*
atm	1.013×10^6	$dynes/cm^2$
Btu	252	cal
Btu/ft^2-hr-(deg F/in.)	3.445×10^{-4}	cal/cm-sec-deg C
Btu/ft^2-hr-deg F	1.355×10^{-4}	cal/cm^2-sec-deg C
Btu/min	17.58	watts
cal	4.186	joules
ft-lb(f)	1.356×10^7	ergs
ft^2/hr	0.2581	cm^2/sec
ft-lb(f)/sec	1.3558	watts
HP	745.6	watts
HP-hr	2.685×10^6	joules
joules	10^7	ergs
kg/m^2	9.678×10^{-5}	atm
KW-hr	3.6×10^6	joules

lb(f)	4.45×10^5	dynes
lb(f)-sec/ft^2	478.8	poise
mm Hg	1333.2	dynes/cm^2
poundals	13.83×10^3	dynes
poise	1	dyne-sec/cm^2
psi	68950	dynes/cm^2
psi	0.068	atm
watts	1	joules/sec
watts/cm^2-deg C	0.239	cal/cm^2-sec-deg C

appendix D

General References

The following books are standard texts and reference works in high polymer chemistry, physics, technology, and in certain engineering sciences related to polymer processing. They provide a more detailed treatment of many subjects that, of necessity, are discussed only briefly in this book.

1. *Mechanical Behavior of High Polymers*, Turner Alfrey, Jr., Interscience Publishers, 1948.
2. *Processing of Thermoplastic Materials*, E. C. Bernhardt (ed)., Reinhold Publishing Corp., 1959.
3. *Transport Phenomena*, R. B. Bird, W. E. Stewart and E. N. Lightfoot, John Wiley and Sons, 1960.
4. *Conduction of Heat in Solids*, second edition, H. S. Carslaw and J. G. Jaeger, Oxford University Press, 1959.
5. *Principles of Chemical Equilibrium*, Kenneth Denbigh, Cambridge University Press, 1955.
6. *Rheology—Theory and Applications*, 3 vols., F. R. Eirich (ed.), Academic Press, 1956.
7. *Viscoelastic Properties of Polymers*, John D. Ferry, John Wiley and Sons, 1961.
8. *Principles of Polymer Chemistry*, Paul J. Flory, Cornell University Press, 1953.
9. *Heat Transfer*, 2 vols., Max Jakob, John Wiley and Sons, 1949.
10. *Polythene*, A. Renfrew and Phillip Morgan (eds.), Interscience Publishers, 1957.
11. *Deformation, Strain and Flow*, Markus Reiner, Interscience Publishers, 1960.
12. *Polymer Processes*, Calvin E. Schildknecht (ed.), Interscience Publishers, 1956.
13. *High Polymers—Theory and Practice*, A. X. Schmidt and C. A. Marlies, McGraw-Hill, 1948.

Author Index

Subject Index